W9-BBE-525

THE OPENING OF VISION

The Author

David Michael Levin is currently Professor in the Department of Philosophy, Northwestern University, Illinois. He has also taught at the University of Connecticut and at the Massachusetts Institute of Technology. Dr Levin has published widely in various periodicals. His previous books are *Reason and Evidence in Husserl's Phenomenology* (Northwestern University Press, 1970), *The Body's Recollection of Being* (Routledge & Kegan Paul, 1985), and an anthology, *Pathologies of the Modern Self: Postmodern Studies on Narcissism, Schizophrenia and Depression* (New York University Press, 1987).

THE OPENING OF VISION

NIHILISM AND THE POSTMODERN SITUATION

DAVID MICHAEL LEVIN

Professor of Philosophy, Northwestern University

Routledge

New York and London

First published in 1988 by
Routledge, Chapman & Hall Inc.
29 West 35th Street, New York, NY 10001 and

Routledge
11 New Fetter Lane, London EC4P 4EE

Set in Palatino, 10 on 11pt
by Columns of Reading
and printed in Great Britain by
Biddles Ltd, Guildford and Kings Lynn

Library of Congress Cataloging in Publication Data

Levin, David Michael, 1939-
The opening of vision.

Bibliography: p.
Includes index.
1. Nihilism (Philosophy) 2. Civilization, Modern—
1950- . I. Title.
B828.3.L39 1988 149'.8 87-20623

British Library CIP Data also available

ISBN 0-415-00412-8 (c)
0-415-00173-0 (pb)

For Francisco Lassiter,
Friend

Can we preclude the possibility of a meaningless emancipation? . . . Is it possible that one day an emancipated human race could encounter itself within an expanded space of . . . will and yet be robbed of the light in which it is capable of interpreting its life as something good?

Jürgen Habermas, 'Walter Benjamin',
Philosophical-Political Profiles

I see the eyes but not the tears
This is my affliction

T. S. Eliot

Where there is no vision the people perish

Proverbs 29:18

Contents

Acknowledgements

Before I thank others, I want to give very special thanks to my parents, Helen and Benedict, for it was in the familiar and congenial spirit of their home at Lake Waramaug, in Litchfield County, Connecticut, that I wrote the first and final versions of this book. As always, their cheerful warmth and many kindnesses helped me immeasurably; but their questions and comments on the problems I often carried with me to their table were also of great benefit. I also want to express my gratitude to a few very special friends: to Cisco Lassiter, who accompanied me on some of my earliest vision quests and shared many experiences of visionary openings; to Roger, my brother, with whom, often until night turned into dawn, I debated different interpretations of these quests and from whom I learned so much of what I know about psychopathology and the spiritual dimension of psychological processes; to Samuel Todes, my colleague at Northwestern, who encouraged me to think out my initially very unclear feeling, my sense, that crying is somehow the source of vision; to Eugene Gendlin, who teaches at the University of Chicago, and without whom I might never have learned how to work phenomenologically, and therefore creatively, with my own gift of experience; to David Wood at the University of Warwick, who read an earlier draft of this book and helped me with its publication; to Calvin Schrag, of Purdue University, who for many years has unflaggingly encouraged and supported me in the course of my work; to Herbert Guenther, whose unequalled scholarship in the field of Tibetan Buddhism was a very inspiring and trustworthy guide as I began to study and practice the Tibetan Vajrayana; to Tarthang Tulku and Namkhai Norbu, my Tibetan teachers, the first, of ris-med Buddhism and the Nyingmapa tradition, and the second, of Dzogs-chen.

I also appreciate the help and cooperation of Tim Staggs,

who patiently lived through with me the proofreading and indexing stages of my earlier book, *The Body's Recollection of Being*, as well as a very difficult stage in the reworking of this book. Audrey Thiel and Marina Rosiene worked very hard on the typing of my sometimes almost indecipherable manuscript. For their sense of humour, their supernatural patience with me, and their supererogatory perseverance, I am deeply indebted. I give them my heartfelt thanks.

Finally, I would like to express my gratitude to the administration at Northwestern University for some very generous research grants and sabbaticals.

I am grateful to Harcourt Brace Jovanovich, Inc. for permission to quote from 'Eyes that last I saw in tears', in *Collected Poems 1909-1962* by T. S. Eliot, copyright 1936 by Harcourt Brace Jovanovich, Inc.; copyright © 1963, 1964 by T. S. Eliot. Reprinted by permission of the publisher. I am also grateful to Faber and Faber Limited, who hold non-exclusive rights to this text in the English language throughout the world excluding the United States of America and throughout the British Commonwealth excluding Canada.

Quotations from Nancy Henley, *Body Politics*, are reprinted by permission of Simon & Schuster, Inc., holder of a 1977 copyright.

I am grateful to Doubleday & Company, Inc. for their permission to quote from 'In a Dark Time', a poem by Theodore Roethke published in *The Collected Poems of Theodore Roethke* (Doubleday, 1966). I am also grateful to Faber and Faber Limited, who hold non-exclusive rights in the English language throughout the world, excluding the United States of America and Canada.

For the right to quote from Jean Paris, *Painting and Linguistics*, I am grateful to the College of Fine Arts, Carnegie-Mellon University, which published this text in 1975.

For permission to quote from Sigmund Freud, *Civilization and Its Discontents*, I am grateful to Doubleday & Company, Inc.

Quotations from Pierre Bourdieu, *Distinction: A Social Critique of the Judgement of Taste* (copyright 1984) have been reprinted by permission of Harvard University Press, holding world rights in English for the United States and Canada.

Basic Books, Inc. has granted permission to quote from *The Renaissance Rediscovery of Linear Perspective* by Samuel Y. Edgerton, Jr. Copyright © 1975 by Samuel Y. Edgerton, Jr. Published by Basic Books, Inc.

By permission of Harper & Row, Publishers, Inc., I have quoted approximately 1,318 words, passim, from *Nietzsche*, vol. II: *The Eternal Recurrence of the Same*, by Martin Heidegger. Copyright © 1984 by Harper & Row, Inc. Analysis copyright © 1984 by David Farrell Krell.

By permission of Harper & Row, Inc. I have quoted approximately 1,350 words, passim, from *Being and Time*, by Martin Heidegger. Copyright © 1962 by SCM Press Ltd. and Basil Blackwell, Ltd. Reprinted by permission of Harper & Row, Inc.

For reprinting rights covering the British Commonwealth, I am grateful to Basil Blackwell, Ltd., presently holding the copyrights to *Being and Time*, by Martin Heidegger, in the British Commonwealth.

By permission of Harper & Row, Inc., Publishers, I have quoted approximately 2,100 words, passim, from *The Question Concerning Technology and Other Essays*, by Martin Heidegger, translated and with an Introduction by William Lovitt. English translation copyright © 1977 by Harper & Row, Publishers, Inc.

For permission to quote from the English translation by Charles Seibert of the *Heraclitus Seminar 1966/67* by Martin Heidegger and Eugen Fink, I am grateful to the University of Alabama Press, which published the text in English translation in 1979. I am also grateful to Vittorio Klostermann, publisher and copyright-holder of the German-language edition from which Charles Seibert made the translation.

For permission to use the line drawing of the gold ring of Isopata, recovered near Knossos, I am grateful to the University of California Press, which holds rights to *Gods and Goddesses of Old Europe*, by Marija Gimbutas (Copyright © 1974), for the United States and Canada, and also to Thames & Hudson, Ltd, holder of rights for the rest of the world market. The drawing first came to my attention in R. Gordon Wasson, *The Wondrous Mushroom: Mycolatry in Mesoamerica* (New York: McGraw-Hill, 1980).

Finally, I am grateful to the Rijksmuseum of Amsterdam

for permission to use a photograph of *Faust in His Study*, an etching made by Rembrandt in 1592.

Introduction

The Postmodern Situation

In his work on Nietzsche, the 'first' of the 'postmodern' thinkers, Heidegger writes: 'That period we call modern . . . is defined by the fact that man becomes the center and measure of all beings.'[1] The 'modern' world begins, in the story I want to tell about the West, with the Renaissance. It is to the people of the Renaissance that we owe the beginnings of modern science and technology, an unprecedented expansion of trade and commerce, the glorious vision of humanism, and a mighty challenge to the mediaeval authority of faith, announced in the name of a self-validating rationality. The spirit of the Renaissance continued for two hundred years, eventually coming to light in the rationality of a mechanical vision: Hobbes, Descartes. In Descartes, humanism assumed a distinctively subjective character, proclaiming its triumph even in the passage where it demonstrates the existence of God. The triumph of Cartesian subjectivity is to be seen in the objective rationality it empowered and imposed. Thus, it is in the seventeenth century that the machine appears as paradigm, partly in a dream of power, partly in a vision of divine glory which was slow to die out. There is ambiguity and paradox in the proof Descartes set down for the existence of God: what the proof really celebrated was the power of human reason, the priority of 'Man' before God, the independence, self-determination, and self-affirmation of the subject. The gaze of these moderns finally turned away from heaven, away from the sky; vision returned, but with a power stolen from its god, to the projects of its life-world.

The Enlightenment saw the perfection of this subjectivity in a social revolution whose progress was grounded in objective reason. But the subjective freedom it won turned brutish and competitive, and a false individualism soon began to inhabit the descendants of the monadic Cartesian ego. In the nineteenth century, people were captivated by a vision of evolutionary progress: they were seeing traces of teleology, a teleology of divine origin, at work in nature. But this vision derived from an egological and essentially anthropocentric vision of reason: reason as instrumental, pragmatic, practical. And people slowly began to lose sight of the difference between reason and power: reason, increasingly asserting itself in self-destructive ways, began to think of itself as the will to truth.

When subjectivity triumphed, it imposed its will on things and brought into being a world ruled by objectivity. But in a world of objectivity, there is no place, no home, for the subject, whose subjectivity – that is to say, experience – is denied value, meaning, and ultimately any truth or reality. This triumph of subjectivity has been self-destructive; we can now see how the subject falls under the spell of its objects; how it becomes subject to the objectivity it set in power. The subject is in danger of losing touch with itself. When reason turned totally instrumental, a function solely of power, it legitimated the construction of a totalitarian state and engineered a Holocaust. The legacy of humanism is terror.

The triumph of subjectivity is self-destructive, because it has inflated the human ego without developing self-respect, the true basis of agency, and the social character of human vision. Moreover, the triumph of 'Man' necessitated the death of God. But, since God had been the sole source of our values and the origin of all meaningfulness, the death of God only accelerated the spread of a latent culture of nihilism, cancer of the spirit, contagion of despair.

We of today are survivors of the Holocaust, World War Two; we face the dangers of nuclear devastation, ecological catastrophe, world-wide famines. The 'modern' world, which began, in the West, with the Renaissance, is now coming to an end. What defines the position assumed by this book, the position I am calling the 'postmodern situation', is a strong and relatively clear sense of this ending. This sense separates us from modernity; it is critical; it sees the modern world with some understanding of it as an historical whole.

We have seen enough of the modern epoch to see how the humanism of reason, emancipatory, nevertheless produces, reproduces, and even legitimates, conditions of alienation and oppression. In the postmodern situation, which is partly a question of awareness, it is not possible to trust in the old vision of reason – and in the 'Man' of its humanism. We are finding it more and more difficult to live in the light of the modern – that is to say, the traditional – paradigm of knowledge, truth, and reality. We are living in a time of crisis: this is our *Befindlichkeit*, our present plight. We must come together to achieve a different vision.

The modern epoch brought into being a world in which the effects of nihilism are spreading. Now, we can see, today, if we look with care and thought, that nihilism is a rage against Being: 'nihilism' means the destruction of Being: the Being of all beings, including that way of being which we call 'human' and consider to be our own. Thus, in the postmodern situation, we need to achieve, both individually and collectively, a recollection of Being, of its dimensionality. This is possible, however, only if the question of Being can become, for us, a question of character – a question that questions the historical character of our vision.

Drawing on the insights of Nietzsche, whose extraordinary foresight positioned him almost a century ahead of his time, the discourse of postmodern thinking begins with a consciousness of deepening crisis, a consciousness that the nihilism which Nietzsche saw, in signs and symptoms, is now unmistakable, too pervasive to be ignored or interpreted away. It is this consciousness of our situation which separates us from modernity; but postmodern discourse is not only a discourse which takes as its problem the experience of a crisis; it is also a discourse which is itself in crisis: a discourse without grounds, without a subject, without an origin, without any absolute center, without reason. This sense of crisis, and of being irremediably separated, by this crisis, from the questions and answers that have challenged us in the past, figures in the thinking of all those who have let themselves be guided by Nietzsche's vision. It figures in Freud's *Civilization and Its Discontents*, and in Jung's essays on 'Mind and Earth' (1927) and 'The Spiritual Problem of Modern Man' (1928), in the second of which there is a discussion woven around Hölderlin's provocative words, that 'Even danger itself makes the saving power grow also' – words, in fact, which Heidegger also attempts to explore in his 1949 lecture on 'The Turning'.[2] The crisis of modernity therefore figures, of course, in the work of Heidegger, and also in the thought of Horkheimer, Adorno, Marcuse, Habermas and Foucault. This study on vision, indebted as it is to all of these thinkers, is intended to contribute to the ongoing conversation which constitutes our break with modernity.

Vision and the Discourse of Metaphysics

In *Being and Time*, Heidegger asserts that,

> The question of Being does not achieve its true concreteness until we have carried through the process of destroying the ontological tradition.[3]

How can we destroy the ontological tradition, the discourse of metaphysics, without abandoning the question of Being? How can we think of Being without letting ourselves be controlled by the history of metaphysics? This is a question which locates Heidegger's thinking within the discourse of postmodernism.

Heidegger chooses the most difficult path. There are, of course, two easier paths. But he does not avoid the postmodern predicament by defending the ontological tradition against the testimony of recent history; nor does he avoid it by including the question of Being in the metaphysics he wants to destroy. Wisely, I think, he challenges the ontological tradition; but he does not refuse to reflect on the question which set this tradition in motion many centuries ago. He distinguished between the question of Being and the answers of history. In that way, he kept himself open, as a thinker, to the question itself: a question of Being.

But questions mean nothing, or mean indeterminately, outside their position in a discursive context: questions cannot be understood in abstraction from their historical background – in abstraction, for example, from the texts of metaphysics and the world in which they are situated. Thus, as Heidegger knows, we may articulate a distinction between the question of Being and the tradition of metaphysics in which that question has been enunciated; but we cannot totally separate and isolate them. Metaphysics can be deconstructed, but not completely destroyed. We can break with tradition; but the rupture can also be seen as a continuity, as an affirmation of continuity. We should struggle to release the question of Being from its metaphysical history – a history of reification and totalization, egocentrism and logocentrism. But we must understand that this struggle is never finished, never final. All we can do is maintain the vigilance of a critical spirit repeatedly questioning itself.

In *Being and Time,* however, Heidegger perpetuates the very tradition he sets out to destroy when he continues to assume the paradigmatic function of vision in the formulation of ontology and the conditions of knowledge. Thus, for example, he writes that,

> In giving an existential significance to 'sight', we have merely drawn upon the peculiar feature of seeing, that it lets entities which are accessible to it be encountered unconcealedly in themselves. Of course, every 'sense' does this within that domain of discovery which is genuinely its own. But from the beginning onwards the tradition of philosophy has been oriented primarily towards 'seeing' as a way of access to beings and to Being. To keep the connection with this tradition, we may formalize 'sight' and 'seeing' enough to obtain therewith a universal term for characterizing any access to entities or to being, as access in general.[4]

We need, as I shall argue, to see this connection with tradition, this hegemony, concretely problematized.

It seems to me that the thrust of Heidegger's critique of metaphysics brings forth serious questions regarding the complicity of vision – vision elevated to the position of paradigm for knowledge and rationality – in the historical domination of our 'universal' metaphysics. Since vision assumes primacy in the hierarchy of the metaphysical tradition, the problematizing of vision and its ontology is one of the tasks which the deconstruction of metaphysics presently requires.

However, the question of Being does not enter history only through the medium of philosophical discourse; in fact, it is first set in motion within the life-world of our experience, being that which *opens our eyes* to the field of possibilities we call 'existence'. The history of Being in the discourse of metaphysics refers back to our historical experience – as individuals, as collectivities – with vision and its ontology.

What is in question in the deconstruction of metaphysics must be, therefore, the *opening* of vision as a 'way of access to beings and to Being'. This question will be understood, here, as demanding a discourse on the nature and character of our normal everyday vision: the vision of *das Man*, of anyone-and-everyone. Such a discourse needs to be spelled out in the hermeneutical language of phenomenological psychology, for experiential concreteness will call our attention to matters otherwise disregarded and can accordingly dispell some persistent forms of mystification and delusion. But our discourse must also bring to light the historical character of our vision: the hidden violence, the hidden nihilism. Therefore it must diagnose hermeneutically, and in an experientially familiar language, the closure-to-Being which underlies the historical character of our vision – its character, in particular, as suffering and affliction, as cultural psychopathology. Our discourse must be a 'speech of suffering,' telling the truth about its social production and reproduction in the world of our vision.

The life-world of this vision is rapidly being gathered, today, into a frontal ontology. Our epoch may close with the domination of the image. The Being of beings is now something to be confronted, placed (*gestellt*) directly in front – that is to say, in a position which gives priority to the demand for optimum clarity, certainty and control. The Being of beings is now something to be mastered, its presence in the lighting reduced to constant availability. In the succession of historical epochs, vision and its ontology have changed. The story of their historical changes belongs, ontologically, to the graphing, the writing of the light in a hermeneutics of concrete illuminations. We are beings of vision. We are beings of light: the lighting in which, and by grace of which, vision takes place. I am concerned about our future in this light. We need to look again at the ontology, the dimensions of the life, our vision has inscribed in its field. The history of Being is inscribed in a light that must no longer be occluded from our thinking. The story of vision and its light must be written into our critical history of Being.

Our task, here, is a necessary one, because the forgetful history of revision continues, and the visibility and invisibility of Being, the dimensionality of visible beings, will otherwise be erased, little by little, from the texts we are reading and writing.

In his 'Letter on Humanism', Heidegger writes:

> Perhaps what is distinctive about this world-epoch consists in the closure of the dimension of the hale [*des Heilens*]. Perhaps that is the sole malignancy.[5]

Perhaps. But what follows? For Heidegger, this situation calls us to a task of thought: 'Thinking conducts historical eksistence . . . into the realm of the upsurgence of the healing.'[6] But can thinking accomplish this when it continues to function metaphysically – that is to say, only theoretically, in a disembodied state, and in abstraction from the practices of everyday life? We need a visionary thinking which is at work in our *experiences* with vision: diagnostic, critical, attentive to closures and even the smallest opportunities for some opening.

'Each field is a dimensionality', as Merleau-Ponty says, 'and Being is dimensionality itself.'[7] When 'vision' takes place, a *Gestalt* makes its appearance: there (*da*), in the clearing (*Lichtung*) for that place of our being (*Sein*). The *Gestalt* is a structural event in which a figure differentiates itself from a surrounding ground. Being is the dimensionality in which the *Gestalten* of vision are grounded. But this dimensionality is now being subject to the most extreme negation. Today, the *Gestalt* is enframed and enframing; it takes place under the structural principle Heidegger calls *das Ge-stell*. *Das Ge-stell*, enframing, the taking-place of a distinctive positing and positioning, has taken possession of the visionary encounter most characteristic of our modern history, excluding the depth of the field, denying what cannot be reified and totalized. In the epoch of *das Ge-stell*, our present epoch, the epoch in which nihilism rages, we are tempted to enframe the very Being of beings.

Our Capacities and their Development

'The essence of thinking', according to Heidegger, 'is the understanding of Being in the possibilities of its development. . . .'[8] Vision is a capacity of our being. As such, vision is an achievement involving a process of development. There is no reason to suppose that this development does not, or could not continue – for a lifetime. If, in our adulthood, vision is ruled over by an ego-logical subject, what could vision become when it is committed to overcoming this rule? Since we are visionary beings, this question is addressed to us, questioning our being. In *Being and Time*, Heidegger observes that,

> Dasein's ways of behavior, its capacities, powers, possibilities, and vicissitudes, have been studied with varying extent in philosophical psychology, in anthropology, ethics, and 'political science', in poetry, biography, and in the writing of history. . . . But the question remains whether these interpretations of Dasein have been carried through with a primordial existentiality comparable to whatever existentiell primordiality they [Dasein's capacities and possibilities] may have possessed. Neither of these excludes the other, but they do not necessarily go together.[9]

Since individual and society are interactively and interdependently co-emergent, the development of our inborn capacity for vision, whatever that may be, is of significance not only for every individual, but also for society as a whole. The visionary life around which a society is gathered reflects and amplifies the character of the vision developed within each one of its individual members; but conversely, the conditions of society as a whole bear in many decisive ways, some of them oppressive and destructive, on the development of individual predispositions and capacities. Thus we must concern ourselves with both dimensions of visionary life: with the individual and with society. As we reflect on how individual vision, an organ of our 'potentiality-for-being', can be developed, we must also reflect, simultaneously, on how the collective visions of our society, the visions encoded in our culture, can be changed in corresponding ways.

Heidegger questions the primordiality of the prevailing interpretations of Dasein's capacities; and he questions the radicality of his own interpretation as well. The significance of this questioning is related to the fact that individual and society are interactive and interdependent. The interpretation we shall spell out in this study must likewise be questioned – questioned again and again. Primordiality is inevitably a relative matter. Nevertheless, if we strive for primordiality, rather than give up on it, there is at least a possibility that we may move beyond the limits of prevailing roles and habits – beyond the prevailing historical conditions.

In his work on Nietzsche, Heidegger clarified the 'moment of vision', the *Augenblick*, about which he wrote in *Being and Time*:

> . . . thinking in terms of the moment . . . implies that we transpose ourselves to the temporality of independent action and decision, glancing ahead at what is assigned us as our task and back at what is given us as our endowment.[10]

Vision is part of our endowment. This endowment is not only a biological program; it is also an existential capacity, a potentiality-for-being, already corporeally schematized, already laid down in the layout, the *legein*, of the flesh. Our endowment is already inscribed in the structuring of our moodedness, our inwrought predispositions. Our visionary endowment, and the world into which we are cast, is our *Befindlichkeit*. But when we 'find' ourselves in this condition, there is an existential question to be answered by the character of our life. This is the question which 'assigns us our task'. This task is an historical task, determined in relation to the history of Being. It is a task for society as a whole. However, it is also a task for us as individuals: a task of individuation; a task which can only be achieved if we are committed, as individuals, to developing our potentiality-for-being – and to doing this in ongoing responsiveness to the question of Being as it figures in our historical experience. Thus Heidegger sought to define the task in relation to the 'overcoming of nihilism'. But we should note that he refers us to our experience of living in a condition of need:

> This implies that we transpose ourselves to the condition of need that arises with nihilism. . . . Our needy condition itself is nothing other than what our transposition to the moment [of vision] opens up to us.[11]

A relative test of interpretive 'primordiality' is suggested here, and it is this: an interpretation is 'primordial' when, in relation to our experience of 'the condition of need', something is seen, seen differently; and something is opened up. With echoes, perhaps, of Spinoza, Nietzsche called this 'affirmation of life' and the 'enhancement' of a 'feeling of power': the power of a moral agent.

We, who call ourselves 'human beings', 'mortals', and 'rational animals' – who are we? And what will become of us? This book is part of a much larger project. The project is an attempt to articulate the emergence of an ontological body of understanding in response to the need of our time. This study was conceived and written as one of four volumes contributing to the ontological project. One of the volumes is already in print: *The Body's Recollection of Being*. *The Listening Self* will follow after this. The first volume is focused on our experiences with gesture and motility: on our embodiment as a whole. The third is of course an analysis of our hearing. The present volume, concerned with our capacity for vision, our visionary being, continues my work on this project.

Like the other volumes, this one is a textual interweaving of seven principal discourses. I will briefly identify them, proposing an order which I continue to find suggestive, but which we could certainly reorganize, and probably should, in other significant ways. The *first* of these, then, owing much to the thought of Merleau-Ponty, is a critical reflection on vision as a perceptual capacity: a capacity for the channelling of our ontological perceptiveness. The *second* is a diagnostic interpretation formulated in the terms of phenomenological psychology, in which I shall question the developmental achievement, the character, of the modern ego-logical subject. The main influences in this discourse are Freud and Jung; Freud, because of his uncanny understanding of adaptive and pathological processes in the constitution of the ego, and Jung, because of his understanding of the Self and his heroic attempt to go beyond a psychology of the socialized ego. The *third*, dependent on the work of Nietzsche and Heidegger, spells out a postmodern interpretation of 'modernity' – our present historical situation – focused on the experience of nihilism: on our experience, therefore, of need, and of suffering and the production of suffering. The *fourth*, also inspired by Nietzsche and Heidegger, is a hermeneutical discourse on the metaphysical history of Being. In this discourse, I will be attempting to articulate the history of Being in a way that could perhaps open it up to another 'beginning'. The *fifth* is a discourse intended to contribute to a practical understanding of our political economy – the character of the life we are living in the practices and institutions of our body politic. Our visionary being is a product of the prevailing economy; but it is also a place where the vision ruling this economy is engaged in its own reproduction. Merleau-Ponty again figures prominently in this discourse, but there are other decisive influences to be named here: Marx, Marcuse, Horkheimer, Adorno, Habermas, Foucault.

A *sixth* discourse might be called 'philosophical anthropology', since it introduces knowledges drawn from work in cultural anthropology and registers my philosophical reflections on the significance of these knowledges for the basic questions of our project. I consider cross-cultural anthropology to be extremely important for our project, because it expands our theoretical sense of the range of possibilities that constitute our being human; and it enables us to envision opportunities for practical action we otherwise would not have considered. In the societies of earlier epochs and other cultures, visions of Being very different from ours have engaged mortal lives. In some of these visions, there is a pre-ontological understanding of Being which mythopoetic narratives in the culture have kept alive, intact, and closer to everyday consciousness. But in the Western 'world', this understanding is an excluded knowledge. The attempt to integrate systems of knowledge our tradition has excluded can be emancipatory, so long as this attempt learns from their differences, and does not suppress them. Philosophical anthropology can serve a critical function here, since it includes previously excluded knowledges, excluded ontologies, within the traditional discourse of Western philosophy. It can become a critical reflection on our age, our cultural life as a whole – but only if the inclusion is open to the value in the differences. I hope to show here how anthropology can open us to different ways of being, different visions.

There is also, finally, a *seventh* textual dimension: a discourse which I shall call a meditation on our visionary being as a spiritual vocation. Our religious traditions have been entrusted with great wisdom and so we shall draw strength and inspiration from their texts and practices. But I do not wish to reduce the life of the spirit to any of its historical forms, the various religions of the world. We need to keep the channels of our visionary being open to what Merleau-Ponty calls 'the immemorial depth of the visible': the dimensionality of existence within which what our tradition calls the 'spirit' lives.[12] This is a dimensionality which corresponds to our 'ultimate concerns'. Its truth comes to light through the visions of our mythopoetic imagination. In the realm of the spirit, our capacities are stretched: sometimes even beyond their breaking point.

Practices of the Self

If we think of vision as a capacity to be developed and a task to be achieved, then we are also thinking of vision as 'practical activity', understood in the sense Heidegger explicates in his work on Nietzsche:

> The essence of thinking, experienced in this way, that is, experienced on the basis of Being, is not defined by being set off against willing and feeling. Therefore, it should not be proclaimed purely theoretical as opposed to practical activity and thus restricted in its essential importance for the essence of man.[13]

It could be said that our study is concerned with what John Welwood describes, in his 'Principles of Inner Work', as 'the process of growing beyond limited views of self towards a greater vision and realization of what it is to be human.'[14] This process *is* 'inner work', but not in the metaphysical sense of 'inner', which overlooks the intertwining of inner and outer: *this* 'inner work' is also a form of 'practical activity'. Vision is a social practice, and it needs to be understood as such.

This conception of self-development (self-realization, self-fulfillment, self-determination) proposes an interpretation of the self which contests the authority and hegemony of metaphysics, and according to which the 'self' is essentially a Cartesian substance, a fixed identity, essentially isolated and disembodied, an ego-logical 'thing' encapsulated in a machine of corruptible matter. Sharing our very different conception, Joel Kovel, author of *The Age of Desire*, argues that,

> We are thus in a position to assess the radical, i.e., transcendent, quality of particular praxes. Clearly, a radical act need not be an explicitly political one, even though its universalizing quality can be consummated only at the level of all society, indeed, for the entire globe. However, as history has yielded a fragmented society, so may it be undone, i.e., transcended, at the level of a fragment. As personal life is a principal one of these fragments, the question of transcendence may validly be asked of it.[15]

What Kovel is calling 'the question of transcendence' is a question which calls upon the self to question itself with regard for its visionary capacity, and for the roles, routines and practices in which this capacity is channelled in response to the task. In our time, this questioning does not bring to light a Cartesian self, a self of reason completely purged of body and feeling, a self without shadows, a self totally transparent to itself, totally knowing of itself, totally self-possessed, totally certain of itself. On the contrary, the self which responds to this question is an embodied, historical self feeling, today, very empty, very much alone, very unsure of itself: a self in fragments, a self in the fury of Being. It is this historical experience of ourselves which forms the basis, the starting point, for our attempt to set in motion a practice of the self very different from Descartes', and indeed very different from the modern: a practice which assigns to our visionary capacity an historical task in response to our need.

The significance of Foucault's final work lies in the fact that it recognizes the importance of relating questions of truth and ontology to the formation of character in a history of practices of the self, self-forming activities.[16] I submit that our visionary capacity as visionary beings is a question that should refer us to character and its potentiality for development. In particular, since nihilism is the negation of Being, our present historical situation calls for a development in which our relatedness to Being, our ontological response-ability, becomes for us a question that demands our attention, our care, and our thought – a question, in other words, in relation to which our character is to be formed. And this is because a self whose character is formed *in this way* simultaneously develops its own ontological potential and initiates historical changes with an emancipatory effect.

Our project in this book is to be understood, therefore, as a contribution to the contemporary postmodern discourse of critical social theory. About this discourse, Raymond Geuss had this to say:

> A critical theory is a very complicated conceptual object; it is addressed to a particular group of agents in a particular society and aims at being their 'self-consciousness' in a process of successful emancipation and enlightenment. A process of emancipation and enlightenment is a transition from an initial state of bondage, delusion, and frustration to a final state of freedom, knowledge, and satisfaction.[17]

Except for its assumption of a 'final state', this conception of critical theory expresses very well the concerns of our project in the present book. However, whereas the authors of critical social theory do not consider the history of Being when they interpret our present historical situation, we shall. For, we cannot respond adequately and effectively to the 'malignancy' of nihilism, nor can we reflect properly, today, on the formation of character, which is, after all, a question of our being the kind of being we call 'human', without letting ourselves be addressed, be claimed, be moved by the history of Being: by the raging of nihilism, by the closure of dimensionality, by the fury of Being.

In justifying the emphasis, in this study, on the 'psychological' character of our reflections, I would like to call attention to an analysis in Adorno's *Minima Moralia*. The passage we shall read says very well how I think phenomenological psychology contributes to critical social theory:

> . . . in an individualistic society, the general not only realizes itself through the interplay of particulars, but society is essentially the substance of the individual.
>
> For this reason, social analysis can learn incomparably more from individual experience than Hegel conceded, while conversely the large historical categories, after all that has meanwhile been perpetrated with their help, are no longer above suspicion of fraud. . . . [T]he individual has

> gained as much in richness, differentiation, and vigour as, on the other hand, the socialization of society has enfeebled and undermined him. In the period of his decay, the individual's experience of himself and what he encounters contributes once more to knowledge. . . . In face of the totalitarian unison with which the eradication of difference is proclaimed as a purpose in itself, even part of the social force of liberation may have temporarily withdrawn to the individual sphere. If critical theory lingers there, it is not only with a bad conscience.[18]

As Merleau-Ponty has very forcefully argued, it is not enough for philosophers 'to create or express an idea; they must also awaken the experiences which will make their idea take root in the consciousness of others'.[19] I share this conviction; however, Nietzsche has helped me to see the nihilism at work in our present historical situation as incapacitating and destructive: damaging to our sense of ourselves as agents of historical change.[20] Nihilism and false subjectivity are inherently connected. There is also a connection, therefore, between the raging of nihilism and the epidemiology of narcissistic character disorders in a culture of extreme narcissism.[21] Consequently, I want to engage critical social theory in a sustained reflection on our capacity for vision, and on the social conditions which limit, disorganize, and predispose us to misinterpret it. In order to accomplish this task, however, social theory must collaborate with a hermeneutically critical interpretation of the history of Being represented in the discourse of metaphysics, and also with a phenomenological psychology which will put us in touch with our historical experience and enable us to understand it without being misled – without being incapacitated – by the prevailing ideological interpretation, presently circulating in our culture.

There are many modes of relating to the self, many interpretations. The modern self, which we must attempt to overcome, is a Cartesian self, a self with an essentially fixed identity: a timeless self, without real history. What we are struggling to envision, then, is rather a self which lives with a continuously *changing* identity: a self open to changes in itself; a self which changes in response to changes in the world; a self capable of changing the conditions of its world according to need. I am not what I am and I am what I am not. 'Self-development', 'self-realization', 'self-determination' and 'self-fulfillment' are words with their own histories, words which have had many different meanings – some of them self-defeating and incapacitating, some of them too entangled in the dualisms of metaphysics, which separate and oppose self and society, experience and world, inner life and external reality, what is given and what is created.

We would consider it unthinkable to condemn science and forbid its practice because of the nuclear catastrophe it has made possible. And yet, there are many thinkers, today, who refuse to consider, albeit in a deconstructive way, questions concerning 'psychological experience' and the development of the self, and who argue against them because these questions have been bound up in a destructive tradition of humanism. We should be more discerning, more careful, around the tradition of humanism, resisting what has been, for some 'post-modern' intellectuals, an irresistible temptation to abolish the discourse of humanism in its entirety and reject all efforts to contribute historically different, and more emancipatory, more developed and more self-developing 'practices of the self'.

I would like to see us attempt to *retrieve* the heart of humanism, even as we acknowledge its inhumanity, its injustice, its irrationality, its violence, its reigns of terror. For this 'same' humanism – 'same' in a sense which is not that of an essentially fixed, non-historical identity – has also changed the world for the better. In this, its deeper sense, 'humanism' is a tradition of caring for the deepest and the best in us; it is a tradition caring for the 'humanity' in us. And this means caring for the development of our potentiality-for-being-human: our capacity, our predisposition, our response-ability as human beings. It is we who have failed in our caring; the caring has not failed us. Despite a long history of errors enacted in its name, despite its complicity in violence, humanism nevertheless represents our highest care for the potential inherent in our gift of being. This, I think, is the position where Merleau-Ponty and Habermas have chosen to take their stand.

The great vision of humanism needs, certainly, to be questioned in the light of a history of domination and oppression to which it has contributed its ideology; but it cannot be identified with, cannot be reduced to, the destructiveness of this history. We are justified in our contempt for 'practices of the self' which are little more than glorifications of narcissism, phantasies of omnipotence and omniscience, acts of self-indulgence, displays of childish regression, forms of social indifference and political irresponsibility. After all, the modernity we are rejecting once itself revolted against 'practices of the self' – the practices of mediaeval Christianity, for example, which required mortifications of the flesh and extreme forms of self-examination, self-restraint and self-punishment. But we must be careful not to cut ourselves off from practices which would enable us to continue our individual and collective development as human beings. Our 'humanity' is a capacity whose character is given shape by the roles, routines, and practices within which we live.

When Foucault was interviewed by Rabinow in April, 1983, he was asked the following question:

> . . . isn't the Greek concern with the self just an early version of our self-absorption, which many consider a central problem in our contemporary society?

The question clearly shows its confused ideological position, but I cannot tell whether or not it surreptitiously manipulated the answer:

> . . . in a culture to which we owe a certain number of our most important constant moral elements, there was a practice of the self, a conception of the self, very different from our own present culture of the self. In the Californian cult of the self, one is supposed to discover the true self, to separate it from that which might obscure or alienate it, to decipher its truth thanks to psychology or psychoanalytic science, which is supposed to be able to tell you what your true self is. Therefore I would say that this ancient culture of the self and the California cult of the self are diametrically opposed.[22]

Although unquestionably true of certain 'practices of the self', this is a very sweeping generalization: what Foucault sees in his metaphorical state of 'California' is only a caricature of practices attempting to encourage a more authentic form of 'subjectivity'. The modern self, the self which appears in the metaphysical texts of modernity, is a self deeply divided, a self in which reason is split off from feeling, from sensibility, and from the innate wisdom of the body. It is also a self moved by the will to dominate. Even our self-knowledge must assume this character: the self exists in self-mastery and in the self-possession of immediately certain knowledge.

If we are ever to develop ourselves beyond this historical self, much experimentation is necessary. Because his attention was captured by the more degenerate practices of the self, which tend, unfortunately, to be more visible, Foucault was blind to the existence of practices which are genuinely individuating, emancipatory, and socially deconstructive. But it would take more than keen observation to see such practices. One would also need a theoretical understanding different from that which prevails, even today, in Freudian psychoanalysis. For, as Eugene Gendlin has argued very persuasively, psychoanalysis is committed to a psychology of the ego, and it therefore can envision no development beyond the social adaptation of an ego-logical subjectivity.[23] From this standpoint, practices of the self which are not centered in the ego, not bound to the traditional social ego and its linear sense of time and development, can only appear as distinctively regressive, i.e., either infantile or psychotic.

Because Foucault never questioned the Freudian theory of the ego, he could not theoretically envision the possibility of practices of the self in which a postmodern character could begin to emerge. I think it is significant, though, that, in what turned out to be the last phase of his work, Foucault reached the conclusion, which he formulated in 'Why Study Power? The Question of the Subject', that 'We have to promote new forms of subjectivity.'[24] I think he is right; and that is why I conceive this study on vision as a contribution to our thinking about practices of the self in a postmodern world.

THE FUTURE OF HUMANISM

According to Foucault,

> humanism is based on the desire to change the ideological system without altering institutions; and

> reformers wish to change the institutions without touching the ideological system.[25]

If this is 'humanism', then Foucault is right to repudiate it. Our ideological system needs to be changed; but the institutions and practices which support this system also need to be questioned and changed. The fact that this study is formulated for the most part in the language of phenomenological psychology should not be taken to imply that I embrace the kind of humanism against which Foucault is struggling. I am arguing not only that vision is a capacity, but also that it is a 'practice of the self'. For me, this means that our visionary being is a capacity the development of which is inseparable from the institutions of our life-world, and that, insofar as our potentiality-for-being is an *ontological* potentiality which calls for a development that *exceeds* the ontical forms of everyday life – the historical structure, for example, of subject and object, the historical development of our visionary being will require fundamental changes in the practices and institutions which significantly determine its character.

Heidegger works with a conception of 'humanism' very different from Foucault's.[26] Heidegger mainly concentrates on that aspect or dimension of humanism which Foucault would call an 'ideological system'; but his critical writings on technology and its transformation of our life-world make it clear that Heidegger certainly sees the pervasiveness of 'humanism' in its organizing of our practices and institutions. Considered, then, as an ideological system, 'humanism' represents, for Heidegger, the vision of modernity: a vision for which 'Man' is the absolute centre and measure of all being; a vision which has seen the arising of omnipotent 'Man' and the 'death' of God; a vision for which we human beings, heirs to God's power, become the 'lords of the earth'. For Heidegger, looking at the modern world in relation to the history of Being, the tradition of humanism continues a metaphysics of the will, of power, of power as the will to master and dominate: a metaphysics which can see the Being of beings only in terms of will, power. Heidegger formulates this conception, this understanding of the tradition of humanism on the basis of his reading of Nietzsche. He reads this conception of the tradition in Nietzsche's writings, locating it, for example, in the vision of Zarathustra.

Heidegger believes that we can now see, in the technocratic world of advanced technology, the concealed character of this will, and he argues that its part in modern history has been aggressive, violent, and extremely destructive. Thus he fears that, in the end, this will could turn even against itself in a paroxysm of total self-destructiveness. Since, for Heidegger, this self-destructiveness of the will is the essence of nihilism, it becomes apparent that nihilism is now circulating, like a fatal toxin, in the very heart of humanism.

So Heidegger sees the destructive character of humanism: sees its inhumanity, its violence, its complicity in the terror of an instrumentalized reason, its responsibility for reification and totalization. However, he looks into the danger more deeply than does Nietzsche, and therefore also sees within it a 'saving power'. Instead of abandoning the tradition of humanism altogether, Heidegger attempts to make visible the still unbroken spirit which survives within this tradition: something from which we could perhaps draw inspiration, something to provide hints of wisdom for our response to nihilism. This spirit, which he can only articulate with the most painful awareness that further thought is imperative, can be evoked, provisionally, in a single word. That word is 'care': the essence of humanism is not will; it consists, rather, in our capacity for caring. We might say that humanism consists in caring for the value, the dignity, of 'Man'. But that, of course, immediately provokes us to consider what this means, if it does not mean the willfulness of the Nietzschean 'lord', the type of human proclaimed in Nietzsche's 'humanism' for the future.

Nietzsche's 'humanism', however, the 'humanism' we of today are beginning to see as problematic, was formulated in response to the danger of nihilism. But since Nietzsche, as Heidegger was the first to understand, did not comprehend – and perhaps, at that time, could not have been expected to comprehend – the historically concealed *essence* of nihilism, the 'humanism' he proposed in response could only contribute to the raging of nihilism and deepen the historical crisis. Although he saw more than any of his contemporaries, Nietzsche saw only the signs and symptoms of nihilism; he saw only the historical actuality, not what we might call the 'essence'. Thus, for him, nihilism consists in the fact that, with the 'death' of God, we are bereft of value and meaning. Since God was the source of all our values and the ground of all possible meaning, the 'death' of God leaves us without any values and any ultimate meaning in life. Nietzsche's 'humanism' consequently calls for a new hierarchy of values. It summons us to create our own values and become the ground of our own meanings. It summons us to appropriate for ourselves the image we projected and alienated from ourselves. It summons us to appropriate the power, the omnipotence, of our dying God.

According to Heidegger, Nietzsche's perception of the signs and symptoms is right; but the essence of nihilism is tragically misunderstood. For Heidegger, the essence of nihilism consists in the negation of Being. And this means the negation of openness, of dimensionality: 'the dimensionality which Being is'.[27] Therefore, Nietzsche's 'humanism' is itself a form of nihilism, since the will to power, understood not as a simple 'affirmation of life,' but rather as the will to master and dominate, is an 'assault' upon beings in forgetfulness of Being, their dimensionality. Nietzsche's 'humanism' represents the final, and potentially most self-destructive moment in the history of metaphysics. Heidegger's radicalized conception of a new 'humanism' grounded in 'care' can only be understood in the context of this critique of Nietzsche.

For Heidegger, 'care' is the proper dignity of 'Man'. 'Care', though, is an attitude of openness: in caring, we open to. . . . Thus, it is essentially different from the attitudes involved in the will to power. In caring for Being, we care for ourselves, for our own dimensionality. 'Humanism' must be a form of caring in which caring for ourselves is also caring for Being: keeping the question of Being alive in the vigilant questioning of ourselves.

Let us now see whether we can find some points of agreement, points around which we may be gathered for further thought and conversation. The points are very general and need to be specified more concretely. We may want to develop them in different ways; but at least we may proceed with some shared discursive ground. Human beings are said to 'exist'. 'To exist' is to be thrown into the openness of Being, cast into an open field of existential possibilities. (This historical moment appears, archetypally, in the myths of Paradise and exile, which represent, for Western culture, an archaic version of our humanism.) To exist is to stand in – to be in – this openness: to *live* in a world open to that dimensionality:

> The 'being' of the *Da*, and only it, has the fundamental character of ek-sistence, that is, of an ecstatic inherence in the truth of Being. The ecstatic essence of man consists in ek-sistence. . . .[28]

If the 'ecstatic essence of man' consists in 'ek-sistence', then we may say that 'humanism' consists in a vigilant, self-critical attitude of caring in relation to the individual and self-individuating realization of this 'essence', and in relation to the various conditions of the world in which such an 'essence' is to be most deeply, most satisfyingly fulfilled. Constant vigilance in questioning, focused on the world of our daily life, is necessary for this practice of caring, because, as Heidegger says,

> if one understands humanism in general as a concern that man become free for his humanity and find his worth in it, then humanism differs according to one's conception of the 'freedom' and 'nature' of man. So too are there various paths towards the realization of such conceptions.[29]

Heidegger's 'humanism' is radically open: it places, it situates, it releases us, as human beings, in an openness-to-Being which is radically decentering, radically unsettling. The ego, the *ego cogito* of modern metaphysics, cannot let itself be open to (the question of) Being without being decentred, cast out, in a kind of exile, into the dimensionality of a wider, more open field. Nor can 'Man' remain standing as the sole measure and ground, in a sense which tolerates false pride, intolerance of difference, neglect of the ecology of the earth, and totalitarianism. Heidegger's new version of 'humanism' is therefore in opposition to the tradition of 'humanism', and it is, in this sense, a kind of 'anti-humanism':

> However different these traditional forms of humanism may be in purpose and in principle, in the mode and means of their respective realizations, and in the form of their teaching, they nonetheless all agree in this, that the *humanitas* of *homo humanus* is determined with regard to *an already established interpretation* of nature, history, world, and the ground of the world, that is, of beings as a whole.[30]

And he continues, arguing that,

> Every [version of] humanism [in our tradition] is either grounded in a metaphysics or is itself made to be one. Every determination of the essence of man which *already presupposes an interpretation of beings* without asking about the truth of Being, whether knowingly or not, is metaphysical. . . . Accordingly, every [conception of] humanism [coming out of our tradition] remains metaphysical. In defining the humanity of man, humanism not only does not ask about the relation of Being to the essence of man; because of its metaphysical origin, humanism even impedes the question by neither recognizing nor understanding it.[31]

The *Dasein*, the human being, is that being, the only being, for whom to be is an open question. Thus Heidegger sees the 'humanity' of human being to consist essentially in being and remaining open to the question of Being.. Humanism must care for our humanity by keeping us open to this questioning; it must care for our humanity as an endowment to be developed in relation to this question. Humanism must always continue to question our interpretations in the light of our historical experience with Being. Humanism must, at the very least, take care that the dimensionality of Being within which we live, the world within which our human being unfolds in its time, is not closed off in ways that even our present experience tells us are destructive.

We need to be able to see how a concealed nihilism has always been at work in the humanism of our tradition, and how its present visibility, its visibility at this time and juncture, challenges the vision of humanity around which many people, communities and nations have for a long time been gathered. For both Nietzsche and Heidegger, the nihilism promoted by the modern vision of reason makes it necessary for us to begin a radical critique of the history of metaphysics, with special attention, perhaps, to the 'subject' of perception, desire, action, and knowledge. Our study, here, will attempt to continue their project. But we shall be particularly attentive, in our reading of Heidegger, to opportunities for translating this critique into questions that concern our visionary being as a practice of the self, a practice through which our humanity, the dimensions of our character as human beings, may be ontologically developed.

Postmodernism begins when the nihilism of the modern world is seriously perceived, and when the vision of reason that brought this world into being is no longer permitted to rule unchallenged. But I think Heidegger is right in saying that only a recollection of Being can protect the dimensionality of our being. What that means is that, if our tradition of humanism is to take care of our humanity, it must become, in a way that I hope this book can make visible, a recollection of Being.

STANDPOINTS AND VIEWPOINTS

In *The Body's Recollection of Being,* in two chapters called 'Taking the Measure in Stride' and 'The Ground and its Poetizing', I proposed an interpretation of our relationship to the ground which involves working with our body of experience, and with an emerging body of understanding, in order to think the possibility of a ground, a principle of the ground, not totally determined by our tradition of metaphysics. Since this tradition has been rejecting of the body, and therefore the body's experience of being grounded, even when it has not actually excluded it from its discourse, this interpretation would seem to promise some way of moving beyond the metaphysical enframing of the ground which takes place in the modern world.

Here I would like to take up the problematizing of standpoints and viewpoints: another question which positions our thinking very precariously on the boundary between modernity and the dawn of a different age. The problematizing of standpoints and viewpoints – in brief, the question of relativism – is an essential part of the postmodern critique of humanism and its vision of rationality. In its most extreme form, relativism does not deconstruct dogmatic positions; it destroys: it undermines all standpoints and calls all viewpoints blind. It becomes totally self-defeating; it offers no alternative to despair, to nihilism.

In *The Marriage of Heaven and Hell,* William Blake affirmed his belief that,

> If the doors of perception were cleansed, every thing would appear to man as it is, infinite./For man has

> closed himself up, till he sees all things thro' narrow chinks of his cavern.[32]

Why do we find it so difficult, today, to believe in this possibility, this opening of vision? Are we compelled to continue seeing things through narrow chinks?

Merleau-Ponty writes that, 'With the first vision, . . . there is initiation, . . . the opening of a dimension that can never again be closed.'[33] If this dimension has been opened up, why does it seem to be closing? What accounts for the closure?

In *What Is Called Thinking?*, Heidegger observes that 'The Being of beings is the most [pervasively] apparent; and yet, we normally do not see it – and if we do, only with difficulty.'[34] This formulates the problematic of visionary being in another way. But we should note that difficulty is one thing and impossibility quite another. In 'The Question Concerning Technology', Heidegger asks: 'Might not an adequate look into what enframing [*das Ge-stell*] is, as a destining of revealing, bring the upsurgence of the saving power into appearance?'[35] Perhaps. But first, before we can begin to see an answer, we need to know something more about an 'adequate look'. We need to give our visionary capacity the gift of thought. Heidegger begins this project:

> Therefore we must consider now . . . in what respect the saving power does most profoundly take root and thence thrive even where the extreme danger lies – in the holding sway of enframing. In order to consider this, it is necessary, as a last step upon our way, to look with yet clearer eyes into the danger.[36]

Heidegger might have learned some of this courage from Nietzsche, who wrote, in his essay 'On the Uses and Disadvantages of History for Life', that,

> It is only through such truthfulness that the distress, the inner misery, of modern man will come to light, and that, in place of that anxious concealment . . . a culture [will emerge] which corresponds to real needs and does not . . . deceive itself as to those needs. . . .[37]

Since our needs are constituted, are brought forth, in response to the danger we experience, Heidegger sought to put his thinking in touch with these needs – real needs, but also ontological, needs constitutive, that is, of our very being, and consequently deeply concealed. He attempted to articulate, in opposition to the prevailing metaphysical understanding of Being, an understanding he wanted to call 'primordial': an experience with Being in its more open dimensionality; an experience, therefore, from out of which our deepest needs would come. But, as he soon realized, he could not avoid the skeptical problematic. Thus, glancing back over the earlier pages, he asks, in *Being and Time*:

> Are we entitled to the claim that in characterizing Dasein ontologically as care we have given a primordial interpretation of this entity? By what criterion is the existential analytic of Dasein to be assessed in regard to its primordiality, or the lack of it? What, indeed, do we mean by the 'primordiality' of an ontological interpretation?[38]

Heidegger confronted this problem and let it suggest a methodological principle, a principle, in fact, which he announced much earlier in the book:

> In any investigation in this field, where 'the thing itself is deeply veiled', one must take pains not to overestimate the results. For in such an inquiry one is constantly compelled to face the possibility of disclosing an even more primordial and more universal horizon from which we may draw the answer to the question, 'What is Being?'[39]

The 'primordial' is only provisional; not final, not absolute. By avoiding ontological dogmatism, Heidegger believes he can avoid the extreme skeptical abrogation of his claim. This move is reminiscent, in some respects, of Kant's well-known architectonic strategy in the first *Critique*: in order to protect the rational claims he wants to make, he repudiates the wild pretensions of speculative metaphysics. Does this strategy succeed in protecting the ontological claim in Heidegger's interpretation? I think it does.

Heidegger emphasizes, however, that, in order for this strategy to work, the interpretation must be spelled out very concretely, so that it will really engage, really touch our daily lives. A 'primordial' interpretation can claim to be true – in this existentially relative sense, of course – only if it makes an ontological *difference* in our lives:

> 'Existence' means a potentiality-for-Being – but one which is authentic. As long as the existential structure of an authentic potentiality-for-being has not been brought into the idea of existence, the foresight by which an existential interpretation is guided will lack primordiality.[40]

So Heidegger avoids the dogmatic rationalism of metaphysics, the viewpoint which claims to see things *sub specie aeternitatis*, sharing what Nietzsche describes as a 'profound aversion to reposing once and for all in any sort of totalized view of the world'.[41] But if we relinquish the possibility of a totalized view, are we then cornered in the vicious relativism of our subjectivity? Are dogmatism and extreme relativism our only alternatives? Can everything we think and say about our present historical situation be reduced to the narrowness of viewpoint constitutive of that very situation? 'Yet how are we to catch sight', Nietzsche asks, 'of the perspective itself?'[42]

Heidegger's discussion of Nietzsche's idea of an 'eternal recurrence' may usefully be read as a sustained meditation on methodology, on perspectivism, relativism, and foundationalism:

> In the second edition of *The Gay Science*, published in 1887, Nietzsche writes (number 374): 'We cannot see around our own corner'. Here man is grasped and is designated as a veritable Little Jack Horner [*als Eckensteher*]. Thus we find a clear expression of the fact that everything that is accessible in any way is encompassed within a particular corner, a clear expression and acknowledgement of the fact that the humanization [i.e., relativization] of all things is unavoidable in every single step that thought takes. . . . [T]he intention to put out of action all humanizing tendencies in our thoughts on the world's essence cannot endure side by side with acknowledgement of mankind's Little Jack Horner essence [*Eckensteherwesen des Menschen*]. If this

> particular intention is held to be practicable, then man would have to get a grip on the world's essence from a location outside of every corner; he would have to occupy something like a standpoint of standpointlessness.[43]

In the next paragraph, however, Heidegger revisions the entire problematic, turning a fault into an exercise of virtue. He neither overlooks the problem nor indulges himself in wishful speculation; rather, he accepts the relativity and regards it as an 'assignment' of his responsibility as a thinker:

> . . . we still have scholars today who busy themselves with philosophy and who consider freedom-from-every-standpoint not to be a standpoint, as though such freedom did not depend upon those very standpoints. These curious attempts to flee from one's own shadow we may leave to themselves, since discussion of them yields no tangible results. Yet we must heed one thing: this standpoint of freedom-from-standpoints is of the opinion that it has overcome the onesidedness and bias of prior philosophy, which always was, and is, defined by its standpoints. However, the standpoint of standpointlessness represents no overcoming. In truth it is the extreme consequence, affirmation, and final stage of that opinion concerning philosophy which locates all philosophy extrinsically in standpoints that are ultimately right in front of us, standpoints whose onesidedness we can try to bring into equilibrium. We do not alleviate the ostensible damage and danger which we fear in the fact that philosophy is located in a particular place – such location being the essential and indispensable legacy of every philosophy – by denying and repudiating the fact; we alleviate the danger only by thinking through and grasping the indigenous character of philosophy . . . , posing anew the question concerning the essence of truth and the essence of human Dasein, and by elaborating a radically new response to that question.[44]

Since we cannot avoid taking a stand and assuming a

viewpoint, we must attempt to formulate a clearsighted recognition of this fact. It is necessary to recognize that we cannot think from a position which is somehow not really a position. To be everywhere is to be nowhere. To be is to be somewhere: to be *da*: to be positioned. In this sense, we are indeed cornered. But does this mean that we are framed, that we can know nothing? If we must know the totality – the whole in the sense of 'totality' – in order to know the corner where we happen to be, then we would not be able to know even our own little corner. Heidegger sees no point in such an extreme requirement for knowledge. (Wittgenstein, with whom I also agree, sees questions of truth and knowledge to be meaningful only in relation to their functioning in the various practices of everyday life.) For Heidegger, we can at least try to 'know which corner is defined in its place'.[45] In other words, we need to become more familiar with, and more knowledgeable about, our own location, our present position: we need to question it, explore it, actively but critically engage with it. In *living* our place, we can try to understand it better. We are limited to our corner, limited in regard to what we can know and understand of it; nevertheless, we can endeavour to familiarize ourselves with its contours and perimeters; we can develop our sense of what is not working, not right, not good: we can deepen, and render more articulate, our intuitive sense of what is most needful in our time, and of what needs to be changed. And we can then commit ourselves to the project, the task, which seems most appropriate, always striving to maintain an open, experimental attitude. Thus we can also, at the same time, learn more about the world *outside* our corner, for, as Heidegger says,

> the essentiality of the corner is defined by the originality and the breadth in which being as a whole is experienced and grasped – with a view to its sole decisive aspect, that of Being.[46]

We cannot escape the fact that our thinking will always be some form of 'humanism'. After all, we, the ones who are thinking, are (called) 'human beings'. But it does not follow from this that our thinking must posit our historical interpretation of the 'human' as the absolute centre, measure, and ground of all beings. Nor does it follow that we must reify and totalize the understanding we think we have achieved. Heidegger concedes that we cannot completely overcome, cannot entirely break with, our tradition of humanism. As he says, there is no way around the fact that 'the essential definition of human beings is executed by human beings.'[47] To this extent, even the most revolutionary thinking will always be a continuation of 'humanism'. Nevertheless, the question is: what is continued and what left behind or changed? More specifically, 'the question remains as to whether the essential definition of human being humanizes or dehumanizes it.'[48] In other words,

> the definition of human being always and everywhere remains an affair of human beings and to that extent is human; but it may be that the definition itself, its truth, elevates human being beyond itself and thereby dehumanizes, in that way ascribing even to the human execution of the essential definition of man a different essence.[49]

Both 'humanism' and 'dehumanizing' are ambiguous words; they may each refer to regressive and progressive change: either a reproduction of the prevailing nihilism or a movement of amelioration. 'Humanism' must be purged of its oppressive rationality, of all that 'dehumanizes' us by making us less than what we are capable of becoming in fulfillment of our individual and collective potential. But this means that 'humanism' must become a vigilant questioning of all dogmatic interpretations of 'the human', of this way or kind of being; of the being we call 'human'. In this sense, it may be said that our humanism must become *more* 'dehumanizing', shaking us loose from socially institutionalized and culturally sedimented conceptions of what it is to be 'human' – of who we *are* as 'human beings'. If we are to be free for the task of engaging ourselves in a progressive process of historical change, we must commit ourselves to being 'dehumanized', separating ourselves as much as possible from the negative aspects of the prevailing interpretation of our 'humanity'.

The problematic of relativism – the relativism we now see inherent in our tradition of humanism, for example – obliges us to recognize our limitations; but our limitations need to be understood, also, as opportunities for questioning, reinterpreting, experimenting, and self-overcoming. Thus, I concur with what Merleau-Ponty has to say in *Humanism and Terror*: '. . . the reign of universal reason is problematic, ... reason like liberty has to be made in a world not predestined to it. . . .'[50] When we understand, as he says, that 'the present and the future are not the object of a science but [the product] of construction or action',[51] then we see that we have no viable alternative but to take responsibility for our corner, our situation, despite its ambiguity, despite its paradox; and that we must make our decisions, make our commitments, take our stands, despite the risks and the anxieties – equally moral and epistemological – which they cause in us. Of course, we should pay attention to these anxieties, for they can keep us honest, vigilant, tolerant of difference, open-minded. If we can remain open, experimental, then it is less likely that our limitations will mislead us into the kind of violence, the kind of inhumanity, which inheres in the dogmatic humanisms of our past. 'Once humanism attempts to fulfill itself with any [totalized] consistency, it becomes transformed into its opposite, namely, into violence'.[52]

Modernity universalized an historically rooted rationality and understanding, but without reflexive understanding of this. We now understand, we who live with a postmodern consciousness, that our understanding of ourselves and our world *is* an historically situated, culturally relative understanding, and we have therefore turned what the tradition of humanism intolerantly thought of as 'universal' into a thoroughly situated understanding. Taking a step in the right direction, we have relativized and particularized the 'universality' we inherited from our cultural past. But this particularization, rooting us in our particular world, has assumed an extreme form which is ultimately no more tenable than its equally extreme opposite. Whether self-defeating and incapacitating, as in liberalism, or self-glorifying and totalitarian, as in National Socialism, it seems to call, now, for an historically different, differently achieved universality. The question is: How can we universalize a particular understanding, the understanding which arises in our corner, without imposing this interpretation on those who inhabit a different particularity?

I think that Habermas is right in arguing that universalization must take place in a context where there is free and open *communication* among the different positions: that universality is a vision whose realization is to be debated;

not assumed, not imposed. The vision must be one which gathers us into its embrace through an enriching sharing of standpoints and viewpoints, through gestures of respect and reciprocity, through an effort to see other points of view and consider other positions.

Since we cannot avoid being politically engaged in the events of our historical situation and cannot avoid co-existing with a multiplicity of different viewpoints, we need to think how we can take part in the life of our world without repeating the tragic errancy of traditional humanism. This means that we must acknowledge our finitude, our particularity, our situated perspective: the reality of error. This in turn would make it possible for us to be more open to criticism, more open to challenges, more open to difference, more open to communication. No matter how hard we try, we cannot avoid all error; but we can at least strive to avoid the more fundamental error of not being really open to the constant questioning of our position in relation to other points of view.

It will be our concern, in this book, to spell out some critical reflections on the world of our visionary being. Our task, then, is to rethink the prevailing vision of humanism . . . and the dehumanizing inhumanity, the culturally epidemic psychopathology, which rages in our vision. Our vision, here, like the other visions of humanism before it, is ineluctably only a viewpoint. If we can remember this, then we may perhaps begin to avoid some of the reifications and totalizations at work in Western metaphysics. Critical thinking is imperative; but so also is a basic trust in our reflectively lived-in experience: how we experience our historical situation. This has to be our starting point.

The experiential focus of this book is not meant to be 'subjective', working merely with our 'inwardness'. We cannot avoid referring to our 'experience' any more than we, being the kind of beings we are, can avoid the humanism in all our visions. But I would like, as much as possible, to disentangle the word 'experience' from its traditional metaphysical interpretation and retrieve its creative, enabling, life-affirming potential. I submit that what we need to do is say very clearly, as carefully, as thoughtfully, and as accurately as we can, just how things seem to us from the standpoint and viewpoint of our corner. This, the saying of our *Befindlichkeit*, is the task I assign to our hermeneutical phenomenology, whose method trains us in the discipline of critical perception: the meticulous, exacting discipline of attention to our life-world, to what we are experiencing in the course of our daily lives. It will become apparent that the use we shall make of this method is a critical one – that we shall be drawing upon the inherently critical power of a method which teaches us to see what previously was not to be seen. In particular, we shall give hermeneutical phenomenology a diagnostic, and therefore potentially emancipatory function, using its powers of awareness to bring to light, through the channels of our experience, the sufferings and needs of our time.

It is surely a matter of significance, a matter not easily dismissed, that Heidegger thought it necessary to call attention to our historical sufferings and needs, and that he chose to put them at the very heart of his writings on the history of metaphysics: in his *Introduction to Metaphysics*, for example, and in 'The Overcoming of Metaphysics'. In the discourse on Being which preceded him, these historical experiences were denied recognition.

Speaking personally, and in a way normally excluded by our patriarchal, logocentric tradition of philosophical discourse, I want to say, here, that I experience – that I see – tremendous suffering and need in the world around me. I see it everywhere, daily, in thousand of shapes and forms, thousands of different dimensions. And I deeply *feel* the suffering and need which I see. Often, I see *from out of* this feeling, so that what I see *with* that feeling I could not have seen otherwise, without its informative presence. This more feeling vision changes my relation to the world, changes my participation in the world, and changes too, therefore, the world itself.

I would like to close this section, this preview of our project, with a question from *Humanism and Terror*, in which Merleau-Ponty urges us to remember our finitude of place. He asks: 'What if it were the very essence of history to impute to us responsibilities which are never entirely ours?'[53]

Pre-ontological Understanding

We are historical beings. This means that our understanding of life, and our vision of reason, are historically situated. It also means that our visionary capacity, our visionary endowment, is historically conditioned. But how deeply into our being does this historicity penetrate? Are we *totally* determined by our historical situation?

In *Being and Time,* Heidegger struggled with these questions. Since he believed that our endowment consists in existential possibilities, in a potential-for-being, he saw that our historicity constitutes a problematic which the tradition of humanism did not address. Now, in *Being and Time,* Heidegger argued that,

> The projecting of the understanding has its own possibility – that of developing itself [*sich auszubilden*]. This development of the understanding we call 'interpretation'. . . . In interpretation, understanding does not become something different. It becomes itself. . . . Nor is interpretation the acquiring of information about what is understood; it is rather the working-out of possibilities [already] projected in [our] understanding.[54]

In this final section of the Introduction, I would like us to reflect together on the project Heidegger is proposing here: the working-out of possibilities already projected in the understanding we 'have' of ourselves and our world.

According to Heidegger, 'The destruction of the history of ontology is essentially bound up with the way the question of Being is formulated. . . .'[55] I concur. That is why we will be attempting, in this study, to reformulate the question of Being by defining, as our problematic, an historical relationship between ontology and vision. By examining our experience of the existential realization and fulfillment of our inborn capacity for vision and turning this questioning not only in the direction of the individual, but also in the direction of society and culture, we shall be in a stronger position to accomplish the deconstruction of our prevailing ontology. When the deconstruction of our traditional ontology is articulated as a question addressed to us in terms of our visionary being, it acquires the power of a project for our living, engaging us in our practices, our roles, our habits and routines.

I want to argue that the persuasiveness of extreme historical relativism, the problematic with which Heidegger struggled, but without success, in *Being and Time,* cannot be seriously contested without bringing into the debate the matter of our embodiment. In other words, I want to argue that, if we continue to interpret our understanding of Being in the Cartesian language of consciousness, i.e., in keeping with the intellectualism of our metaphysical tradition, then we cannot adequately defend this ontological understanding against an extreme version of historical relativism. Intellectualism turns the question of historical determinism into a dilemma: either our understanding of Being is totally determined by history, or else it is totally free. Thus I want to suggest that, if we interpret this understanding in terms of our embodiment, a much stronger defense can be spelled out.

It will be recalled that, for Heidegger, our 'endowment' is an unrealized potential which constitutes, for us, an historical task. In *Being and Time,* he writes that,

> We understand this task as one in which, by taking the question of Being as our clew, we are to destroy the traditional content of ancient ontology until we arrive at those primordial experiences in which we achieved our first ways of determining the nature of Being. . . .[56]

Now, what Heidegger is proposing is a *phenomenological* task, and 'primordiality' is therefore a heuristic concept, the meaning of which is *relative* to the depth of our interpretive work with life-world experience. Heidegger is not arguing, here, that we should be searching for a metaphysical origin. What he is saying is that, when we have made contact with experiences we are compelled, after appropriately critical reflection, to call 'primordial', or 'inaugurative', our retrieval of their relatedness to Being makes it possible for us to contribute to the deconstruction of traditional ontology with a much clearer sense of its failings, of how it has 'failed' us. (See Chapter Two, where I argue that the 'child of joy' symbolizes 'those primordial experiences in which we achieved our first ways of determining the nature of Being'.) For the Heidegger who wrote *Being and Time,* the hermeneutical return to 'primordial experiences' meant spelling out, always only provisionally, 'the basic structures of Dasein . . . with explicit orientation towards the problem of Being itself. . . .'[57] In other words, Heidegger begins where it seemed, prior to the *Kehre,* that he must: with what appear to be the 'essential structures' of Dasein's existence. Of course, it is clear to Heidegger from the very beginning that the average 'commonsensical' understanding of our tradition keeps Dasein

> from providing its own guidance, whether in inquiry or in choosing. This holds true – and by no means least – for that understanding which is rooted in Dasein's ownmost being, and for the possibility of developing it – namely, for ontological understanding.[58]

We need, therefore, to work out an interpretation of 'this average understanding of Being' which will bring to light its fundamental deficiencies, its participation in our collective historical 'pathology', and allow us to 'develop the concept of Being' from out of a deeper, more primordial level of experience.[59] If we begin our ontological inquiry with a critical examination of our ontical experience, that is because, as Heidegger says, 'Dasein is ontically "closest" to itself and ontologically farthest. . . .'[60] To which he adds the clause: 'but pre-ontologically it is surely not a stranger'.

The last clause in this passage merits some discussion. Although Heidegger sometimes uses the term 'pre-ontological' to cover *all* that precedes our ontological understanding of Being, i.e., *both* our socially normalized understanding (the 'ontical' in what we might call a 'restricted' sense) *and also* our undeveloped, unacknowledged, and more primordial understanding (the 'pre-ontological' in a correspondingly restricted sense), here, in this passage, he clearly separates out these two equally non-ontological understandings. In terms, then, of our more differentiated schema, we may say that the ontological understanding, as the relatively clear realization and fulfillment of our visionary being, is phenomenologically farthest, whilst the pre-ontological, as that gift (*Es gibt*) of understanding which must be retrieved before we can accede to a genuinely ontological relationship to Being, is somewhat closer; and the ontical (mis)understanding is naturally closest. We begin our ontological meditations with the ontical, the everyday experience of everyone-and-anyone, because, despite its errancy, it is what we can most immediately recognize and understand. But we can get closer to an ontological understanding, and to an ontologically attuned vision, only by making contact with, and retrieving, a more primordial experience of Being: that which we shall be calling 'the pre-ontological', meaning to distinguish it from the ontical, which also, of course, is 'pre-ontological', in the sense that it, too, precedes the phase of our fulfillment in the ontological.

In *Being and Time,* Heidegger contends that

> the question of Being is nothing other than the radicalization of an essential tendency of being which belongs to Dasein itself – the pre-ontological understanding of Being.[61]

Because of the pre-ontological gift of our primordial nature, Heidegger maintains that

> What we seek when we inquire into Being is not something entirely unfamiliar, even if at first we cannot grasp it at all.[62]

In other words:

> Inquiry, as a kind of seeking, must be guided beforehand by what is sought. So the meaning of Being must already be available to us in some way. As we have intimated, we always conduct our activities in an understanding of Being. Out of this understanding arise both the explicit question of the meaning of Being and the tendency that leads us towards its conception. We do not *know* what 'Being' means. But even if we ask, 'What *is* Being?', we keep within an understanding of the 'is', though we are unable to fix conceptually what that 'is' signifies. . . . *But this vague average understanding of Being is still a fact.*[63]

'We do not *know* what "Being" means.' Most of us live, most of the time, within a world experienced in terms of a 'vague, average understanding'. But this understanding – of our 'human nature', our 'endowment', and the potentialities-for-living which it grants to us – is not the only possible understanding. For Heidegger, we are gifted with a *pre-ontological understanding of Being* which contests the prevailing historical understanding. And he therefore appeals to our capacity to get in touch with this understanding and think from out of this contact, in order to gather us into the task of resistance and deconstruction. But, in Heidegger, we will look in vain for any recognition of the body's role in carrying our pre-ontological understanding of Being. Heidegger does not see that the human body is the primordial bearer of this wisdom.

Thanks to our 'gift' (the *Es gibt*) of a corporeal nature always already inherently attuned by Being, that is to say, thanks to our *Befindlichkeit* as *Gestimmtsein*, we 'have' an understanding of Being which is *neither* the normal understanding of anyone-and-everyone *nor* the understanding which Heidegger calls 'ontological'. By grace of our embodiment, we enjoy the possibility of access to an understanding of being which is more primordial, and more radical, than that which constitutes our belongingness to the world of social construction.

For Heidegger, 'Dasein always understanding itself in terms of its existence – in terms of a possibility of itself: to be itself or not be itself.'[64] And it is because of this understanding, however deficient, that he wants to assert that:

> The task of an existential analytic of Dasein has been delineated in advance, as regards both its possibility and its necessity, in Dasein's ontical constitution.[65]

Heidegger says only that the task has been 'delineated in advance'. I want to add that there is a 'pre-ontological understanding of Being' which has been corporeally schematized for us in the *a priori* gift and throw (*Geworfenheit*) of our bodily being. Heidegger's account cannot see this corporeal schematism; his interpretation of the 'advance delineation' does not succeed, here, in breaking away from a very traditional metaphysical formalism, blind, as always, to the gift of our embodiment. His formalism makes it difficult for him to see that, by grace of our embodiment, a primordial gift (*Es gibt*), we are able to carry 'within' us a *felt sense* of 'the meaning of Being' which cannot be identified with the socially constituted (mis)understandings of anyone-and-everyone.

In the present study, Dasein's 'ontical constitution', its 'ontical structure', will be interpreted by reference to an inherent, and therefore primordially operative pre-ontological understanding of our ontological capacity for vision. The familiar character of normal vision points back to an existential structure of possibilities, a pre-ontological understanding, constitutive of our visionary being.[66] Our critical task therefore consists in a process of recollection which begins in the body, retrieving that pre-ontological understanding of Being which, whenever it is recognized as such, is always to be 'found' already schematized for us corporeally, and therefore only in a rudimentary and provisional way, deeply inwrought into the structural flesh of our visionary-being-in-the-world. It must be emphasized, however, that what is 'found' and 'retrieved' is not to be conceptualized metaphysically, i.e., as a complete, fully developed, conceptually clear understanding, a doctrine inscribed like a stencil upon our submissive bodies.

Now, in *Being and Time*, Heidegger argues that Dasein

> is ontically distinguished by the fact that, in its very being, that being is an issue for it. But in that case, this [questioning] is a constitutive state of Dasein's being, and this implies that Dasein, in its being, has a relationship towards that being – a relationship which itself is one of being. And this means further that there is some way in which Dasein understands itself in its being, and that to some degree it does so explicitly. . . . *Understanding of Being is itself a definite characteristic of Dasein's being*. Dasein is ontically distinctive in that it *is* ontological.[67]

Continuing this argument, Heidegger makes a point of distinguishing, here, between (i) our 'being ontological', in the sense that we are always already in relatedness to Being by grace of our pre-ontological understanding, and (ii) our purely theoretical activity of 'developing an ontology'. This distinction is significant:

> So if we should reserve the term 'ontology' for that theoretical inquiry which is explicitly devoted to the meaning of entities, then what we have had in mind in speaking of Dasein's 'being ontological' is to be designated as something 'pre-ontological'. It does not signify simply 'being-ontical', however, but rather 'being in such a way that one has an understanding of Being.'[68]

We shall be interpreting this pre-ontological understanding of Being as a 'gift of nature' always carried by the body of experience, and as a 'bodily felt sense' already implicit in our familiar visionary experience. Is this interpretive strategy justified? We must keep in mind Heidegger's observation that,

> ontologically, moodedness [*Gestimmtsein*] is a primordial kind of being for Dasein, in which Dasein is disclosed to itself *prior* to all cognition and volition, and *beyond* their range of disclosure.[69]

Before any clear theoretical articulation of an ontological understanding, long before the beginning of the discourse on ontology, we are already enjoying, by grace of the 'flesh', the field of Being in which our historically distinctive visionary being arises, a deeply felt sense of the meaning of Being and our incarnate relationship to this meaning. This 'felt sense' is our pre-ontological attunement: our fundamental *Befindlichkeit*, a gift of nature, an ultimately inexplicable cast of the die.

By grace of this initial attunement, we are never completely 'in the dark'. But of course this felt sense *calls for* a deep commitment to questioning and exploring its implicit potential: it needs to be recognized, made explicit, conceptually articulate and clear; but it also needs to be taken up by our existential 'practices' and integrated into our daily life, so that, as a capacity for vision, a way of being human, it can continue to develop and unfold.[70] It is only when we attempt to integrate this bodily felt sense of Being, this pre-ontological understanding, into our daily life that we can *continue* to question this sense, and our understanding of this sense, whilst *also* calling into question, from that deep, primordial, and very radical viewpoint, the way we *normally* live – the way of life, and the understanding of it, which are characteristic of anyone-and-everyone.

In the works which followed *Being and Time*, in the works which followed the *Kehre*, Heidegger no longer refers us to a pre-ontological understanding. This, though understandable, is unfortunate. It is understandable, because, as we know, he felt it necessary in his later thinking to abandon the entire 'analytic of Dasein', a project which he had come to see as dogmatic, i.e., as continuing the metaphysical tradition, in its interpretation of (our experience with) Being.

But, as I read him, Heidegger's conviction that he had no alternative derives from the fact that he did not appreciate the human body as an organ of Being: as the organismic bearer of a pre-ontological understanding; as the primal medium into which this pre-understanding of Being is always first inscribed. Consequently, it seemed to him that the pre-ontological understanding which he affirmed in the 'analytic' of *Being and Time* would be totally determined by the metaphysics of our historical world. But by overlooking the natural body, the wild body of metaphorical existence, he could not see the sense in which our pre-ontological understanding of Being is transcendent, is not totally reducible to the understandings imposed by our historical life. He should not have abandoned this pre-understanding, because, after the *Kehre*, the turning point, the reversal in his thinking, he was looking for a way to interpret his experience of our historical need. As the very word he uses in his essays on the history of metaphysics says, he turned towards this need, this *Not-wendigkeit*, but found it difficult to escape an extreme form of historical relativism, and could not settle the question of ideological delusion. In relation to the body of understanding, the body to the emergence of which his own texts contributed, it must therefore be said that Heidegger remained, despite himself, a faithful son of the patriarchal, logocentric tradition.

Heidegger attempted to 'overcome' the nihilism of Western metaphysics by retelling its history from a critical standpoint and making of this retelling a recollection of Being. I think we should continue this project; but I see a need to make it possible for this recollection to happen 'within', and by grace of, the body of our experience. Thus, in *The Body's Recollection of Being*, I began the task of 'performing' the recollection through the hermeneutical medium of our embodiment. Here, in the present volume, I propose that we continue this project, recollecting the dimensionality of Being in our capacity as visionary beings, beings endowed, by grace of our embodiment, with a primordial vision of Being as a whole.

Because of the functional, developmental importance our project recognizes in the pre-ontological understanding of Being, and because of the crucial role we ascribe to it in the task which lies ahead, I would like to conclude this Introduction with a brief recapitulation of the conception with which we shall be working.

We human beings always find ourselves, as Heidegger says, already thrown, already cast, into an historical world not of our own making. We are inherently historical; as soon as we become aware of history, we find that we have already been shaped by its forces, positioned in an historical corner long before we are old enough to realize it. We can never exist outside history.

However, when we reflect more deeply, and 'look into ourselves', as the tradition would say, we see that we are always endowed with, and already predisposed by, a general existential attunement: an attunement by, and to, the Being of beings as a whole. It is this attunement which constitutes our primordial *Befindlichkeit*; and it is this attunement which we are calling a pre-understanding – a pre-ontological understanding – of Being. This 'understanding' is ours by grace of the fact that we are biological beings, embodied beings, for our bodies are already attuned by Being, attuned from the very beginning. In other words, our bodies are from the outset already unconsciously related to Being. Because we are bodily beings, we have received a primordial understanding of Being: an understanding already encoded in, and carried by, our bodily nature. This understanding, though initially rudimentary and undeveloped, is the most precious dimension of our 'endowment'. It is a potential-for-being we can develop. It is an implicate order deeply encoded within the primordial structures of experience, and it informs all our subsequent world-oriented attunements, the organs of our perceptiveness and responsiveness to the world.

This biological preprogramming is *not* a timeless, eternal, immutable structure; but it *is* transhistorical: a condition which, though it never appears except *in* history, is yet never totally determined by historical conditions alone. It is unknowable, therefore, except insofar as it enters into history; nevertheless, it compels us to recognize that it surpasses our historical knowledge. And we do at least know this much: that it *limits* what we know, what we can think and understand, even about ourselves, and what we can become, what we can make of ourselves. Our pre-ontological understanding of Being is, then, historically relative; and yet, . . . it is not totally determined by this relation to history. It is a 'primordial' and 'transhistorical' understanding, but always channeled through the specific historical conditions into which we are thrown. Insofar as our bodies transcend history, there is also a pre-ontological understanding of Being which transcends the impact of history.

Our pre-ontological understanding is determined from two directions: one is our bodily nature, our animal biology; the other is the historical world in which we live and act. But each of these both limits and enables the determinism of the other: our biology is an endowment, a layout of dispositions and potentialities, which call us to an historical task at the same time that they set limits we cannot transcend, limits on what is historically possible; by the same token, our situatedness in history limits and channels the biological predispositions which it selects for actualization. These two sources of conditioning *intersect* in the lived body, the body of experience, where they become factors which organize the field in which individual and collective freedom, freedom and action, are made possible.

As we can see, however, after we have followed Merleau-Ponty into the intertwining, the chiasm of the flesh, what Heidegger does not work out is the phenomenology, the reflective dynamics, of this intersection in the body of felt experience. Between our 'animal nature' and our distinctive 'rationality', between the body as a machine of desires and the mind as pure consciousness, there is a body of felt experience, actively creating or developing itself in the process of self-examination, self-reflection, and self-understanding. Without *this* body to inform him, Heidegger cannot easily defend his interpretation of our transhistorical pre-understanding of Being against the problem of historical relativism: relativism therefore puts him, as he was forced to acknowledge, in a very difficult corner.

What makes the body of felt experience so very significant is the fact that it is transhistorically informed, always already informed, by a pre-ontological understanding of Being. As Eugene Gendlin argues in his commentary on Heidegger's *What Is A Thing?*: 'A context of meaning [which includes possibilities of action] is projected by the way we are feelingly in our situations'.[71] Thus, I should 'attempt to live from out of my own authentically felt senses; but I can do this only by explicating and elaborating the historically *given* senses I actually already feel and live'.[72] Being thus informed, the body can speak, if we are prepared to listen, with an 'authority' which must be respected. We let it speak in a process Gendlin calls 'explication', drawing on 'what was already comprehended in feeling'. But, as he points out, 'what was already comprehended' is not 'in' the feeling, the felt sense, like a sentence in a book, sap in a tree, or a gift concealed in its wrapping. Explication is a hermeneutical interaction, simultaneously discovering and creating: 'a process of further drawing out and further creating, which, when authentic, expresses my directly felt "thrownness" and creatively develops what I am, i.e., my felt [sense and way of] being in my situations.'[73] (For a detailed demonstration of the body's self-creative capacity when it is lived as a hermeneutical process, see Gendlin's book on *Focusing*.)

Our pre-ontological understanding of Being is the precious gift which the lived body, and only the lived body, can give to thinking. Instead of abandoning his conception of a pre-ontological understanding, Heidegger should rather have spelled out its grounding in the self-developing body of felt experience. But, of course, this would have required of him a willingness to listen to the body and give to it the gift of thought. Indeed, working hermeneutically with the body for the sake of its further historical development, working, in particular, with the body's felt sense of being, of being in relation to Being as a whole, would have required of him a much more radical break with the logocentric tradition than he was prepared to make.[74] Our project in this book, however, depends on our willingness to make precisely this break.

THE EMERGING BODY OF UNDERSTANDING

(5) *The Ontological Body*
This is a hermeneutical body, because (i) it is accessible only through hermeneutical phenomenology and (ii) it is itself hermeneutical, i.e., disclosive of the presencing of Being.

(4) *The Transpersonal Body*
This is our ancestral body, the ancient body of our collective unconscious, that dimension of our bodily being through which we experience our connectedness with all sentient beings, our participation in nature's organic processes, and the cessation of our total identification with the conventional time and space of the socialized ego. Religions use ceremonies and rituals to schematize and bring forth such a body.

(3) *The Ego-logical Body*
This is the civil body, socially constituted in the economy of a body politic. It is personal and interpersonal, and consists in masks, roles, habits, routines and social practices. It is formed through child-rearing practices, education, and participation in social structures. It is the body of *das Man*: the body of anyone-and-everyone.

(2) *The Prepersonal Body*
This body is pre-civil and pre-egological. It is the body of the infant and child: a body which adults still carry within them, however split off it may be; a body which adults can retrieve through memory or a relaxation of defenses, letting it take part in life involuntarily and spontaneously.

(1) *The Primordial Body*
This is the wild body, the dreambody, the animal body, the body of nature, the vegetative body rooted in the earth. This body can only be invoked with the language of metaphors, symbols, stories, legends, fairy-tales, myths, poetry and dreams. This body is both pre-egological and pre-ontological. It carries around with it a dark, implicate pre-understanding of Being: a subsidiary guardian awareness of the meaningfulness of Being.

Interpretation of the Schema

- Development from stage (1) to stage (3) is normal, and is typically 'completed' when the child becomes an adult. Stages (4) and (5), however, represent stages of individual development that require special effort, commitment, and maturity. Stages (1) and (2) are basically biological. Stage (3) is distinctively cultural – a body produced in and by society. The ego-logical body is the body shaped according to the ego's image of itself. But stages (4) and (5) go beyond what society requires. We might call them 'spiritual' stages: stages frequently of concern to religious traditions.
- Normal development (stages 1-3) is always, more or less, a linear progression; but the process of self-development beyond (3) is not: it is essentially hermeneutical, involving a return, a turning *into* the body of experience, to retrieve a present sense of the earlier stages. Beyond (3), it is necessary to go 'backwards' in order to move 'forwards'. Stage (3) is the moment when, for the first time, this return and retrieval become existentially possible.
- Although the development from (1) to (3) is more or less linear, the earlier stages are never lost, never left entirely behind. Even when split off and largely unconscious, with effort they can be sensed and brought to bear on present living. Though *aufgehoben*, sublated, sublimated, they can always be reintegrated into present living, so that their unique sense of being-in-the-world can take part in, and contribute to, further steps of self-development.
- 'Body' and 'self' develop simultaneously and, in the normal case, in the closest interdependence. Just as the ego can develop beyond itself, so can the body. If the self is to grow beyond its identification with the socialized ego, it must work on developing a body not totally determined by the ego's self-image.
- Bodies (1) through (4) are all, in effect, pre-ontological, since they all precede the ontological stage of development. Nevertheless, in general, we will restrict the term 'pre-ontological' to stages (1) and (2), which are also in fact pre-ontic, in that they precede the fully *ontic* stage, which is (3).

- Bodies (1) and (2) are 'transhistorical' in the sense that they are biological, and therefore not totally determined by historical conditions, even though they are never to be found apart from, or outside of, a social history. Stage (4) is 'transhistorical' too, but in a different sense: it is certainly not a biological endowment; on the contrary, it is a difficult spiritual achievement; nevertheless, the body which has achieved stage (4) enjoys a different understanding, a different sense, of being-in-time. It is 'transhistorical' because there is a sense of living in the wholeness of time, a sense of not being totally measured by the passage of time in a linear progression from past to future. The transhistorical body of stage (4) is an accomplishment that requires a hermeneutical retrieval of the experience with being-in-time carried by the body of stages (1) and (2). This early experience with time is normally suppressed, split off, *aufgehoben*, in the development from (2) to (3).
- When I say that the ontological body is hermeneutical in the sense that it is 'disclosive of the presencing of Being', I mean to be referring to the achievement of a homology (*homologein*) between the comportment of the body as laying-down and gathering (the meanings Heidegger gives to the Greek *legein, logos*) and the presencing of Being as *itself* a primordial laying-down and gathering (*Legein, Logos*) of the visionary field. This homology takes place in the world of everyday living, i.e., the disclosing of Being takes place in the world, in the course of our daily life. Thus, for example, our vision becomes the organ of an 'ontological body' insofar as the character of our gaze is one which is open and responsive to, and shows appropriate care for, the dimensionality of the things it sees; maintains a vigilant openness to the invisible, the unseen; and cares for the invisible as much as for that which is, or can be, made visible.

CHAPTER 1

Das Ge-stell: The Empire of Everyday Seeing

Opening Conversation

(1) . . . A twofold frame of body and mind./
The state to which I now allude was one/
In which the eye was master of the heart,/
When that which is in every stage of life/
The most despotic of our senses gained/
Such strength in me as often held my mind/
In absolute dominion.

William Wordsworth, *The Prelude*[1]

(2) But none of them owns the landscape. There is a property in the horizon which no man has but he whose eye can integrate all the parts, that is, the poet.

Ralph Waldo Emerson, 'Nature'[2]

(3) . . . the still covetous vision of things . . .

Heidegger, 'What Are Poets For?'[3]

(4) Innumerable, pitiless, passionless eyes/
Cold fires yet with power to burn and brand/
His nothingness into man.

William Blake (The subject is Isaac Newton)[4]

(5) In these [existential moods of everydayness], Dasein becomes blind to itself, the environment with which it is concerned *veils* itself, the circumspection of concern gets led astray. States-of-mind are so far from being reflected upon, that precisely what they do is to assail Dasein in its unreflecting devotion to the 'world' with which it is occupied and on which it expends itself.

Heidegger, *Being and Time*[5]

(6) But to look in order to know, to show in order to teach, is not this a tacit form of violence, all the more abusive for its silence, upon a sick body that demands to be comforted, not displayed? Can pain be a spectacle?

Michel Foucault, *The Birth of the Clinic: An Archeology of Medical Perception*[6]

(7) The correlation of consciousness with masculinity culminates in the development of science, as an attempt by the masculine spirit to emancipate itself from the power of the unconscious. Wherever science appears it breaks up the original character of the world, which was filled with unconscious projections. Thus, stripped of projection, the world becomes objective, a scientific construction of the mind. In contrast to the original unconscious and the illusory world corresponding to it, this objective world is now *viewed as* the only reality. In this way, under the continual tutelage of the discriminative, masculine spirit, ever searching for laws and principles, the 'reality principle' comes to be represented by men.

Erich Neumann, *The Origins and History of Consciousness*[7]

(8) This dominance of masculinity, which is of crucial importance for the position of the female in patriarchal societies, determines the spiritual development of Western humanity.

Neumann, *The Origins and History of Consciousness*[8]

Part I

A Certain Blindness

In his *Philosophical Investigations*, Wittgenstein has his own way of registering his dissatisfaction with metaphysics. The therapy he recommends for this peculiar pathology is communicated with simple force: 'Don't think, but look.'[9] This, however, it is much easier to say than to practice. Be this as it may, I am quite sure that in life itself, and not only in metaphysics, thoughtful looking is exceedingly uncommon. And it might well be argued, in fact, that such comportment is becoming increasingly infrequent as enframing takes control of our world.

In Akenfield, an English village where the process of modernization, and therefore enframing, have only recently, and only very slowly, begun to change the way of life familiar to the older villagers, there are still a few people around – smithies, bell-ringers and saddlers, for example – who have observed the historical changes and can speak with an unexpected eloquence about the difference they see and feel:

> The old village people communed with nature but the youngsters don't do this. . . . The old people think deeply. They are great observers. They will walk and see everything. They didn't move far, so their eyes are trained to see the fine detail of a small place. They'll say, 'The beans are a bit higher on the stalk this year. . . .' I help to run the school farm but I'd never notice things like that. The old men can describe exactly how the ploughing turns over in a particular field. They recognize a beauty and it is this which they really worship. Not with words – with their eyes. Will these boys be like this when they are old? I'm just not sure. Nobody is trying to bring it out in them. Nobody says to them, 'This is heritage'. Somebody should be saying to them, 'Let's go and look. . . .'[10]

The villager concludes:

> The young are different today. They have common communications with the world. The old look inward at things we cannot see. The young have a common image.[11]

But there is in any case, i.e., in all of us, a certain blindness: a blindness characteristic of what Heidegger calls 'the everyday perception of mortals'.[12] Eyes which can see may yet be blind; and Oedipus, blinded, discovers a new capacity for vision – the oldest and the deepest. This fact may be striking, but it is not, as it seems, a paradox. For vision is the gift of an existential capacity: a gift we are free to abuse or deny as well as cultivate. What we do with our natural endowment – how we respond to the gift of nature – constitutes the *character* of our vision. Whether, and how, we take up our visionary project: that is the measure, the test, of our character, our development of self.

'Character', says Emerson, 'is this moral order seen through the medium of an individual nature.'[13] But, as character reflects, and makes visible, the quality of the social order in which it inheres, so, conversely, the social order is a field of visibility, which reflects – makes visible – the quality of character that predominantly inhabits it. Moreover, the realizing of our capacity for vision, and above all its visible flowering, can take place only within, and only with the sustained encouragement of, the institutions of the social order, which itself comes into being *as* an articulation of the visible and the invisible. The field of visibility is, and unfolds in, a world of sociality – a world of history, tradition, and political institutions. Vision is nature's gift of a possible *adventure* in the social, or cultural order. It is not only an opportunity for individuation; it is also at the same time a project of responsibility for the social order as a whole. Are we responding, then, responding appropriately, to the visible need for a far-sighted vision of real community? Are we contributing to the institution of a new social order, in which the creative, visionary potential inscribed within the heart of each and every one of us will be greeted with visible kindness, and wisely encouraged to make itself visible? 'Where there is no vision, the people perish.'[14]

For the most part, our vision conforms, conforms to the gaze of a social order which reflects and multiplies our fears, ignorance and passions, and which extinguishes many of the sparks that might otherwise kindle some effort of vision:

> In no case is a *Dasein*, untouched and unseduced by this [normalizing] way in which things have been interpreted, set before the open region of a 'world-in-itself', so that it just beholds what it encounters. The dominance of the public way . . . has already been decisive even for the possibilities of having a mood. . . . The 'they' prescribes one's state-of-mind, and determines what and how one 'sees'.[15]

If, in the ease of its conformity, our vision lets itself become too deeply inscribed in the light of 'public reason' and 'common sense', it is failing to cultivate its capacity. We are responsible for such cultivation because our eyes have *already seen and responded to* the lighting of this potential and cannot entirely forget the claim which transpired.

In 'Der Anfang des abendländischen Denkens', Heidegger once again takes up the question of vision and blindness.[16] The subject of his discussion is the two stories one hears told about Herakleitos as an object of the 'curious' gaze. In the one story, a group of people – spectators – come to see the thinker 'thinking' and are disappointed to find him simply warming himself by a stove. Seeing the people gaping, Herakleitos is reported to have said, 'Even here the gods are present'. We may suppose that, for his visitors, this shining presence was not very visible. Only Herakleitos saw it. In the second story, Herakleitos is found playing with a group of children in the temple of Artemis:

> The resting places of the thinker show each time precisely the opposite of what one expects [to see]. At the oven one is disappointed [*enttäuscht*]; in the precincts of the temple, one is astonished [*überrascht*]. Of the ways and spirit [*Wesen*] of the thinker, or indeed of his thinking, nothing shines forth [*nichts kommt zum Vorschein*]; at least not immediately; at least not for the merely gaping eye of the crowd [*das bloss gaffende Auge der Menge*]. This eye sees only what immediately strikes it: the visibly striking and that alone which is pleasing to it [Heidegger benefits, here, from the meaningful resonance of his native language: *das Augenfällige und das nur ihm Gefällige*]. This eye of the crowd is not inclined to notice what the strikingly visible [*der Augenschein*] indicates beyond itself. The eye of the crowd is in general not practiced in following that which such a showing [*Zeigen*] actually shows. The eye of the crowd is *blind* to the ciphers [*Zeichen*]. What extends beyond the most strikingly visible is regarded by the *polloi* . . . as sheer phantasy [*Phantasterei*] and invention. The crowd, by contrast, adheres to that which it presumes to be 'real' and obviously given. But, by the same token, the eye of the crowd has no regard [*Blick*] for the unexceptional, in which alone the most precious ciphers are concealed. The oven makes manifest

> [*zeigt auf*] the sheer presence of the bread and the fire, and in the fire, its glow and brightness. The 'reasonable person' sees just an oven.

I would like to connect this statement with Nietzsche's criticism of the 'reasonable man'. (His criticism is quoted in the Preface.) In his essay on 'The Uses and Disadvantages of History for Life', Nietzsche asserts that 'there are things he does not see which even a child sees.' Now, old Herakleitos, called 'The Obscure', is *like* a child, a child of nature (*Physis*); and, like a child, and like *Physis* itself, of which he speaks obscuringly, he likes to play with hiding and concealment. His eyes can see what is hidden by the 'reality' of the many. He has not forgotten, not rejected, the vision of his childhood. Is it really surprising that he could see the enchantment of Being where others did not?

As infants, closer to nature than to society, we began life with eyes opened by enchantment. As adults, we tend to conform to the crowd, seeing only *its* reality. This has always been true; not only nowadays, but also in earlier epochs – in the time of Herakleitos. Even then. But there is, nevertheless, a difference between the past and the present. For, the culture of blindness has reproduced and multiplied its blindness. In the present epoch, the blindness of the crowd, a perfectly normal blindness in which we all participate, is steering the machines of our technological rationality and poses an especially perilous and fateful danger to our visionary being. Our everyday vision has brought forth a science and technology which have profoundly changed our world. The historical changes, themselves brought about in a certain blindness, have in turn affected our vision and our capacity for vision. More and more, we are alienated from an exceedingly precious gift of nature: the experience of Being we could *begin* to retrieve and cultivate by making contact, like Herakleitos, with the visionary being of our childhood.

Deep spiritual wisdom, deep enlightenment, lies hidden in the visionary being with which we are gifted in the innateness of childhood. For Herakleitos, the sheer presence of light, the play of light, of Being, is a vision of enchantment: and this is what he sees. But it is much harder for us, living in a society, a world, a reality brought forth from out of so much blindness, to see in that way. The historically given culture in which we live is itself more closed off from the presence of Being and its enchantment of vision. Earlier epochs in our civilization were, like the earlier stages of individual development always are, more open to the enchantment of Being. Thus, we moderns leave behind, along with our childhood, a vision more akin to the older visions of Being which predate the history of our metaphysics. But the two stories about Herakleitos suggest that, if we are ever to break out of the metaphysical determination of our vision, we will need to retrieve, and then cultivate, the sense of visionary being we enjoyed, all too briefly, and for the most part, unconsciously, in our childhood. Herakleitos embodies this paradox.

In *The Dawn of Tantra*, Herbert Guenther argues that, in the developmental experience of the individual, the process of 'transformation' which we call 'growing up' is actually one of 'growing narrowness and frozenness'.[17] In *Being and Time*, Heidegger describes a similar process, and connects its unfolding in the individual with the history of Being: according to his reading of metaphysics, there has been a progressive closure in the dimensionality of Being, and it involves a certain restriction (*Einschränkung*), a certain closure, of our vision and its field of visibility.[18] Later, in his essays on science and technology, Heidegger uses the word 'enframing' (*Gestell*) to designate the character of this historical process as he experiences it in its modern phase. But in the phenomenology of his earlier work, he spells out its 'universal' psychology: in the beginning phase, an experience of wonderment (*thaumazein*); then, gradually, the emergence of curiosity (*Neugier*). In wonder, we stand open to the enchantment of Being and are disposed to let beings be. In curiosity, however, we fall into the structure, the pathological pattern, of the ego-object relationship, and we lose contact with the more open dimensionality of the existential field. The loss occurs because curiosity is the arousing of desire, and it is desire which brings about the more restricted form of vision, which is more interested in the objects brought forth by the lighting of the field than it is open to the enchantment in the presence of the lighting itself. Today, more than ever, desire, technologically manufactured desire, the masculine will to power, is what dominates our vision.

Heidegger points out that,

> The basic state of sight shows itself in a peculiar

> tendency of Being which belongs to everydayness – the tendency towards 'seeing'. We designate this tendency by the term 'curiosity', which characteristically is not confined to seeing, but expresses the tendency towards a peculiar way of letting the world be encountered by us in perception.[19]

This tendency is an inveterate tendency or predisposition to live in a certain 'normal' blindness. We know, from the stories, that Herakleitos experienced its prevalence even in his own culture; but in our culture of today, this tendency is reinforced, for the effects of blindness accumulate, and the world in which we live, a world dominated by enframing, was brought into being with the power of its blindness.

Can this inveterate tendency be reversed? Can we encourage a radically different tendency? In thinking through such possibilities, we shall find that our visionary being is at stake. Our destiny may depend on our response to the historical challenge.

In his study on Anaximander, Heidegger sets the tone for a revolutionary response. Upsetting a powerful assumption, he 'reminds' us that,

> Seeing is [primordially] determined, not by the eye, but by the lighting of Being. Presence within the lighting articulates all the human senses.[20]

If we could actually realize this reversal in our lives together, our vision would become a path of liberation. It would light our way through history.

Part II

The Yielding of the Visible

'Yielding' is a marvellous word; it will facilitate our access into the primordial dimension of our visionary experience. 'Yielding' means giving up and giving over; but it also means allowing, accepting and receiving. In other words, it speaks of the unity or identity of the two, the two as one. That which is yielding, then, is a oneness, a harmony of opposites, which can become two-fold: a giving and a receiving, a gift and the reception of a gift. When the dimension of yielding is entered through our awareness, giving and receiving take place in a different mood, a different spirit, a different space. The giving gives without holding back; it gives, in the end, itself: such giving, giving up even itself, becomes selfless. And it is open to receiving whatever may be given to it. Correspondingly, the reception is one in which the deepest, most thoughtful attention is given to the gift. Even the givenness of the gift, the phenomenon of its givenness, is itself received, given the utmost thought, the utmost attention, the utmost awareness. And this giving of thought is a reception which reciprocates the gift.

The givenness of the given and the conditions of our receptivity are matters which have traditionally figured very prominently in the history of philosophical discourse about perception and knowledge. However, the authoritative texts in this tradition have never given thought to the character, the felt quality, the 'mood' of our reception. Nowhere do these texts consider the givenness of the given as a gift and question our receptivity in this light. Here, in this study, we shall remedy the neglect of our tradition; and we shall do this in a very radical way, so that the two poles, the giving and the receiving, can be experienced in the dimension of their oneness.

'Yielding' names the original 'ecstasy' of giving and receiving – the elemental give and take – which illuminates the vital flesh of the field of visibility. The subject of perception is no more an isolated, self-sufficient individual than is 'the subject of history', about which Merleau-Ponty states that, 'There is an exchange between generalized existence and individual existence, each receiving and giving.'[21]

In his *Phenomenology of Perception*, Merleau-Ponty argues, against the tradition, that,

> Sense experience, thus detached from the affective and motor functions, becomes the mere reception of a quality, and physiologists thought they could follow, from the point of reception to the nervous centres, the projection [i.e., the stamp, the imprint, the impression] of the external world in the living body.[22]

Let us give thought to our experience with receptivity in the realm of vision. We undoubtedly fail to recognize our familiar experience, as we live it, in the physiologist's description of 'the mere reception of a quality'. On the other hand, as we listen to Heidegger's description of a 'greeting', we may feel no less removed:

> The genuine greeting offers to the one greeted the harmony of its own nature.[23]

I am proposing that we interpret Heidegger's words to refer to the experience of perception. This reference is not explicitly proposed by Heidegger himself, but it is an extension which, far from doing violence to the text, significantly specifies and enriches it.

Vision is a form of sensibility: it is our susceptibility to the yielding of illuminations and reflections which inflect or accent the field of the visible; it is a way of *being affected* by what is given. The eyes, it is said, are 'receptive' organs. But *how* – in what spirit, and with what character of feeling – do they receive? How do they greet and welcome? What is the mood, the *Stimmung*, of their relationship to the given? If the beings we behold are given unbidden to our eyes, they must be 'gifts' of light, gifts yielded up by the visible, gifts our eyes could enjoy. But a gift cannot be enjoyed unless it is properly received, accepted with a co-responding gift of appreciation. And it cannot be received or accepted so long as our vision reflects the fact that we are neglecting or denying its being as a gift. What is involved in receiving the gift of the visible? What is involved in giving that gift an appropriate reception? Could it be that, in order to develop our capacity for vision, we need first to return to the dimension of our experience where the giving and the receiving are not split asunder, but rather are felt as one? Could it be that we need to move beyond the dualism of active giving and passive receiving – a dualism which, in our patriarchal culture, is caught up in the antagonism between the masculine and the feminine – in order to make contact with the field of their primordial wholeness?

Western rationality – we should consider, above all, our philosophy and psychology – has reflected for centuries on the giving, the yielding, of the visible; it now calls it 'stimulation' and 'excitation', leaving behind such older, more mechanistic terms as 'imprint', 'trace', and 'impression'. The gift itself, that which is said to be 'given', is then readily called a 'sense datum' or 'stimulus'. What kind of reception therefore, does our rationality give? What kind of *experience* of the visible world does the giving of such a reception manifest? And what takes place, when we respond with aversion or hostility to the presents which the field of visibility yields and presents? What takes place, when our visual reception is not yielding, when it is so *willful* that it cannot give way, and does not visibly give thanks? What takes place, when our reception is *merely* 'passive'? Is it any wonder that our eyes are so often strained, tired, bored, clouded by tears of loss and mourning? When we reflect on our everyday habits of receiving the presence yielded up, already articulate, by the light of the field of visibility, perhaps we should not be greatly surprised that our experience is so often bereft of visible meaning. Mikel Dufrenne tells us that, in the field of visibility, there where vision takes place, an 'apotheosis of the sensuous' erupts into the light, 'appearing in its glory'.[24] If this 'apotheosis' is of the essence, then we need to give much more thought to the way we characteristically *receive* the yielding presence of the visible.

But the act of receiving is always also the act of giving: we *give* a reception to that which comes forth into our field of vision. Thus we say that perception is always giving: a mode of *giving attention to. . . .* Perception gives as well as receives. But how giving, how yielding, is it? How yielding are we, for example, in the way we behold, and thus give attention to, that which the light of visibility yields, and gives over, to our vision? And what is the *character* of our way of giving attention? We need to give much more thought to the way we give our attention to that which shines forth in the field of light.

Speaking from the depth of his experiencing of vision's ontological dimension, Padmasambhava, Tibet's most beloved teacher, joyfully declared:

> Whatever I behold is my Wish-fulfilling Gem.[25]

And he adds, for our benefit:

> I contemplate the Three Jewels *without averting my gaze.*[26]

Padmasambhava's teachings of compassion (*circa* A.D. 775) merit our consideration here, in that they provide some concrete experiential clarifications for Heidegger's assertion that,

> it is Being itself, whose truth will be given over to man when he has overcome himself as [ego-logical] subject, and that means when he no longer [aversively or aggressively] re-presents that which is as an object.[27]

In this passage, Heidegger names the two main concerns of this section, viz., the process of re-presentation and the nature of the subject-object relationship. We now turn to this problematic in order to understand how enframing takes hold of the yielding of the visible and tightens its willful grasp.

To begin, suppose we ask ourselves this question: What kind of 'presence' would we expect to correspond to the visible hostility and, in general, the negative energy, of our vision? In *Psychology as a Human Science*, Amadeo Giorgi maintains that,

> one should be concerned about proper and improper attitudes toward objects and the world, rather than merely objective and subjective attitudes.[28]

And he argues that,

> objects are merely one type of 'presence' to consciousness. For example, we are aware of the spaces between objects, and they are not object-like; we are aware of silences between musical notes and they are not object-like; we are directly aware of other persons in ways that are object-like. Thus, 'presence of objects' is merely one type of presence. But because this type is so obvious [and so tempting], it has received exhaustive analyses, and even worse, it has dominated our thinking so much that other types of presence are understood primarily in terms of object-like presences.[29]

Thus, as he says, 'there are many other presences that are waiting to be classified, if we know how to look!'[30]

Furthermore, 'each type of presence would have its specified attitude in which it could be known maximally and most accurately.'[31]

Paul Valéry reminds us that 'To see is to forget the name of the thing one sees.' In *Russian Formalist Criticism*, Viktor Shklovsky articulates a similar awareness: 'After we see an object several times, we begin to recognize it. The object is *in front of us*, and we know about it; but we do not see it.'[32] There is a seeing which is not a seeing. Most of the time, an ingrained tendency to live by habit and in the projections of desire blocks or obscures our vision: we look, and in that sense see; yet the experience is dull, stale, unfulfilling. We may even be seeing what we want most of all to see; but even in this case, and perhaps precisely because of the intensity of the desire, the experience may seem flat and lifeless. As Merleau-Ponty points out, 'Visual experience . . . pushes objectification further than tactile experience.'[33] Further, too, than hearing. (This point is argued in my forthcoming book on *The Listening Self*.) In fact, the field of visibility yields itself much more readily than do all the other fields of sense to the kind of structuring process which willfully *re-presents* whatever presents itself, so that every presence manifesting in the field of vision is essentially reduced to the ontology of a mere thing.

Heidegger argues that this process of objectification – this enframing – is particularly prevalent in the modern world, where visual experience and its extensions have been dominant. Enframing is a historical manifestation of the 'will to power'. Perhaps its pervasive determination of our world is related to the domination of vision, the most reifying of all our perceptual modalities, as the paradigm for knowledge. The split in our vision between giving and receiving, a split between active and passive modes of being, confirms Heidegger's sense that 'willing determines the nature of modern man'.[34] Later, I shall argue that the character of this determination needs to be understood historically in relation to the domination of patriarchal consciousness and the suppression of the 'feminine principle' of wisdom.

According to Heidegger, this willfulness is responsible for the process of re-presentation. Therefore he insists that,

> The first step [forward] . . . is the step back from the thinking that merely re-presents to the thinking that responds and recalls.[35]

Thinking which 'merely re-presents', however, is *metaphysical* thinking: '. . . starting from what is present, [metaphysical thinking] re-presents it in its presence.'[36] Our particular concern, of course, is with *vision*, insofar as it either embodies such thinking or is capable of *freeing* itself from that body of thinking. We therefore continue our study by focusing on the enframed *vision* of metaphysical thinking.

The re-presenting of the visible in its absolute otherness is a process which takes place as the field of visibility begins to take shape in and around a subject-object structure. The problem with this structure is that it tends to polarize into a situation of opposition, or conflict, and to condense into a rope of pain drawn between two knots of being: the ego-subject which 'sees' and the object which, held tightly in its grasp, is 'seen'. The objectification which re-presents the visible in its absolute otherness is, however, a reflection of the *ego-logical need for security* – a need which reappears in the madness of metaphysics as 'the quest for certainty'. The ego is *always* – and of necessity – attached to the issue of certitude.[37] For the ego – as Freud conclusively demonstrates in *The Ego and the Id* and *Inhibition, Symptom and Anxiety* – is basically nothing more, nothing other than, the product of anxiety – a structure, or system, of defense which, once it is established, at once becomes the *source* of continued anxiety, since the very rigidity of the system intensifies the need for defense. (I would maintain that the global armaments race clearly repeats this same process and visibly reflects the insanity of its essential nature.) To be sure, the infant and child require strong defenses; their experience, otherwise, would be painfully overwhelming. And even the adult needs some measure of defense. Our capacity for openness is always finite. Indeed, at every level of human existence, the field of visibility will undergo an articulation which is also at the same time a delimiting of its luminosity. But our finitude is not what we are attempting to overcome. The problem we feel a need to address, rather, is the tendency for the ego-object structure to become *pathological*, i.e., to be *experienced* as rigid, restrictive, and, in general, as an unwholesome, unsatisfactory situation.

The negative polarization is under way once 'lust' and greed narrow down the focus of vision; under way, also, once extreme anxiety and insecurity inhibit or restrict the receptivity of our response. It is important, therefore, to bear in mind the inherently *aversive and aggressive character* of re-presentation. As the prefix itself informs us, re-presentation is repetition: a process of delaying, or deferring, that which visibly presences. It is a way of *positing at a distance,* so that vision can 'again' take up what presences – but this time on ego's terms. The more restricted and defensive our response, the more the being we have encountered will be *confined* within the limitations of our response. This results in a *reciprocal* subjugation, for the more we *oppose* the visible, the more we *object* to its *own* way of presencing, and insist on viewing it through eyes of anxiety, or eyes that are defensive and *therefore* hostile, the more the visible will negatively conform to that set-up, that enframing – and it will give us *very little* of itself.

Moreover, when we feel threatened by what we have encountered, either because we are *attracted* and fear its loss or because we want to *avoid* that being and are constantly distressed by its 'threat' of presence, the visible field gets caught up in the web of our anxiety, as we ourselves get caught: the whole field, in fact, is transformed, and we within it, so that nothing escapes reduction to the being of an object. Thus, every subject eventually becomes an object, subject to the objectification it produced, since it inevitably suffers the *reflection* of all that destructiveness into which it enframed the visible.

In 'The Age of the World Picture', Heidegger diagnoses the pathology of the process:

> That which is, is no longer that which presences; it is rather that which, in re-presenting, is first set over against, [is] that which stands fixedly over against, [and] which has the character of object [*das Gegenständige*]. Re-presenting is making-stand-over-against, an objectifying that presses forward and masters.[38]

Why, then, should it be surprising that what presences for us in the field of the visible has *its own way* of resisting that pressure – its own way of objecting to that mastering violence? (The word 'master' is significant: it accurately accuses the patriarchy.)[39] Why are we surprised by the *dullness* of our visible world? Why are we perplexed when we cast our eyes into the mirror of the visible and end up with feelings of frustration and disappointment? How could our situation be otherwise?

We need to focus on how our vision yields to attraction and aversion, the two basic patterns of egological attachment, and on how these patterns of response initiate various negative cycles of experience in the realm of vision.[40] We need to give thought, for example, to the ways in which we *stereotype* the visible; and to the ways we avoid the challenge of finding *a fresh response* even to that which is familiar.[41] The re-presentation of a visible presence, fore-closing an experience of its fullness, cannot possibly bring the kind and quality of visual fulfillment we might otherwise enjoy. (This is a problem which Husserl's studies on intentional fulfillment in perception do not attempt to investigate.) For a vision closed by the need to re-present the visible life literally *loses its colour*: whatever our gaze reaches appears to be dull and inert; paltry: not glowing and vibrant. The field of vision will also seem to be lacking in *depth*, and therefore incapable of yielding any visible presence of great significance. (The 'meaningfulness' of the visible is in part a function of its visible depth, while depth is, in turn, a function – at least in part – of shadows and reflections. But the gaze which is tightly enframed will be incapable of the gentle, hovering, balanced attention, and the playfulness, which would elicit that aspect of the visible field.)[42]

In the 'Es gibt', which means 'there is', Heidegger's ear of thought hears 'it gives', the long-silent *truth* of perception. In its field of visibility, Being yields, gives up beings in gifts given to our eyes. The visible deeply *objects* to our habitual objectification; it will not fully give itself, will not wholly yield itself, to our desire. The most extreme evidence in which this is visible appears when we engage in an exercise of intensive staring: 'a fixed staring at something that is purely present-at-hand [*vorhanden*).'[43] In German, the word which we translate as 're-presentation' is *Vorstellung*. Now, this word signifies a gesture of setting down (*stellen*) in front (*vor*), a gesture which corresponds to the 'frontal' ontology of our modern, nihilistic world. I submit that the concealed essence of 're-presentation' begins to appear through this interpretation, and that it is, in a word, *staring*. If not literally true, this analysis is at least true to the 'emotional essence' of our modern visionary experience.

In the authoritative texts of modern philosophy, vision is always re-presented by a straight line of sight, sharply focused, absolutely clear, and fixed on its object, something positioned directly in front. It is true that, in his *Essay*

Towards a New Theory of Vision, Berkeley does not overlook the encompassing *presence* of the field of light.[44] And yet, his paradigm of vision is not essentially distinguishable from a stare. His theory makes questionable assumptions about clarity and focus. And it assumes that what principally *moves* our eyes is the desire to know, and that knowledge is mastery and control. In his theory, vision is already being prepared for its functioning in a technological world. But Berkeley is only an example. A careful reading of Descartes would encounter a similar blindness in his scattering of propositions concerned with vision. How can he believe, for example, that what his vision sees, when he looks out the window, are only hats and coats?

What happens when we stare intensely at something? Instead of clear and distinct perception, blurring and confusion; instead of fulfillment, the eyes lose their sight, veiled in tears; instead of stability and fixation at the far end of the gaze, we find a chaos of shifting, jerking forms, as the object of focus violently tears itself away from the hold of the gaze. The more intensely we struggle to deny, suppress, or retard the natural processes of change; the more we refuse to accept impermanence; the more we impose a metaphysics of constantly available substances, the more violently our world will make visible our self-destruction.

Staring is an attempt to dominate; but, in the end, it always compels us to see spontaneous, uncontrollable changes in the field of visibility: changes that occur whether we will them or not. Thus, if we fix our eyes upon the corner of a building and persist in staring at it, we will find, sooner or later, that the whole building has·begun to move about, shifting in peculiar and 'unnatural' ways. Likewise, if we stare at a fixed point on a distant hill, that hill will start to stretch; and it will rotate 'endlessly' along the horizon.[45] As a *Vor-stellung*, the stare is that possibility of vision which begins to dominate our being in the modern epoch, and which manifests in an extreme form the inherent nihilism in *das Ge-stell*. When our vision is dominated by the *Ge-stell*, it tends to become a stare, fixing its objects before it: it implements the *Vor-stellen*.

But, if we can begin a recollection of the yielding in which, and from out of which, vision takes place, the hold of the stare, the hold of 're-presentation', could at long last be broken.

Part III

The Horizon's Embrace

Opening Conversation

(1) . . . when a person starts on the discovery of the absolute by the light of reason only and without any assistance of sense, and perseveres until by pure intelligence he arrives at the perception of the absolutely good, he at last finds himself at the end of the intellectual world, as in the case of sight at the end of the visible.

Plato, *The Republic*[46]

(2) The understanding of Being already moves in a horizon that is everywhere illuminated, giving luminous brightness.

Heidegger, *The Basic Problems of Phenomenology*[47]

(3) And from this source we have derived philosophy, than which no greater good ever was or will be given by the gods to mortal men. This is the greatest boon of sight, and of the lesser benefits, why should I speak?

Plato, *Timaeus*[48]

(4) Two things fill the mind with ever new and increasing admiration and awe, the oftener and more steadily we reflect on them: the starry heavens above me and the moral law within me. I do not merely conjecture them and seek them as though obscured in darkness or in the transcendent region beyond my horizon: I see them before me, and I associate them directly with the consciousness of my own existence.

Kant, *Critique of Practical Reason*[49]

(5) Only a horizon ringed about by myths can unify a culture.

Nietzsche, *The Birth of Tragedy from the Spirit of Music*[50]

(6) Cities give not the human senses room enough. We go out daily and nightly to feed the eyes on the horizon, and require so much scope, just as we need water for our bath.

Ralph Waldo Emerson, 'Nature'[51]

(7) . . . it is the magical lights of the horizon, and the blue sky for the background, which *save* all our works of art. . . .

Emerson, 'Nature'[52]

(8) The eye is the first circle; the horizon which it forms is the second; and throughout nature this primary figure is repeated without end.

Emerson, 'Circles'[53]

(9) My present field of consciousness is a centre surrounded by a fringe that shades insensibly into a subconscious more. . . . What we conceptually identify ourselves with and say we are thinking of at any time is the centre; but our full self is the whole field, with all those indefinitely radiating subconscious possibilities of increase that we can only feel without conceiving and can hardly begin to analyze.

William James, *A Pluralistic Universe*[54]

(10) My eyes, however, strong or weak they may be, can only see a certain distance, and it is within the space encompassed by this distance that I live and move; the line of this horizon constitutes my immediate fate, in great things and small, from which I cannot escape. Around every being there is described a similar concentric circle, which has a mid-point and is peculiar to him. Our ears enclose us within a comparable circle, and so does our sense of touch. Now, it is by these horizons, within which each of us encloses his senses as if behind prison walls, that we *measure* the world, we say that this is near and that far, this is big and that small, this is hard and that soft; this measuring we call sensation – and it is all of it an error!

Nietzsche, *Daybreak*[55]

(11) . . . every elevation of man brings with it the overcoming of narrower interpretations . . . every strengthening and increase of power opens up new perspectives and means believing in new horizons.

Nietzsche, *The Will to Power*[56]

(12) . . . one is constantly compelled to face the possibility of disclosing an even more primordial and more universal horizon from which we may draw the answer to the question 'What is "Being"?'

Heidegger, *Being and Time*[57]

(13) . . . what lets the horizon be what it is has not yet been encountered at all.

Heidegger, *Gelassenheit*[58]

Here is an exchange of words between Hamm and Clov, characters in Beckett's *Endgame*:

> *Hamm*: And the horizon? Nothing on the horizon?
> *Clov*: What in God's name could there be on the horizon?[59]

Perhaps the advent of nihilism is related to the absence of the horizon *as* horizon: as that which both confines us within its limits and holds us open to the limitless beyond; as that which provides wholeness, but never totality; as that which encourages hope, but mocks our human arrogance.

It would be difficult nowadays, if not in a sense impossible, to say what Kant said, only two centuries ago. Not even his firm tone of voice, relatively untroubled by the thought of nihilism, or by the technological concealment of horizon and starry heavens, could now be repeated. I want to take this fact to be a measure of our historical distance from Kant's time: a measure, therefore, of the technological enframing which differentiates our time from his, and which conceals from us, in an historically new way, the dimensionality of the ontological difference in the field of its visibility. Is there a *necessary* connection between the visibility of sky and horizon, and the human experience of moral law? As we fall into lawlessness, chaos and nihilism, I find this question increasingly pressing. Must we first be denied an experience of horizon and the heavens in order to realize the meaning of the connection Kant makes? Technology conceals. Technology reveals.

In his 'Letter on Humanism', Heidegger gives thought to what he calls 'the fundamental experience of the oblivion of Being.'[60] In this section, we shall attempt to understand this in terms of our characteristic experience of vision's horizon. This undertaking is suggested, I believe, by Heidegger himself, who has given thought to our experience of the horizon in at least four major studies. To begin with, let me name the ones of which I am aware: 'The Age of the World Picture', 'The Word of Nietzsche: "God is dead" ', 'What are Poets For?' and *Gelassenheit* in its entirety.

So that we may begin with a good sense of the problem, let us listen to Descartes' *Second Rule of Cognition*. It reads as follows:

> It is advisable to attend only to the circle of objects for which our natural faculties suffice to perceive with certainty and without doubt.[61]

I take this rule to recommend that we overlook, or, in effect, that we forget, the omnipresent horizon, which makes visible the finitude of our vision while gathering the field of visibility into the unity of its vast embrace. Certainty, mastery, and a clearly defined order, must take precedence, therefore, over the *phenomenon* of truth. Nothing, I think, could be more alarmingly explicit.

Althusser describes the invisible – that depth of the visible which the horizon preserves – as a 'prohibited sight'. And he contends that,

> To see this invisible . . . requires something quite different from a sharp or attentive eye: it takes an educated eye, a revised, renewed way of looking, itself produced by the effect of a 'change of terrain' reflected back upon the act of seeing.[62]

For Descartes, the deep invisible into which the horizon withdraws is invisible: it simply does not exist. Vision must resist enchantment, must resist being drawn into the region of the horizon. But if its encompassing presence should somehow become visible, disrupting the sharp line of sight and reminding us of our finitude, and indeed of the vulnerability, the uncertainty of our visible world, then Descartes' *Second Rule* would promptly call us to task, strictly prohibiting any revision of the rational objective and any deviation of the gaze from the line of duty. Awareness of the horizon is not only anarchic; it also demands of the gaze a radically different modality of engagement: presence, rather than re-presentation; diffuse attunement, rather than concentrated focus. In effect, Descartes' disciplinary regulations define a mode of vision which prepares for the modern political economy. For Descartes, the gaze is fulfilled by what can only be called its 'possession' of the object. In this economy, things are sub-stances, standing under the masterful gaze, transfixed and possessed. The gaze must always dominate; uncertainty must be completely ruled out. There must be nothing hidden, nothing hiding, nothing beyond the point of focus which appears with the event of the gaze. Moreover, the gaze will take nothing for granted: just as 'rigorous' discourse should be without unacknowledged assumptions, so the gaze should depend on nothing outside itself.

This forgetting of the horizon, this deliberate, willful suppression of its embracing presence, both opening and closing the field, is an essential characteristic of the enframed, enframing mode of vision. It has been happening, so slowly that it is almost imperceptible, since the beginning of the modern epoch.

Heidegger tells us that,

> Enframing blocks off the shining forth and holding sway of truth.[63]

Let us take the (so-called) metaphor seriously and translate this reflection into a phenomenological observation regarding our experience with vision in the modern epoch. If we do not interpret enframing experientially, as, for example, something which actually *happens* in the field of our vision, we may miss an exceptional opportunity to work with ourselves and prepare for a vision which could participate in the advent of a different historical existence. For the same reason, it is of the greatest importance that we attempt to think of truth as the circumambient light or illumination of the visual field. Heidegger would accordingly be suggesting that enframing blocks off the shining forth and holding sway of the light by blocking off our guardian awareness of the field and its horizon. Enframing requires total visibility and constant surveillance; otherwise, its control is not secure. When truth is thought phenomenologically, as unconcealment (*alētheia*), as lighting (*phainesthai*), it is an event which cannot be abstracted from a field and its horizon. The horizon is that region of strangeness where concealment and unconcealment happen.

We shall return, in our final chapter, to the problematic of truth. For the time being, I would like to continue our meditation on the horizon, making use of a text written by St. Augustine. Although what concerns him is our relationship to the light of God, it can serve as a parable – an analogy – for a process which also takes place in the everyday world of vision. The passage we shall consider is taken from his work *On Free Choice of the Will*, and reads thus:

> For turning, as it were, their backs to Thee, they are *fixed down* upon the works of the flesh, as it were in their own shade, and yet what even there delights them, they still have from *the encompassing radiance* [i.e., the horizon] of Thy Light. But the shade being loved, weakens the mind's eye, and makes it unequal to bear Thy countenance. Wherefore a man becomes more and more darkened, while he prefers to follow what, at each stage, is more bearable to his increasing weakness. Whence he begins to be unable to see that which, in the highest degree, *is* [and which presences through the horizon].[64]

Now, to be sure, as Heidegger says,

> Horizon and transcendence . . . are experienced and determined only relative to objects and our [way of] re-presenting them.[65]

To what extent is this the historical character of a pathological experience narrowed and blinded by the 'will to power' which prevails in our time? The question is complicated. Merleau-Ponty points out that, because of its opening,

> The horizon . . . is what guarantees the identity of the object throughout the exploration [of the gaze].[66]

In an epoch closed to the presence of the horizon, we might

therefore expect to see substantial changes in the identity of what Merleau-Ponty is calling, here, 'the object'. According to Heidegger's interpretation of the history of Western metaphysics, such changes have indeed taken place. The reduction of 'things' to 'objects of reckoning' is perhaps the most fundamental, and most disturbing, of these changes. The very being of things – what Heidegger calls their 'truth' or 'disclosedness' – can no longer be guaranteed. Without a circumspective awareness of the horizon, a 'guardian awareness' guarding its presence, it is questionable whether any stable or coherent perception of things in the truth of their being could possibly take place. Perception (*Wahrnehmung*) can only be perception, i.e., perceive the truth (*das Wahre*) or perceive *with* truth, insofar as it takes up (*nehmen*), and becomes, a guardian awareness: in German, a *Gewähren*. We need the presence of the horizon. When the horizon is forgotten, the space of enchantment, the clearing of light which allows vision to open and beings to presence as they are, gets diminished: diminished, finally, as we are now seeing, to a point where the historical psychopathology begins to insist on recognition.

Violating the 'preserve' (*Wahrnis, Wahren*) of the visible, enframing is a mode of perception (*Wahrnehmung*) which reduces the horizon to a collection of objects available for total, comprehensive control. Merleau-Ponty must have sensed the historical danger, for in his essay on 'Interrogation and Intuition', he argues, against the metaphysicians of modern thought, that what they call 'the immediate' represents what in truth 'is at the horizon and must be thought as such.'[67] And he adds that, 'it is only by remaining at a distance that it remains itself.' Taking Merleau-Ponty's thinking another step by connecting it with the critical interpretation of the history of metaphysics suggested by Nietzsche and Heidegger, I would like to argue that the 'error' which appears, through their interpretive schema, in the history of our metaphysics is a reflection of our visionary being; and in particular, I would like to argue that the doctrine of 'the immediate', which makes its appearance during the modern history of metaphysics, corresponds to an actual psychopathology which begins to afflict our visionary being at the dawning of our modern civilization: a psychopathology, carried by the body, which involves the negation of the horizon in our lives and in our perception, and which we should now, therefore, begin to interpret as a significant historical manifestation of nihilism, since the nothingness of the horizon, our indifference to the ontological difference it makes in our world, corresponds to the diminishing of our visionary being and to the oblivion into which Being itself has withdrawn.

What Merleau-Ponty says about the metaphysician's 'immediacy' is also true of the visible horizon:

> [it] recedes in the measure that philosophy [or the enframing vision it reflects] wishes to approach it and fuse into it.[68]

In 'The Intertwining – the Chiasm,' Merleau-Ponty repeats his calls for thinking to acknowledge the open being of the horizon. He understands that the reflections and illuminations of our thinking are intimately intertwined with our sense of what vision could be for us, and that, since Being presences as a field of visibility, we bear a singular historical responsibility for the guardian character, the truthfulness, of our vision. The Question of Being, the *Seinsfrage*, is a question addressed to our visionary being. Merleau-Ponty therefore contests the metaphysical interpretation which imposes its will on the phenomenology of our experience with the horizon:

> When Husserl spoke of the horizon of things, . . . it is necessary to take the term seriously. No more than are the sky or the earth, is the horizon a collection of things held together, or a class name, or a logical possibility of conception, or a system of 'potentialities of consciousness': it is a new type of being, a being by porosity, pregnancy, or generality, and the one before whom the horizon opens is caught up, included within it. . . .[69]

According to the Teacher in Heidegger's *Gelassenheit*,

> What is evident of the horizon . . . is but the side facing us of an openness which surrounds us; an openness which is filled with views of the appearance of what, to our re-presenting, are objects.[70]

The horizon we face visibly encloses our field of vision

within the expanse of its protective embrace. Within the given limits of this enclosure, a clearing, there is (*es gibt*) an openness which is receptive to the taking place of mortal vision. In making a clearing, the horizon sets limits; yet it also makes visible the presence of an invisible field of Being opening out beyond itself. The receding of the horizon should remind us that the clearing and its illumination are a gift (*Es gibt*) to our vision:

> We say that we look into the horizon. Therefore the field of vision is something open; but its openness is not due to our looking.[71]

The Scholar then contributes the reflection that,

> Likewise, it is not we who place the appearance of objects, which the view within a field of vision offers us, into this openness.[72]

It is precisely because the seer – let us say a woman – is forever mindful of the protection of vision, the gift of the horizon, that, as Heidegger states,

> The seer speaks [to us] from the preserve [*Wahr*] of what is present.[73]

She speaks 'from the preserve' in two senses: first of all, she speaks in order to preserve the truth of vision; secondly, she speaks from a place where her vision, her perception, her taking up (*nehmen*) of that which is given to perception (*Wahrnehmung*) for its preserving (*Wahren*), is always and already an appropriately thoughtful reception of the gift. The seer speaks in a way that makes visible the gift to our vision. The vision of the seer turns our gaze in the direction of the horizon: there where the gift, the *Es gibt*, can be seen. In 'The Anaximander Fragment', Heidegger invites us to see what is happening in that enchanted region:

> [T]he region of the open expanse, in which everything present arrives and in which the presencing to one another of beings which linger awhile, is unfolded and delimited.[74]

At the horizon, the yielding of the light is recollected. The vision of the seer is always attuned by a guardian awareness to the presence – and that means, the way of presencing – of the horizon. This 'way of presencing' is essentially different from that of the object, and my intention, in this book, is therefore to *specify the character of this way* in terms of a phenomenological psychology which interprets the character of this way in relation to our self-development as visionary beings.

Our capacity to experience the presence of the horizon without seeing it as another object is related to a more general capacity to perceive the background *as* background and the whole in its wholeness, letting them be what and as they are. In the modern epoch, the epoch of science and technology, these capacities, which, of all our capacities, are the most deeply attuned by their ontological awareness of Being as a whole, are very poorly developed out of their initial, more primitive pre-ontological stage. Since, in the modern epoch, the will to power, functioning as a tendency to dominate and control, determines all our experience, these particular capacities are increasingly being suppressed in favor of tendencies more compatible with the calculative rationality of our political economy.

The wholeness of what I am calling a 'whole' is essentially different from the totalization of a 'totality'. A totality can be mastered, dominated, controlled; it can be grasped and possessed; it can be fixed and secured; it can be known with certainty; it is absolutely complete. A whole has its own completeness, but this completeness remains open. A sonata, a painting, a photograph, a sunset, a summer shower, a conversation, an individual's psychoanalysis, a gesture: each of these has a beginning and an end, a certain unity, coherence, and completeness. Yet this wholeness does not preclude their continuity, and they are open to further enrichment or development, different completions. The difference between a whole and a totality is an ontological difference which cannot be understood by a reductively calculative rationality; it can only be understood aesthetically, that is to say, in an experience grounded in our sensibility, our capacity for feeling. We need, to begin with, a familiarity with the feeling of wholeness; we need to consult our deepest *sense* of wholeness.

The significance of *das Ge-stell*, the enframing characteristic of the modern epoch, is manifest in the historically prevailing formation of the *Gestalt*. More particularly, it is the dominant formation, the dominant *Gestaltung,* taking place in the field of our vision. In the modern epoch, the perceptual structuring of figure-ground is pervasively determined by our closure to Being – our closure to 'dimensionality itself'. (In his 'Working Notes',[75] Merleau-Ponty writes that 'Each [perceptual] field is a dimensionality, and Being is dimensionality itself.') This closure in the *Gestaltungen* happening in our field of vision is an affliction, a pathology: a 'closure', as Heidegger says in his 'Letter on Humanism', in the dimension of the hale'. And it is of the greatest significance for the understanding of our present historical situation.

Among the 'Working Notes' published in *The Visible and the Invisible,* we find this remark: 'the Gestalt contains the key to the problem of the mind.'[76] I would add that it also contains a key to the interpretation of *das Ge-stell,* for it enables us to think about enframing as a historical appropriation of our visionary being. This historical interpretation permits us, in turn, to relate our experience with vision to the advent of nihilism.

In a note written first of all for himself, Merleau-Ponty says:

> understand perception as differentiation, forgetting as undifferentiation . . . disarticulation which makes there be no longer a *separation* (*écart*), a *relief*. This is the night of forgetting. Understand that the 'to be conscious' = to have a figure on a ground, and that it disappears by disarticulation – the figure-ground distinction introduces a third term between the 'subject' and the 'object'. It is *that separation* (*écart*) first of all that is the perceptual *meaning*.[77]

The 'night of forgetting' will be our subject for meditation in the penultimate chapter. The *écart* will be considered in the very next chapter, where I shall interpret it as the *Riss,* and relate it to the ontological difference. Here I want mainly to call attention to the differentiation constitutive of *Gestalt* process in perception, for, in this regard, enframing effects a distinctive kind of disarticulation: a historically distinctive forgetfulness of the difference between figure and ground.

Merleau-Ponty is certainly accurate when he claims that, in the realm of everyday vision,

> it is necessary to put the surroundings in abeyance
> the better to see the object, and to lose in
> background what one gains in focal figure.[78]

But I should like to question the assumed universality and necessity of this characterization. From a more primordial, more radical point of view, namely the ontological, the processes of figure-ground structuration typical of our present historical form of everyday experience must be called into question. What Merleau-Ponty is describing is a matter of 'dimensionality'. And, as he himself has argued, Being is a question of dimensionality. Consequently, we must take care to observe the fundamental difference between the dimensionality of perceptual processes characteristic of vision in its present form of everydayness and the dimensionality of such processes in those precarious 'moments of vision' enlightened by a guardian awareness of Being as a whole, presencing in the field of visibility.

Merleau-Ponty's assumption, made during a time of greater orthodoxy (for his *Phenomenology of Perception* is a work belonging to a much earlier period than that of the 'Working Notes' we have been considering), belongs to a metaphysical tradition unconsciously shared by the authors of *Gestalt Therapy: Excitement and Growth in the Human Personality*, who state that,

> For the Gestalt to be unified and bright – a so-called
> 'strong Gestalt' – all this varied background must
> become progressively empty and unattractive.[79]

A 'good gestalt', they argue, is a 'unified figure against an empty ground.' But what is the norm, the standard, the point of reference for this judgment? Are we destined, as visionary beings, to remain stuck within a pattern of affliction, forever divided in our being between the extremes of loss and gain, strong and weak, empty and full, dull and attractive? Are we destined to perpetuate the 'dull suffering', as Heidegger calls it, which inhabits the blind and forgetful vision of our everydayness, repeating the suppression of the ground, the field, the clearing, the opening, embracing horizon – the dimensionality of Being – which now prevails in this epoch of enframing? The questioning of norm and point of reference should not be a matter of indifference to us, for we need to consider whether or not this figure-ground structuration, implicating, as it must, a distinctive ontological attitude towards the perceptual *ground*, constitutes an experience with vision beyond which we can grow, both as individuals and as a historical culture. Are we being asked to consider an 'ontical' description of perceptual fact, a description in the language of phenomenological psychology recording our everyday experience? Or are we considering a description with 'ontological' or 'transcendental' authority? If the description is not meant in the latter sense, then the recognition that the *Gestaltung* in question constitutes the basis for enframing, and consequently the historical form of suffering we presently experience in the field of our vision, would encourage us to give thought to our capacity for self-development as visionary beings and to the historical opportunities this kind of growth, this kind of 'self-overcoming', might bring to light.

As a sage, the 'seer' in Heidegger's study of 'The Anaximander Fragment' is one who knows from his own experience that our vision is capable of outgrowing the painful exchange between loss (in grounding awareness) and gain (in instrumental attention) which is characteristic of the various historical forms of *Gestalt* process with which we are familiar in the ontological forgetfulness of everyday life. In *Phenomenology of Perception*, Merleau-Ponty has not yet understood what he understands in 'The Intertwining – The Chiasm' and in the 'Working Notes', published many years later in *The Visible and the Invisible*: that a different gaze, a different way of seeing, is actually possible, despite great difficulty, and despite the risk of estrangement, and that we do not have to lose touch with the ground of our vision as we go about our lives. There are other ways for us to experience the interplay, the yielding, between figure and ground, withdrawing and bringing forth, concealment and unconcealment: ways which are historically available to us at this time, and in which that interplay could be experienced without forgetting, as in the vision ruled by the *Gestell*, the presence of the ground.

The presencing of the horizon, and more generally, the presencing of the visual field in which our vision takes place, call for the development of our *capacity* to perceive. Just what this means, what this involves, defines the task we have set for ourselves in this book.

Now, since it is the early Greek thinkers to whom we look

for exemplary teachings concerning vision, we should not bring this section to a close without taking note of Heidegger's long discussion of the experience of the horizon as it has been communicated to us through Greek philosophy.

Consider, first, Protagoras, who seems to have declared – in words whose meaning may indeed be less obvious than we initially supposed – that the human being is the 'measure' of all things. Heidegger asserts that, for Protagoras,

> The ego tarries within the *horizon* of the unconcealment that is meted out to it always as this particular unconcealment. Accordingly, it apprehends *everything that presences within this horizon* as something that is. The apprehending of what presences is *grounded* in *this tarrying within the horizon of unconcealment*. Through its tarrying in company with what presences, the belongingness of the I in the midst of what presences *is*. This belonging to what presences in the open *fixes the boundaries* between that which presences and that which absents itself. From out of these boundaries [of ontological difference] man receives and keeps safe the [horizon's] *measure* of that which presences and that which absents. Through *man's being limited* [by the horizon] to that which, *at any particular time*, is unconcealed, there is given to him *the measure that always confines a self to this or that. Man does not*, from out of *some detached I-ness*, [himself] *set forth* the measure to which everything that is, in its being, *must* accommodate itself. Man, who possesses the Greeks' fundamental relationship to that which is and to its unconcealment, is *metron* (measure), [but *only*] in [the sense, and to the degree,] that *he accepts restriction . . . to the horizon of unconcealment* that is limited after the manner of I; and *he consequently acknowledges the concealedness of what he is and the insusceptibility of the latter's presencing or absenting to any* [merely human] *decision*, and to a like degree, *acknowledges the insusceptibility to decision of the visible aspect of that which endures as present.*[80]

And Heidegger comments on this:

> The fundamental metaphysical position of Protagoras is only a *narrowing down*, but that [still] means nonetheless a *preserving*, of the fundamental position of Heraclitus and Parmenides.[81]

Clarifying even further where Protagoras stands, in contrast, on the one hand, with the two Greek thinkers who preceded him, and, on the other hand, with the thinkers of our modern epoch, Heidegger writes:

> It is one thing to *preserve* the horizon of unconcealment that is *limited* at any given time through the apprehending of what presences (man as *metron*). It is another to proceed into the *unlimited sphere of possible objectification*, through the reckoning up of the representable that is accessible to every man and binding for all.[82]

When *we* listen to Protagoras, what we *hear* is a proclamation of our power – our will to power. Being is (to be) measured by *human* being. But Heidegger suggests another way of hearing him: a way that suspends, or puts into brackets, the *noise* of our will to power, and lets us at least begin to hear the ring of a very different saying. Perhaps what Protagoras meant is that, where human beings are the measure, the measure is mortally limited, and that we need therefore to recognize the presence of a measure which is *not* of our own making: a measure *by reference to which* our merely human measure may itself be measured. Without such a measure, *our* measuring merely makes visible the magnitude of our insolence. Insolence, however, leads by the most fateful necessity to a time of self-destruction. As we will measure, so are we measured.

Our mortality is bounded by the horizon, whose awesome beauty is sufficient to establish its truth as our measure and limit. It is at the horizon that our earth, the realm of mortals, is *most visibly touched* by the closedness of the overarching sky, realm of the gods who embody our ideals. The horizon marks an intersection, a contact; but it also articulates a visible difference, an ontological difference that we may powerfully contest, but will never abolish. The horizon, *laying down* this difference while *gathering* us into the depths of its meaningfulness, is, in truth, a visible disclosure, a phenomenal manifestation, of the elemental

Logos, which reminds us of our limits while demonstrating, at the very same time, the unlimited meaningfulness of those limits.

The horizon is that gift of measurement by reference to which our mortal lives on earth may enjoy a sense of wholeness: circumscribing mortal dwelling under the *archē* of the sky, the horizon is that which unifies and integrates the meaning of our existence. It is at the horizon that the overarching sky grants to us a healing vision, a sense of human life in its wholeness; and yet, the meaningfulness it gives is *not* a wholeness which we can totalize, totally comprehending its gift. The horizon teaches us understanding, not knowledge. At the horizon, region of enchantments, the overarching sky *embraces* our existence with an encompassing meaningfulness that is benevolent precisely because it *withdraws* from the reach of our understanding, and *withholds* from us any final disclosure of meaning or purpose. Its gift is a wholeness which is open, rather than closed; a meaningfulness which can be deeply felt, but never reduced to the concepts of our calculative rationality. And whether we see that or not – that is the true measure and bounty of our lives.

Now let us consider Plato. How did Plato experience the horizon of vision? What did he learn, what under-stand, thinking under the gaze of the stars and in sight of the horizon? According to Heidegger:

> the sun and the realm of light are the sphere in which that which is appears according to its visible aspect, or according to its many countenances (ideas). The sun forms and circumscribes the field of vision wherein that which is as such shows itself. 'Horizon' refers to the supersensory world as the world that truly is.[83]

And then he notes, by way of contrast, that, for us moderns,

> The realm that constitutes the supersensory, which, as such, *is* in itself, *no longer stands over man as the authoritative light. The whole field of vision has been swept away*. The whole of that which is as such [and 'which envelops all'], the sea, has been drunk up by man. For man has *risen up* into the I-ness [posture]

> of the *ego cogito*. Through this uprising, all that is, is transformed into object. . . . *The horizon no longer emits light of itself*. It is now nothing but the *point* of view posited in the value-positing of the will to power.[84]

The question of an 'authoritative light' refers us to our experience; its sense is symbolic, or metaphorical, but not without *truth* for our perceptual experience. The experience in question is one which Merleau-Ponty examines at another level. In his *Phenomenology of Perception*, he points out that 'we perceive in conformity with the light, as we think in conformity with other people in verbal communication.'[85] I submit that there is a need for us to think the connection between the perceptual 'conformity' described by Merleau-Ponty and the absence of any 'authoritative light' coming from the supersensory dimension of the visible field. Heidegger's observation suggests that a decisive historical shift has taken place in the field of our vision: the source of the authoritative light in conformity with which we perceive – and live as visionary beings – is no longer a supersensible, invisible dimension of the field of visibility. For Heidegger, this change registers the dawn of the modern world. But Heidegger does not take up for thought the question which would seem to be next, namely: What then *is* the authoritative source of light for our vision in the modern epoch, and in particular, at the present time?

Given the discursive context of Heidegger's observation – a discussion of nihilism in relation to the history of Western metaphysics – this question invites us to take another look at our bearing as visionary beings. Since 'authority' and 'conformity' are in fact primarily political words, I would like to suggest that the historical significance of the 'event' Heidegger notices is one which calls, with great urgency, for a *political* analysis of the everyday world in which our vision presently happens. Such an analysis does not emerge in the texts we have been considering. In Merleau-Ponty, there is no recognition at all that the light in conformity with which our vision takes place is not a 'pure' light, or a light entirely belonging to nature, but rather a condition of visibility determined in many ways by the world our vision has built for itself. And in Heidegger, this recognition is implicit, but it remains, for the most part, unthought.

What happens when the light – the lighting, the level of

illumination – to which our visionary being must conform is pervasively determined by a political economy ruled by *das Gestell* and the technology it has built? What happens to our vision, our vision of ourselves and our world – what has already happened to the world – when the authoritative source of light can no longer be lived and experienced as 'shining forth' into the field from a supersensory, invisible, self-concealing dimensionality? Since it was towards the beauty of this dimensionality that the horizon's embrace inspired us to turn our thought and vision, the historical occlusion of the horizon from the world that our vision has built threatens now, as never before, to cut us off from all sense of the wholeness of Being.

I am concerned by what I see. What I see is that the authoritative source of the light by which we are enabled to see is increasingly restricted to an environment completely controlled by the political economy which advanced technological power dictates. 'Conformity to the light' means one thing when the authoritative source is a supersensory dimension of the political economy. It may not seem important, for example, that our vision, during the day, is increasingly attuned, not by the sun, but by the lighting installed in our buildings, and that, at night, our artificial systems of illumination increasingly deny us any relationship to the lighting of the moon and the stars. But since the very beginnings of human history, the lighting provided by these 'heavenly bodies' has granted us an essential perspective on our mortal, earthbound lives. If we should ever lose this perspective altogether, we are in danger of finding our vision subjected to a pervasive technology of normalization. The extremity of this danger has in fact already demonstrated its character, for intrusive systems of surveillance and supervision continue to multiply.

Part IV

In a Diminished Light

Opening Conversation

(1) The lower worlds 'form in an unbroken flow from the lights which grow steadily dimmer. . . .'

Gershom Scholem, *The Kabbalah and Its Symbolism*[86]

(2) Enframing blocks off the shining-forth and holding-sway of truth.
Heidegger, 'The Question Concerning Technology'[87]

(3) The unconcealment of beings, the brightness granted them, obscures the light of Being.
Heidegger, 'The Anaximander Fragment'[88]

The enframing which determines our relationship to the horizon also conditions, in a still more general way, our experiencing of light. We have already touched on this phenomenon, but here, in this part of the chapter, I would like to clarify the perceptual significance of our interpretation.

We recall, from our reading of Heidegger's essay on Anaximander that,

> Seeing is determined, not by the eye, but by the lighting of Being. Presence within the lighting articulates all the human senses.[89]

The second sentence should dispel any lingering doubts that what Heidegger means by 'the lighting of Being' must be understood phenomenologically in relation to our actual *experience* with vision. Now if, at this point, we turn to the *Phenomenology of Perception*, we will find the help we need in order to understand this 'lighting' as it happens in our field of experience. Distinguishing between 'light' and 'the lighting', Merleau-Ponty retrieves, in its concreteness, the phenomenon which I think Heidegger is adumbrating. He writes that,

> The lighting is neither colour nor, in itself, even light; it is anterior to the distinction between colours and luminosity.[90]

As I suggest we read this passage, it locates the 'ontological difference' within the field of visibility: the lighting, being 'anterior', is constitutive of the ontological dimensionality of our visual field; it is that level or gradient of illumination within which, and by grace of which, we are enabled to perceive that which shines forth in the event of vision. The lighting is that which makes light and vision possible. It thus, as we shall see, is anterior to the splitting of the field into structures with a subject and an object.

Now, for Merleau-Ponty, it seems clear that, because of its primordiality, the lighting does not allow itself to be experienced *as such* in the same way that we perceive the multitude of beings which appear, which presence, by grace of its illumination:

> Lighting and reflection . . . play their part only if they remain in the background as discreet intermediaries, and lead our gaze instead of arresting it.[91]

We must agree that this is true. But the matter is more complicated than this observation suggests, since there are *different ways* for the lighting to remain in the background. Is it necessary that the lighting withdraw into oblivion? Is it necessary that its presence be totally forgotten, totally concealed from our awareness? Is it necessary that its gift – the *Es gibt* of the clearing it makes for our vision – remain unseen? Heidegger tells us that what disturbs him is the fact that, in their everyday perception, mortals 'keep to what is present without considering presencing.'[92] Formulating this point in the more experiential terms of the *Phenomenology of Perception*, I would argue that, because of its obsession with objects, everyday perception is predisposed to lose the elemental contact it originally enjoyed with the presencing of the lighting: as the gaze allows itself to get caught up in relationships of attraction and aversion fixated on the objects it finds 'arresting', it tends to become more and more indifferent to the presencing of that which makes those relationships possible; and it continues to restrict itself to object-relationships, increasingly split off from its initial field-oriented awareness. The initial attunement, eventually forgotten, once opened our eyes, and kept us open, to the phenomenon, the sheer fact, of presencing as such.

In his study on 'Alētheia', Heidegger argues that,

> Mortals are irrevocably bound to the revealing-concealing gathering which lights everything present in its presencing. But they turn from the lighting, and turn *only* toward what is [presently] present, which is what immediately concerns them in their everyday commerce with each other. They believe that this trafficking in what is present *by itself*

> creates for them a sufficient familiarity with it. But it nonetheless remains foreign to them. For they have no inkling of *what they have been entrusted with*: [the recollection of] presencing, which in its lighting first allows what is present to come to appearance. *Logos*, in whose lighting they come and go, remains concealed from them, and forgotten.[93]

What we need to think, today most of all, is the character of a guardian awareness, a mode of perception which would recollect the presence of the elemental background without obliterating the ontological difference between figure and ground, objects of light and a lighting which brings them forth. We need to learn a way of seeing which neither objectifies the lighting nor leaves it in neglect; we need to learn a way of seeing which continues to be attuned, letting the lighting be present in its 'discreet' withdrawal *as* that which withdraws to make things visible. But the very same conditions which give rise to this need also make our attempt to overcome them extremely problematic. The historical conditions of the modern world are not very encouraging; for the most part, they reinforce our natural tendency to see the gift of the lighting as *nothing* of any importance. Even in the world of antiquity, this tendency prevailed, as the stories about Herakleitos attest. But it is only in the modern epoch that the world such restricted vision built could so thoroughly discourage the more fulfilling, and more precarious tendency, by virtue of which we would grow, like the 'seer', in the preserve of the lighting.

According to Heidegger, Western metaphysics is not yet prepared to think this metamorphosis. Thus, our thinking itself needs to undergo a crucial turn, if it is ever to provide us with a guiding vision:

> Metaphysics thinks about beings as beings. Wherever the question is asked what beings are, beings as such are in sight. Metaphysical re-presentation owes this sight to the light of Being. The light itself, i.e., that which such thinking experiences as light, does not come within the range of metaphysical thinking; for metaphysics always re-presents beings only as beings.[94]

Our metaphysics, merely reflecting the dullness of everyday vision, cannot possibly enlighten us.

Thinking gets under way, however, when it takes up the structure of the subject's relationship to its object. In *Being and Time,* Heidegger makes this claim:

> In 'setting down the subject', we dim entities down to focus.[95]

The claim in this formulation can be understood in two senses: as a claim about language, noting the apophatic character of predication in the propositional structure we take for granted; and as a claim about how the subject-object relationship affects our everyday experience with vision. Neither of these claims should be disregarded. It would be particularly unfortunate, though, if the latter interpretation were to be belittled. If we construe the 'dimming down' to be a 'mere' metaphor, its ornamental function would deprive it of all critical significance. But Heidegger's words are intended as a phenomenological description of 'the everyday perception of mortals'. And this means that they are intended in a critical sense. Is it not precisely to ensure that we will not read his words metaphorically that Heidegger adds, at once, that wherever subject and object are involved, 'our seeing gets restricted'? I take this point to exclude readings which prefer mystification to truth.

A brief reflection on the notion of the *lumen naturale* (natural light) might be useful at this point. (We shall return to this matter in the final chapter.) Perhaps the best way to begin would be to consider a remark Heidegger makes in his long work on Nietzsche. In the midst of an analysis of the classical Greek experience of beauty, Heidegger writes:

> they had such an originally mature and luminous knowledge, such a passion for knowledge, that in their luminous state of knowing they had no need of 'aesthetics'.[96]

A much more rigorous formulation of this question, set in the context of his 'recollection' of the history of metaphysics, will be found in 'Moira', Heidegger's study of Parmenides. In this essay, Heidegger writes:

> The essense of *Alētheia* remains veiled. The visibility

> it bestows allows the presencing of what is present to arise as outer appearance [*Aussehen*] (*eidos*) and aspect [*Gesicht*] (*idea*). Consequently, the perceptual relation to the presencing of what is present is defined as 'seeing' (*eidenai*). Stamped with this character of *vision, knowledge and the evidence of knowledge cannot renounce their essential derivation from luminous disclosure,* even where truth has been transformed into the certainty of self-consciousness. *Lumen naturale,* natural light, i.e., the illumination of reason, *already presupposes* the disclosure of the duality [i.e., the ontological difference which separates beings from Being]. The same holds true of the Augustinian and medieval views of light . . . which could only develop under the tutelage of an *Alētheia* already reigning in the destiny of duality.[97]

Mircea Eliade, probably writing without any knowledge of Heidegger's work, offers an anthropological analysis which confirms that interpretation in the wisdom of ancient myth:

> In the Greco-Roman world the sun, having become the 'fire of intelligence', ended by becoming a 'cosmic principle': from a *hierophany,* it turned into an *idea* by a process rather similar to that undergone by various sky gods.[98]

And he ventures the thought-provoking observation that, in subsequent epochs: 'constant rationalization makes it paler still'.[99]

Heidegger, too, sees 'rationalization', particularly in the form of re-presentational thinking, diminishing the visible presence of the 'natural light'. In 'The End of Philosophy', we read:

> Philosophy does speak about the light of reason, but does not heed the lighting of Being. [However,] the *lumen naturale,* the light of reason, needs it [the lighting] in order to illuminate what is present in the lighting.[100]

I submit that, just as metaphysics has participated in the suppression of the lighting of Being, so it has concealed an anterior light of nature, the so-called 'natural light', which is the radiance of spontaneous ontological awareness and the glow of inborn wisdom. Heidegger clearly sees only the first of these concealments; he sees the light of reason concealing the more primordial lighting of Being. But he does not see that the light of *reason* conceals the light of *nature*: the text we have just read follows the metaphysical tradition, which assumes that the light of nature and the light of reason are the same. This assumption, which I contest, reduces to a much dimmer and more abstract light the visible presence of our inborn awareness, our guardian awareness, of Being. We are born with an inborn attunement by, and to, the lighting of Being. The visible presence of this attunement, coming from the 'depths' 'within' us, is a manifestation of the warmth, the vitality, and the joy we experience through this bodily felt awareness. Thus I would argue that the light of metaphysical reason conceals the light of nature within us at the same time that it conceals the lighting of Being. And that is because the two concealments must be thought in their correspondence; or rather, stated in more phenomenological terms, they must be thought in their dynamic attunement. We must attempt to think what Heidegger leaves unthought, namely: that decisive moment in the history of metaphysics when this natural light, in its attunement by, and to, the lighting of Being, is seen – and translated – as the light of reason. For that moment is the moment when our vision finally consented in the enframing of our modern epoch.

What we have been calling 'reason' is, rather, when we are capable of seeing more deeply into the nature of its embodiment, a *natural* light: a pervasive field of luminosity sealed, for a certain time of protection, within the trust of the inwardly reflective human body. The light of reason is therefore continuous with this pervasive lighting of nature, although it is temporarily enclosed within our flesh, as the gift of our natural capacity for awareness and reflection. Depending on the character of our enlightenment, the flesh which seals in this spontaneously reflective luminosity will be more or less radiant, brighter or dimmer. The light of reason is but a pale reflection of the light of nature. How does it happen that Western metaphysics can see no difference? When metaphysics calls the natural light a 'light of reason', its confusion contributes to the diminishing of that light within us which we enjoy by grace of our inherence, as visionary beings, in the field of visibility.

We need to give thought to the possibility that the dimming of luminosity, a process which differentiates the light of reason from the natural light, and the light of nature from the even more primordial lighting of Being, correlates with the capacity of our vision to participate, at the present time, in the giving and receiving which make visible the presencing of Being as a field of luminosity.

James Hillman, a psychologist of Jungian inspiration, helps us to focus our thinking by making the connection between the dimming of luminosity and the establishment of the ego as master of the body-Self:

> The Ego steals its light from the *lumen naturale*, and the Ego expands, not at the expense of primordial darkness . . . , but at the cost of childhood's godlike, dimmer light of wonder.[101]

On my reading of this text, four dimensions of illumination are implied: (i) the primordial darkness, (ii) the lighting of nature, (iii) the inner illumination of childhood vision, godlike, but still dimmer than the lighting of nature, and finally, (iv) the 'stolen' light appropriated by the ego-logical subject. Hillman's interpretation leaves out the lighting of Being, but I take it that he would agree that the borrowed light which the ego raises into consciousness is a light much dimmer than the light of our childhood vision and the elemental light of nature. Since the world of the modern epoch is a world ruled over by the ego, it is in the story of the ego's appropriation of our visionary being that we will find a preliminary answer to the question at the heart of our present concern: Why do we live in a world of diminished light?

In *Being and Time*, Heidegger says this:

> In everydayness, Dasein undergoes a dull 'suffering', can sink away in the dullness of it, and evade it by seeking new ways in which its dispersion in its affairs may be further dispersed. In the moment of vision, indeed, and often just 'for the moment', existence can even gain the mastery over the 'everyday' [i.e., over its inherent suffering], but it can never extinguish it.[102]

Our thinking, here, has been guided by the hypothesis that it could be useful to interpret the 'dullness' Heidegger notes as a historical phenomenon related to the diminished lighting distinctive of our present epoch. But this interpretive project cannot proceed without a corresponding phenomenological psychology, for the 'moment of vision' requires the development of our *capacity* for vision. Since this capacity concerns us as visionary beings, we need to think the 'moment of vision' as a possibility for which, in some way, we could prepare ourselves. Thus considered, our thinking in this book becomes an *existential* process, engaging us in a 'practice of the Self'.

Part V

Technological Eye

Opening Conversation

(1) Things are so strictly related, that according to the skill of the eye, from any one object the parts and properties of any other may be predicted.

Emerson, 'Nature'[103]

(2) . . . sovereignty of the gaze . . . the eye that knows and decides, the eye that governs. . . .

Foucault, *The Birth of the Clinic*[104]

(3) Profound aversion to reposing once and for all in any sort of totalized view of the world.

Nietzsche, *The Will to Power*[105]

(4) . . . the matter into which we are here inquiring, Being as a whole, can never be represented as some thing at hand concerning which someone might make this or that observation. To be transposed to Being as a whole is to submit to certain inalienable conditions.

Heidegger, *Nietzsche*[106]

In his *Meditations on First Philosophy*, Descartes reflects on the 'fact' that, 'when looking from a window and saying I see men who pass in the street, I really do not see them, but infer that what I see is men.' And he asks us: '. . . what do I see from the window but hats and coats which may cover automatic machines?'[107] Since Descartes stands at the beginning of the modern epoch, what he has to say about vision should not be lightly dismissed after his argument about knowledge and truth has been refuted. His way of seeing things inaugurated the epoch of modern science and technology; and it could be argued that the Cartesian gaze is not a philosopher's fiction – that it not only exists, but actually, in today's world, prevails. Only now, the strange character of this gaze is more concealed: more concealed, in part, because it has become so pervasive, so normative.[108]

What I am calling, here, the 'technological eye' views the world theoretically and instrumentally. When Descartes sees a piece of wax taken freshly from the hive as nothing but 'a certain extended thing which is flexible and movable', when he distinguishes the wax as such, quite 'naked', from its external forms and vestments, the character of his gaze is theoretical.[109] Theoretical vision sees all things as being *vorhanden*: present-at-hand in their sheer extantness, present just in regard to their suchness, their substantiality, their being something. When the character of the gaze is more instrumental, what it sees is seen as being *zuhanden*: ready-to-hand, useful, readily available for practical application. Although the theoretical view is a secondary and derivative posture, a capacity which began to evolve only in situations where the instrumental attitude was not sufficient or not possible – suppose, for example, that the handle of my hammer breaks or that the stone I am carefully chipping should happen to shatter from the blows – its historical emergence has contributed in the most fundamental and decisive way to the modern extension and domination of the instrumental world-view.

Now, in *Being and Time*, Heidegger argues that,

> By looking at the world theoretically, *we have already dimmed it down* to the uniformity of what is purely present-at-hand, though admittedly this uniformity comprises a new abundance of things which can be discovered by simply characterizing them.[110]

In the first part of this passage, Heidegger is certainly describing the theoretical-instrumental gaze from an extremely critical position. Since we shall be continuing

Heidegger's project from this critical position, I would like to emphasize at once that, as the second clause points out, the theoretical-instrumental way of looking at the world has also been responsible for bringing forth 'a new abundance of things'. I see no need to praise the technological eye, although I recognize, as does Heidegger, that it has brought forth many things for which we have reason to be thankful. The vision represented in such works as the *Dioptrics* of Descartes, Newton's *Opticks* and Berkeley's *Essay Towards a New Theory of Vision* has contributed significantly to the building of our modern world, and many of the historical possibilities it has brought to light must be reckoned fortunate and beneficial.[111] If, therefore, we take for thought a very critical position, that is because I feel that what more urgently needs our attention at the present time is, rather, the suffering and madness which grow with this kind of vision.

We are different from the other animals: we stand and walk erect. This posture has enabled us to make use of our eyes in new ways, and our vision, our visionary being, has developed correspondingly. Our sightedness, like our upright posture, graces us with inherent nobility.[112] (Hans Jonas has shown this in an important essay called 'The Nobility of Sight'.) Our eyes are organs of pride. Their power to survey, to encompass and comprehend, to fix and stabilize relatively permanent substances, and to gather entities into a totality of simultaneous co-existence, cannot be matched by any other human organs of perception, nor by any other animals. Our eyes, the organs of our visionary being, are graced with a latent virtue, a latent capacity for excellence, an *aretē*, which is the source of their – and finally our – nobility.

The eyes are privileged organs for long-range, comprehensive sensing: surveying what the field gives them to behold, grasping at a distance without need to be in touch or penetrate into the opaque depths of things, they make it possible for us to acquire a superficial, representative knowledge which is already, in a certain rudimentary way, abstracted and idealized. Thus, not surprisingly, our vision became the source of our historical paradigm for knowledge; in this sense, then, it would, I think, be accurate to say that it constitutes an 'incipient science'. (Merleau-Ponty argues, in *Phenomenology of Perception*, that vision is *not* an 'incipient science'. But he is making a different point.) This may seem altogether good. However, inherent in the nature of the gaze, there is an inveterate tendency to develop only one aspect of its primordial ontological potential, viz., its detached, dispassionate, theoretically disinterested power to survey, encompass, and calculate or categorize with one sweep of a glance. The development of this aspect of our vision is reflected in the history of Western metaphysics, for the predominant tendency in our vision, theoretical and instrumental, unquestionably privileges a metaphysics of permanence, constancy, fixity, simultaneous co-presence, substance, and totalization. The world which our science makes visible is a world which reflects back to us the operations of a science which arises with the emergence of theoretical-instrumental vision, and which therefore brings to light what was already implicit, already encoded, in a natural propensity of our vision.

What Heidegger sees, he communicates to us in a language which is sometimes strange at first, and difficult; but it helps to remember that it is intended to be the sharing of actual experience:

> The Open becomes an object, and is thus twisted around toward the human being. Over against the world as the object, man stations himself and sets himself up as the one who deliberately pushes through all this producing.[113]

Heidegger speaks out because he sees us being self-destructive. And I invoke his words because I consider that we are responsible for what we are now seeing in the world around us. We are responsible for the way we see. We are responsible for our vision – and for our lack of it. What we are seeing today – the violence, the destruction, the growing wastelands – is a reflection, a mirror, making visible the historical character of our vision.

In 'Science and Reflection', Heidegger sets out to trace the process by which, beginning with the investigation of nature among the early Greek thinkers, scientific theorizing has progressively, if almost imperceptibly, affected us: affected our capacity for vision and our experience of ourselves as visionary beings, as seers, determining at the same time the being of the thing which is seen and the presencing of that which remains invisible.[114] Our vision, our visionary being, belongs to history; and it is because of this belonging that we need to undertake a recollection which retrieves the past in order to contribute our visionary potential to the making of a future it would be more joyful to behold. Near the end of his life, Husserl, Heidegger's teacher, saw the tragic irony: that the sciences our vision produced, the very sciences which not only originated in, but even now continue to depend upon, the theoretical-instrumental powers always already latent in our everyday habits of seeing, should have turned against us, and against their own origin, denying any value, any validity, in our lived-through experience with vision; reducing our capacity to physiology and a physics of light; separating our vision from its rootedness in our dreams, our needs and hopes; concealing from our view the depth of the field in which visible and invisible intersect and interpenetrate.[115] The sciences have turned against the vision of their own origin, for they accuse this origin of 'subjectivity' and impose on the world our vision inhabits a theoretical-instrumental logic – a *legein*, a layout, a grid – which conceals the lighting and clearing of Being on which all vision, even the theoretical-instrumental itself, very deeply depends. Pain is what happens when the 'objectivity' of science arrogantly proclaims itself to be the sole standard for human vision and the only measure of its experiential field.

'Science and Reflection' cuts through our cultural delusions and obsessions to unmask the way that

> theory entraps objects in order, for the sake of representation, to secure those objects and their coherence in the object-area of a particular science at a particular time.[116]

In order to show this, Heidegger deepens his phenomenology with the insight and 'suspicion' of hermeneutics.

Heidegger notes that, according to its root meaning in the Greek language, our word 'theory' refers to 'the outward look, the aspect, in which something shows itself.'[117] To 'theorize' originally meant (something like) 'to look at something attentively, to look it over, to view it closely.'[118] Thus he concludes that:

> it follows that *theōrein* is . . . to look attentively on the outward appearance wherein what presences

> become visible and, through such sight – seeing – to linger with it.[119]

This sense of what it means for human beings – mortals – to undertake a 'theoretical observation' of nature, a sense which, if we are silent and very attentive, we may still hear in the ancient Greek words, is completely missing today, buried and forgotten long ago under many sedimented layers of errant and narrowed awareness. In the triumph of theoretical-instrumental vision, much has been gained – but much, perhaps too much, has been sacrificed.

For the Greeks of antiquity, the 'theoretical life' (*bios theōretikos*) is:

> the way of life of the beholder, the one who looks upon the pure shining-forth of that which presences.[120]

It involves a

> pure relationship to the outward appearances belonging to whatever presences: those appearances that, in their radiance, concern man in that they bring the presence of the gods to shine forth.[121]

But when this wonderful experience treasured by the word 'theoria' was passed on and entrusted to the Latin word *contemplatio*, an experiential *paradigm shift* of earth-shaking consequence took place:

> there comes to the fore the impulse, already prepared in Greek thinking, of a looking-at which sunders and compartmentalizes.[122]

And there 'comes to the fore', through vision, a 'frontal' ontology. Heidegger's *Introduction to Metaphysics* points out, in this regard, that things are seen as being

> available to everyone, already there, no longer embodying any world – now man does as he pleases with what is available. Every being becomes an object, either to be beheld (view, image) or to be acted upon (product and calculation). The original world-making power, *Physis*, degenerates into a

> prototype to be copied and imitated. . . . The original emerging and staying of energies, the *phainesthai*, appearing in the sense of a world epiphany, becomes a visibility of things that are already there and can be singled out. The eye, the vision, which originally allowed the dynamism of the *phainesthai* to presence, becomes a mere looking-at or a looking-over or a gaping-at. Vision has degenerated into mere optics.[123]

What the Romans – and still later, the mediaeval scholars – experienced by 'contemplation' cannot easily be recognized in what we moderns experience through the notion of 'observation': 'theory as observation', says Heidegger, 'would be an entrapping and securing refining of the real.'[124] 'Contemplatio' still involved a sense of wonder and enchantment; it was not aggressive; its presence could be felt as a meditative quality organizing the region of its enactment. There was a time when the *contemplatio* inaugurated an open space marked out for observation: it was an augur's clearing, a place of disclosure consecrated by the seer's presence, a temple brought into being through the enactment of a vision grounded in supersensory awareness.

In the world of post-Galilean physics, *ratiocinatio* is finally reduced to the act of reckoning: see Hobbes's *Leviathan*, Part I, chapter 4. And vision itself, correspondingly 'rationalized', likewise assumes a strictly calculative function. For Heidegger, our modern, patriarchal science:

> challenges forth the real, specifically through its aiming at objectness. Science sets upon the real. It orders it into place to the end that, at any given time, the real will exhibit itself as an interacting network, i.e., in surveyable series of related causes. The real thus becomes [totally] surveyable and capable of being followed out in its sequences.[125]

By the middle of the fifteenth century, theoretical-instrumental vision was already projecting the standardization and homogeneity of an infinitely extended physical space. *De pictura*, Leon Battista Alberti's treatise on linear perspective in painting, appeared in 1435. According to Samuel Edgerton, scholar of art history, Alberti proposed a technique – and therewith a method of seeing, for 'organizing the visible world . . . into a geometric composition, structured on evenly spaced grid co-ordinates'.[126] The Renaissance 'rediscovery' of linear perspective codified, and at the same time promoted, a new historical style of vision: a new way of looking at the world; a new way of looking at human and inanimate nature. It was this new style, and the succession of changes in our vision which followed, which made it possible for us, beginning perhaps in the seventeenth century, to *see* every event fitted into a scientific ground-plan of nature in which all places would be essentially equivalent, essentially interchangeable.[127]

The negative consequences of this cumulative development in the willfulness of vision can no longer be overlooked: the gods finally fled; the temples, places of contemplation, have long been in ruins; and now – we of today are calling ourselves a generation displaced, disheartened, homeless. We occupy a space of places in which there is no place of the heart to serve as our 'home'. We are a restless people, always on the move. Heidegger tells us that,

> Ordinary perception never perceives place, *topos*, as an abode, . . . as a home to the presencing of what is present.[128]

But it was 'ordinary perception' which brought forth a theoretical-instrumental gaze even more inimical to the spirit of this kind of place. Without a home, the heart grows weary. Our age is not only the epoch of nihilism; it is also a time of chronic depressions.[129]

To make the character of our modern space more visible, Heidegger attempts to reconstruct the 'nature' and 'place', *Physis* and *topos,* of ancient Greek thinking – visionary thinking.[130] *Physis,* once an overwhelming power, a luminous eruption of living energy inexhaustibly appearing and bringing beings into the light, has been progressively 'restricted', he says, to 'an already finished [and therefore inert] space.' The technological eye, organ of the modern will to power, confines *Physis* to 'the rigid measure of this [claustrophobic] space'. *Physis,* life itself, living space, a nature in which life can flourish, becomes a vast realm of death, a realm of dead, or even deadly matter. There are more and more places where life is so threatened that it can no longer survive. The difference between ancient and modern is an ontological difference: a difference rooted in different visions of Being; a difference belonging to two different epochs in the visibility, the unconcealment, of Being. It is not at all essential that Heidegger exactly *replicate* the visionary world of the ancient Greeks: a task which, in any event, is not really possible. What is called for, rather, is an effort: a genuine struggle to see what, and how, the Greeks saw. The effort itself is sufficient, and serves its purpose, if it brings to light for us a *visible difference* between *our* vision and *an other* vision. For it would be through whatever that *difference* opens up that a new historical vision could finally emerge.

Even within science itself, however, there may be some hopeful signs of a visionary theoretical turning point. For example, in the discipline of physics, which has served for almost three centuries as the paradigm of our vision of objectivity, objectification has reached such an extreme point that it has finally begun to cancel itself out. The very refinement of theoretical-instrumental vision now discloses the limits of its objectivity. As Heidegger points out,

> classical physics maintains that nature may be unequivocally and completely calculated in advance, whereas [contemporary] physics admits only of the guaranteeing of an objective coherence that has a statistical character.[131]

And he calls our attention to 'the way in which, in the most recent phase of atomic physics, even the object vanishes. . . .'[132] It is still true, nevertheless, that enframing, as a vision of the 'standing reserve', rules over this new, postmodern physics, for it serves a technology of war and a politics of global domination.

But who knows what some day might come to pass? Perhaps it will be a radical science, and not our exhausted humanism, which brings to light a new truth for the ailing spirit. The Einsteinian revolution, introducing a radical relativity of observational frameworks into our vision of the universe, has disclosed a time and space much closer to our lived experience than the time and space of Newtonian physics, while a revolution in particle physics, recognizing a field of continuity and interaction underlying the observer and the observed, moves the thinking of physics into a kind of sympathy for that equally radical thinking which espouses the hidden unity of all that happens in the field of visibility. The new physics may deepen our insight into that dimension of our visionary being which Merleau-Ponty calls 'the intertwining'.

In 'The Age of the World Picture,' Heidegger in fact implies that there must be some alternative way for science and technology to be *with* nature. That implication of hope can be heard, for example, when he points out that

> Nature, in its objectness for modern physical science, is only one way in which what presences – which from old has been named *physis* – reveals itself and sets itself in position for the refining characteristic of science. Even if physics as an object-area is unitary and self-contained, this objectness can never embrace the fullness of the coming to presence of nature. Scientific re-presentation is never able to encompass the coming to presence of nature; for the objectness of nature is, antecedently, only one way in which nature exhibits itself.[133]

If objectness is only one way, there must have been others. Heidegger looks to the Greeks of antiquity in order to see, through different eyes, a different nature. His gestures of recollection are not futile efforts in the spirit of Romanticism. On the contrary, they are intended to *release* our vision from the grip of its fixations and obsessions: once we see that there have been other ways, we can begin to look with eyes open to a new future – a future as different from the present as the present is different from the past. Heidegger looks back into the past, not for the sake of a futile repetition, but rather to learn from its difference. We are following the project of his recollection so that we can *look forward* to a new vision for our dark times.

John Dewey observes, in his courageous work on *Democracy and Education*, that,

> When nature is treated as a whole, like the earth in its relations, its phenomena fall into their natural relationships of sympathy and association with human life.[134]

If the nature of the earth is to become, once again, a home for our culture, we need to see it in a *civilized* vision. How, then, might we learn to see it with a sense of wholeness? How could we make the world our vision has produced more touchingly visible, so that we begin to feel the pain in our vision? Nothing will change until the pain is acute enough to tear off the veil of our ontological indifference. We must *behold* the rape and plunder, the mindless acts of profanation, by which we steadily change this patient, fertile earth, our home from times beyond memory, into a lunar wasteland too barren for any life. If theoretical-instrumental vision has operated for many centuries in the service of a *masculine* will to power, it may now be time, psychologically speaking, for us to ground our vision in the principle of the *feminine* archetype.

However, as a Tibetan teacher and scholar, Tarthang Tulku, warns us, in his new book, *Knowledge and Freedom*,

> From the viewpoint of *the separate self*, it is very difficult to 'let be', resting in pure observation, without thinking of what we can do as a result of what we are observing. Every situation is an invitation to create another one we view as more desirable. Succeeding situations also have possibilities for improvement, so we repeat the cycle again and again. . . .[135]

It is because Heidegger, too, understands this that, late in his life, he speaks of *Gelassenheit*, letting be, and of a time of awaiting, a time for the cultivation of guardian awareness. And he disavows Nietzsche's 'active nihilism', arguing that Nietzsche himself was caught in the darkness of our time, since the will to power, and its enhancement, can only direct us into an even deeper region of darkness, destruction and despair:

> The will must cast its gaze into a field of vision and first open it up so that, from out of this, possibilities may first of all become apparent that will point the way to an enhancement of power.[136]

This is how Heidegger characterizes Neitzsche's view. He shares with his predecessor the horror at what is seen. But he is alerted by Nietzsche's 'so that . . .' to the continued domination of the (masculine) will to power in Nietzsche's thinking, and therefore sees the master himself falling at times into confusion and perpetuating the cycle of nihilism. So long as our visionary being is dominated by the will to power; so long as our vision is one of mastery and domination; so long as the gaze of theory is an instrument of the will, we have not yet seen beyond the nihilism of our present situation.

The historical power of our sciences is rooted in what Erwin Straus would call the essentially 'gnostic' character of our vision. A 'gnostic' character, for Straus, is analytical, divisive, and particular, in contrast to the 'pathic' character, which unifies in an appreciation of the dimensionality of the field as a whole. Our gnostic science, the successful issue of a theorizing tendency implicit in everyday vision itself, now is in the most visible need of balance. Since we have not yet properly experienced the true 'nobility' of vision, which is a *gift* of nature, we cannot yet envision a science and technology of nature which would correspond to the historical overcoming of the gnostic will to power. But this we do know: that there is, in fact, a still unrealized potential, implicit in our vision, for a 'pathically' constituted mood of relatedness. This potential constitutes a complementary tendency which our thinking must begin to draw out. In the next chapter, therefore, we will give our thought to the crying need for a different vision, a vision in which the two ontological moods – the gnostic and the pathic, the involved as well as the detached, and the compassionate no less than the unmoved – are able to function together in a harmony.

Standing at the window to the modern world, Descartes looks out at men with a mechanical eye, withdrawn from the flesh of the world, immobile, unmoved by all fluctuations of sense and sensibility, functioning according to the laws of a strictly monocular rationality. And he reports what he sees: nothing but hats and coats. The Cartesian gaze has decisively altered our history as visionary beings. It is a gaze, therefore, whose concealed metaphysics we must continue to interrogate, whilst we struggle at the same time to make contact once again – once again, and yet also for the very first time – with the pathic mood, long forgotten, long suppressed, which sees to it that we are firmly rooted in the world. Recollection in the history of metaphysics needs to be accompanied by a process of recollection taking place *in* our lived experience with vision – our concrete experience of Being as visionary beings. Perhaps this is why, when Heidegger speaks to the students in one of Medard Boss's

Zollikon Seminars, he tells them: 'I am only concerned that you open your eyes and do not immediately dim and distort your vision once more with artificial suppositions or theoretical explanations'.[137] Our phenomenological psychology is *not* reductionism and ontological indifference, but an attempt to continue in the direction of this spirit.

When we follow Heidegger, here, and think our modern epoch as the age of the world picture, we must not, however, forget that our visionary propensity for picturing can be traced back into the world of the ancients. As Samuel Edgerton has reminded us, we know from his *Geographia* that Ptolemy (367-283 B.C.) 'intended the whole three-dimensional "earth" to be posited frontally before the eyes in the conventional manner of looking at a picture.'[138]

In concluding this part of the chapter, I would like us to listen to Heidegger's attempt to think beyond both the fatalism of passive despair and the nihilism in the will to power. In 'The Turning', he affirms his conviction that

> Enframing is, though veiled, still glance, and no blind destiny in the sense of a completely ordained fate.[139]

Whether, and how, we may see through the veil of tears and deepen our understanding of this epochal enframing as it gives shape to our 'glance'; whether, and how, we may be moved by what we see: these are the fateful questions which claim us, and to which, in the chapter which follows, we shall attempt to respond with our thought beholden.

Part VI
The Image of Our Suffering

Opening Conversation

(1) But we do not yet hear, we whose hearing and seeing are perishing through radio and film under the rule of technology.

Heidegger, 'The Turning'[139]

(2) As their telescopes and microscopes, their tapes and radios become more sensitive, individuals become

blinder, more hard of hearing, less responsive, and society more opaque, more hopeless, its misdeeds . . . larger and more superhuman than ever before.
Max Horkheimer, *Dawn and Decline*[140]

(3) . . . in industrial society, everyone is reduced to the mode of spectator at the show of his own alienated activity.
Paul Breines, 'From Guru to Spectre: Marcuse and the Implosion of the Movement'[141]

(4) The gaze at the technical object, a passive gaze, attentive only to functioning, interested only in structure (disassembling and reassembling), fascinated by that spectacle without background, complete in its transparent surface. This gaze becomes the prototype of the social act.
Henri Lefebvre, *La Vie Quotidienne dans le Monde Moderne*[142]

(5) . . . a fixed staring at something that is purely present-at-hand . . .
Heidegger, *Being and Time*[143]

(6) . . . Being is equated with constant presence-at-hand.
Heidegger, *Being and Time*[144]

(7) The destruction of the inner life is the penalty man has to pay for having no respect for any life other than his own. The violence that is directed outward, and called technology, he is compelled to inflict on his own psyche.
Horkheimer, *Dawn and Decline*[145]

(8) The concern with violence is mere sensationalism when it is not understood in terms of an analysis of the fabric of the social order which generates violence and its onlookers. For we should not overlook that we are the consumers of violence as much as its producers. . . .
John O'Neill, *Sociology as a Skin Trade: Essays Towards a Reflexive Sociology*[146]

(9) [Hannah] Arendt warns against the tendency to place the human emotions in opposition to 'rationality' when in fact these emotions become 'irrational' only when they sense reason itself is distorted. Indeed, much of what outrages contemporary rationality is nothing but the outrage of a more humane reasonableness driven to expose the sham of established rationality. It is in this context that we must understand . . . the attempt to connect with the humane roots of reason which are progressively destroyed by technological rationality. . . .

O'Neill, *Sociology as a Skin Trade*[147]

(10) . . . it was Marx who first expressed the nature of the social experience of industrialization and class struggle in terms of the alienation of the human sense and the enforced privatization of experience derived from the separation of labour and ownership of the means of production. . . . But from Hegel, Marx understood that the development of human sensibilities under capitalist conditions involved a terrible alienation of the human senses in all individuals, not just the working class. While conceding the expansion of human possibilities through industrial development, Marx attacked the distortion of the sense of reality and value.

O'Neill, *Sociology as a Skin Trade*[148]

(11) From the material and muscular body, continuous with physical reality and capable of performance within physical reality, a reduced and simplified body is abstracted. In its classical and Albertian formulation, this body of perception is monocular, a single eye removed from the rest of the body and suspended in diagrammatic space. Having no direct access to experience of spatial depth, the visual field before it is already two-dimensional, is already a screen or canvas.

Norman Bryson, *Vision and Painting: The Logic of the Gaze*[149]

(12) [The modern] Gaze . . . atomizes the most individual flesh and enumerates its secret bits, [and]

is that fixed, attentive, rather dilated gaze which, from the height of death, has already condemned life.

Michel Foucault, *The Birth of the Clinic*[150]

(13) For Adorno, we must start from where we happen to be historically and culturally, from a particular kind of frustration or suffering experienced by human agents in their attempt to realize some historically specific project of 'the good life'.

Raymond Geuss, *The Idea of Critical Theory: Habermas and the Frankfurt School*[151]

(14) What's effectively needed is a ramified, penetrative perception of the present, one that makes it possible to locate lines of weakness, strong points, positions where the instances of power have secured and implanted themselves.

Foucault, 'Body/Power'[152]

(15) . . . one of the tasks that seems immediate and urgent to me, over and above anything else, is this: that we should indicate and show up, even where they are hidden, all the relationships of political power which actually control the social body and oppress and repress it.

Foucault, 'The Politics of Health in the Eighteenth Century'[153]

Epoch in Perspective

I propose that we begin our reflections on representation and the image by considering a very thought-provoking argument which Jean Paris suggested in a lecture not many years ago at Carnegie-Mellon University. In 'The Misfortunes of the Virgin Mary', Paris analyzes what he describes as 'the successive distortions of the Madonna' in order to demonstrate his theory about the history of painting as a history of human vision.[154] Paris limits his analysis to shifts in the representation of 'symmetry' and shifts in the event he calls 'regard' – that is to say, the look, the gaze, the glance, the beam of vision. Paris's study begins with the representation of God the Father in Byzantine art and concludes with the representation of Madonna and Child in the art of the Renaissance. As the analysis moves through a succession of shifts in the representation of the holy Christian trinity, God, Mary, and Jesus, it uncovers a hidden, previously invisible pattern – a 'deep structure' of transformation in the 'syntax' of painting – which indicates a corresponding pattern of shifts in the cultural history of our visionary being.

Professor Paris does not refer to the work of Heidegger, and I cannot tell, from what he has to say, whether or not he is familiar with this work and has let it influence his thinking. Nevertheless, I am struck by the implicit connection I perceive, and would like to spell it out. According to my reading, then, Jean Paris provides a concrete exemplification in art history which 'confirms' Heidegger's interpretation of the history of Being in terms of a reading of the history of metaphysics. Both tell a story of decline and closure; both tell a story of narrowing vision; and both see in this story the rise to power of willful subjectivity.

The interpretation gets under way with an analysis of The Pantocrator, a thirteenth century mosaic created from the Basilica San Marco in Venice. Paris calls our attention, first of all, to the frontal symmetry of the image: 'His visual beam, His Regard' is projected right in front of Him, perpendicular to the wall.'[155] This symmetry, he says, is to 'assert the Almighty's Authority', for it is an extremely lucid way of 'emphasizing His visual vigilance, His quality of supreme Seer, of "Pantepopte" '. And he explains this analysis as follows:

> If there is no depth in Byzantine mosaics, if the divine space prevents our intrusion by opposing a dazzling wall of gold to our own 'regard', as a supernatural frontier which reveals and at the same time forbids the absolute infinity of the Being, clearly the third dimension is not to be found at the background of the image, but *in front* of it, protruding straightforward as the very Regard of Transcendence itself: *we* are the third dimension, *we* are the picture![156]

According to Paris, this visual relationship brilliantly resolved the theological conflict between the need somehow to represent God and the danger of idolatry, which reduces the dimensionality of His transcendent divinity to a visible object-for-us. But how can the Almighty be painted, how depicted *per se*, and not *for us*? Here, Paris argues, is the genius of the Byzantine answer:

> . . . by inverting the relation between the observer and the observed. By imposing God as an Eye, i.e., not as an object to be looked at, but as a Subject staring at us, from His inaccessible source, as a Regard filling our own world, watching the faithful, inside the Church, so that no one may escape His all-embracing attention.[157]

In other words, they *inverted* our profane, i.e., normal relationship to the image:

> if God is the viewer, *we* are His perspective, and it is logical that [perspectival] lines be *converging to us* [rather than receding, as we moderns have come to expect, towards a vanishing point in the background].[158]

In effect, then, it is *as if* God sees us but we cannot see Him: He is the One-who-sees; we are only the seen. We live in the realm of the visible; God, as the All-seeing, remains in the dimension of the invisible.

Now, what Paris argues is that, 'Right from the beginning, the Virgin Mary obeys the same canons as her husband and son.'[159] Paris invites us to consider the *Virgo Orans*, an early twelfth century mosaic in Ravenna. This mosaic represents 'the primordial Virgin'. Like the mosaic of God the Father, this one also adheres to a very strict, and therefore 'unnatural' frontal symmetry: 'she is standing upright on the gold wall, hands stretched, arms open, eyes straight ahead'.[160] What happened to this frontal posture, and to the divine stare which confronts and overpowers us, is the story Paris then tells:

> The following story of the Virgin simply relies on the *ontological* succession of her images, which ideally retraces the transformation of Byzantine sacred space into Italian profane space – in other words, the gradual emergence of perspective.[161]

(Paris's emphasis on the word 'ontological' means that the story he has to tell does not necessarily follow the merely chronological generation of images.)

In the next ontological phase, exemplified in a sixth century mosaic located in the Basilica of Saint Apollinaris the New, the Virgin is shown *sitting down*, on a throne, surrounded on both sides by angels. This, he asserts, is a momentous historical shift, since,

> up to this point, in the Christian era, standing up, arms open, hands outstretched, the child attached to Her by magic, regardless of the laws of gravity, She has asserted Herself as a purely metaphysical symbol, as a counterpart of the Creator. In such an absolute schema, a seat, however imperial it may be, introduces a third and suspicious element. First of all, it implies *localisation*. . . . Divine Power now appears to be rooted in a single place, depending on a piece of furniture. It is not a Power any more, but a person. At the visual level, . . . the new position humanizes her, suggests a rest, hence a fatigue, incompatible with the immaterial.[162]

Nevertheless, as he says, 'the body remains hieratic, quite symmetrical, and the eyes are still staring at us. . . . Frontal position, geometrical schematization, direct glance: the previous characteristics of the divine are still connected by a deep necessity.'[163]

Paris sees a third phase set in motion when the infant Jesus, always depicted, in the preceding phase, in accordance with the same frontal symmetry we will see in the representations of the Virgin Mother belonging to the second phase, is suddenly *rotated* on the Virgin's knee: 'And, all at once, in disrupting the frontality, he necessarily disrupts the steadiness of his gaze. No longer perpendicular to the wall, his visual beam is now wandering far from us, losing itself in the distance.'[164] (One example Paris mentions is *The Hodigitria*, an anonymous Italian painting of the twelfth or thirteenth century located at present in the Fogg Art Museum. Another is the 'Madonna with Angels' by Cimabue, painted near the end of the thirteenth century.) This simple rotation carried revolutionary implications. First of all, the axis of symmetry was now broken; second, the frontality of Mary and Jesus gave way to a variety of more humanized, more profane and worldly postures; and third, the powerful stare, subjecting us to its mysterious power, was now increasingly deflected, looking elsewhere. Mother and Child were no longer painted according to the requirement that they both look directly at us; often, indeed, neither would be positioned to look directly at us. Thus, by a succession of archetypal shifts,

> From high metaphysics, art falls down to the level of daily psychology: instead of a unified, integrated structure, we now have two rival figures loosely connected by tenderness, boredom, or melancholy – that is, by an external element, a human, too human projection subjectively, capriciously imposed upon the [sacred] scene.[165]

In particular, Paris maintains that,

> In turning her attention away from us, Mary renounces her last supernatural privilege: her Regard. The power that she formerly projected from the gold wall is now merely enclosed in the image. All the so-called 'sacred conversations' take place in a space that is no longer sacred, since it is no longer protected against our own violation of it. Its intermediary character retains some features of the primeval sphere of absolute Being: it does not concern us, we have no part in it; *but* it is offered to our eyes, it is a show. We are now in the position of 'voyeurs' peeping in at a private scene without any risk of being disturbed or discovered, since the two figures ignore us entirely.[166]

Thus, as he says, 'instead of being *objects* for the gods to look at, we become plain *subjects* looking at them.'[167] The story Paris tells is the story of the rise to power of modern subjectivity. And I appreciate his story because it connects this rise to power with historical shifts in our visionary being.

For Professor Paris, the rise of subjectivity announces itself in the emergence of perspective – the 'transformation of a sacred *surface* [deflecting our transgressive look] into a profane *volume*' – a volume, namely, which *invites* us to experience our godlike power to gaze into the depths of the invisible.[168] Now, Paris contends that,

> Perspective . . . is of no concern for the Byzantines. As long as it characterizes our physical world, our binocular vision, imposing it on divine space would amount to sacrilege. God is beyond all laws of optics, and so must be His representation: His flat, abstract silhouette proclaims it aggressively, and all His heavenly relatives are similarly exempted from depth, as they are, say, from death or gravity. . . . But we have also discovered that the consequence, or better, the condition of this aesthetics is that the depth, excluded from the wall, will be projected *in front* of the image, signified by the Divine Regard. This is why this Regard, like the Symmetry that goes with it, constitutes the true perspective of the fresco: *inverted* perspective only from our [modern] viewpoint, *logical* perspective from the transcendent one. As indicated by the few objects that may find their way into the composition, . . . it is toward us that space is converging, since it is at us that God keeps staring. Obviously, then, altering one of these factors is bound to destroy the two-dimensional space.[169]

The 'discovery' of perspective, then, represents *the viewpoint of the modern ego-subject,* whose will to power has finally appropriated visionary being. The stare of the gods becomes *our* stare; the frontal look which once made us confront ourselves gives way to an aggressive look that thrives on confrontations outside itself. Our vision inaugurates a 'frontal' ontology based – among other things – on the strengthening of foveal attention and the repression of peripheral awareness.

Paris concludes his lecture with a comment on the chain of events Alberti set in motion:

> 'At long last', exclaims Alberti, 'perspective makes me see the world as God saw it.' We can take this statement at its face value: when space becomes the endless travel of our eyes, when every character, every form, discovers an utter loneliness, when the world has nothing to offer but absence, then the Divine 'Regard' will be well defeated, so that

> painting, like literature, reversing its whole course, may finally proclaim the victory of Man.[170]

'The victory of Man'? Yes. But Paris is not looking beyond the perspective of the Renaissance and the Enlightenment. When we look beyond, what we will see, as did Nietzsche, Freud, Heidegger and Foucault, is the dark shadow of nihilism – and the *death* of Man. It is perhaps still too soon to tell; but this victory, celebrating the rule of subjectivity, could already be turning into a struggle against defeat.

Death to the Emotions

Let us turn, now, to Francis Cornford, whose scholarly studies on the Orphic origin of science in a vision which became 'theoretical' support our Heideggerian interpretation. According to Cornford,

> The [Orphic] state of mind is that of passionate sympathetic contemplation (*theōria*), in which the spectator is *identified with the suffering god*, dies in his death, and rises again in his new birth.[171]

But the Orphic mode of vision did not endure. It eventually grew dim, faded into the light of archetypal memory, and gave way to other, very different historical modes of visionary being. Philosophical thinking itself participated in, and contributed to, the transformations:

> Pythagoras gave a new meaning to *theōria*; he reinterpreted it as the *passionless contemplation* of rational, unchanging truth, and converted the way of life into a [more restricted] 'pursuit of wisdom' (*philosophia*). The way of life is still also a way of death [and rebirth]; but it now means *death to the emotions* and lusts of this vile body, and a release of the intellect to soar into the untroubled empyrean of theory.[172]

I do not doubt that the Pythagorean modification was a necessary historical condition for the possibility of a distinctively 'scientific' science of nature. But, as we near the end of the twentieth century, the destructive character of this Pythagorean vision, deeply concealed from those who took part in its conception, is rapidly becoming all too visible. If we turn our gaze into the beginning of the next century, what we *see* can hardly be called an 'untroubled empyrean of theory'. And I, for one, am now convinced that the detachment of theoretical-instrumental vision from its *body of felt experience* is finally making itself visible to us as a decisive factor in the historical advent of nihilism. For this detachment of vision from the body of feeling encourages the rise to power of an ego-logical subjectivity whose will to power has lost touch with that primordial ontological attunement we once enjoyed as visionary beings.

The life-threatening violence implicit in the technological operation of theoretical vision became a matter of great concern to Foucault, who wrote *The Birth of the Clinic* in order to make visible the pervasive presence of 'death-bearing perception' within the functioning of medical practice.[173] According to Foucault,

> The most important moral problem raised by the idea of the clinic was the following: by what right can one transform into an object of clinical observation a patient whose poverty has compelled him to seek assistance at the hospital?[174]

Theoretical vision and clinical observation succeeded in bringing forth a 'rational, well-founded body of medical knowledge'.[175] However, at the same time that this way of seeing, this way of making visible, extended the knowledge and care of medicine, it also strengthened and legitimized the power of a 'mood' of vision which Foucault proposes to describe as:

> the absolute eye that cadaverizes life and rediscovers in the corpse the frail nerve of life.[176]

It is appropriate that this eye should be connected to a mechanical, essentially hydraulic heart in the physiology of William Harvey. Since the body as corpse is an object with no 'rational essence' to hide; since this objective condition yields itself totally and without resistance to the 'cold' observation of the strictly 'rational' gaze, it may be said with truth that modern medicine has built the foundation for its temple of healing knowledge on the 'rational' dismemberment of human corpses. Foucault is certainly not accusing this mode of vision as such; certainly not condemning the use of cadavers. Nor am I. What is at stake, here, is rather the need for public awareness: individual awareness of the historical need for the development of a very different gaze and a very different vision. The gaze of theoretical-instrumental reason needs to be reintegrated with a vision of wholeness, a vision of feeling, a vision of life. And because, as John O'Neill puts it, the 'forms of the body politic' are 'existential structures of political experience', I share with Foucault a conviction that it is necessary, today, to make visible the ways in which our present historical mode of visionary being is related to our current political experience.[177] Unless we open our eyes to this need, we may some day see, all too keenly, the consuming death our 'perfectly rational' vision has engineered.

The Age of the Image: Triumph of False Subjectivity

According to Heidegger,

> The fact that whatever is comes into being in and through representedness transforms the age in which this occurs into a new age in contrast with the preceding one.[178]

Indeed, it must be acknowledged, he thinks, that 'what distinguishes the essence of the modern age' is precisely 'the fact that the world becomes picture at all'.[179] Heidegger invites us to consider the Middle Ages:

> . . . never does the Being of that which is consist here in the fact that it is brought before man as the objective, in the fact that it is placed in the realm of man's knowing and of his having disposal, and that it is in being *only in this way*.[180]

If we attempt to 'reconstruct' the experience of the ancient Greeks, we find that, in his words,

> man is the one who is looked upon by that which is. . . . To be beheld by what is . . . – is the essence of man in the great age of the Greeks.[181]

Thus, 'in the age of the Greeks, the world cannot become picture [*Bild*].'[182] 'Yet, on the other hand' – and here Heidegger's hermeneutical phenomenology penetrates insightfully a dimension of their own experience which the Greeks themselves could not have realized – 'that the beingness of whatever is, is defined by Plato as *eidos* [aspect, view] is the presupposition, destined far in advance and long ruling indirectly in concealment, for the world's having to become picture.'[183] The age of the world picture is, I think, the ontological phase which *follows* the emergence of perspective and the 'victory of Man'. In the first age, we are in *God's* picture: the world is a picture seen only by God. In the second age, of Renaissance and Enlightenment, we usurp God's place: the world is pictured, but what the picture represents is what is visible *to us*. God's being is therefore in *our* picture. It is only in the third age, however, that the world in which we live is finally *reduced* to the ontology of the picture – the picture, that is, *for us*.

Heidegger explains that,

> world picture, when understood essentially, does not means a picture *of* the world, but the world conceived and grasped *as* picture. What is, in its entirety, is now taken in such a way that it *first* is in being and *only* is in being to the extent that it is set up by man, who re-presents and sets forth. Whenever we have the world picture, an essential decision takes place regarding what it is, in its entirety. The Being of whatever is, is sought and found in the representedness of the latter.[184]

What, then, is this 'representedness'? Heidegger holds that,

> to represent means to bring what is present at hand [*das Vorhandene*] before [or in front of] oneself as something standing over against oneself, . . . and to force it to remain in this relationship. . . . Therewith, man sets himself up as [the godlike arbitrator of] the setting in which whatever is must henceforth set itself forth, must present itself, i.e., be picture.[185]

In brief, re-presentation imposes, on whatever is, a metaphysics of objectivity. Everything which presences, which is, must present itself for our re-presentation. And only the re-presentation is regarded as real. In this way, re-presentation manifests the most extreme form of subjectivity, for the imposition of this metaphysics of objects is the work of the will to power, which thrives on setting up situations structured in terms of the relationship between a subject and its object. As the etymology of the word *object* (ob-ject, *Gegen-stand*) tells us, this relationship is inherently organized around opposition, conflict, struggle and violence. The 'frontal' ontology established by ontically limited, ontologically indifferent vision is an ontology born of confrontation. (More than once, Heidegger has even used the word 'assault' to describe the character of our re-presentational ontology.)[186] The age of re-presentation, the ontology of the world picture, our age, has consequently become a time of terrible strife and destruction.

The final part of the present chapter – this part – is called 'The Image of Our Suffering'. The preposition in this title allows me to suggest (i) that the being of our suffering, its ontological significance, is now caught in an image, reduced to its (primarily visual) re-presentation, and (ii) that the domination of the image (*Bild*) in the present historical epoch somehow participates in, and is somehow responsible for, the epidemiology and character of our manifest suffering. First of all, then, I would like to show how our blind attraction to the ontology of the image, and our correlative propensity to restrict ourselves within a mode of vision for which the image is predominant, compel us to take part in the production of human suffering. Secondly, I would like to show how the ontology of vision and image which prevails today in what Heidegger calls 'the age of the world picture' is related epidemiologically to one of the most prevalent forms of suffering in evidence at this time, namely, the so-called 'narcissistic character disorders'. Taking it for granted that psychopathology is always related to historical conditions, to society and culture, I will argue that such disorders are to be expected in the epoch of nihilism: in an epoch when the most extreme subjectivity – a false, or sham subjectivity caught up in the production and consumption of images – asserts itself and rules over the world with a vision moved by the will to power.[187]

Let us consider, now, how everyday vision, the prevailing mode of vision, supports the domination of an ontology on account of which we take part in the production of human suffering. Our analysis is set in motion by something which Heidegger says in 'The Turning'. I used his words to open the conversation that begins this part, Part VI, of the chapter. What he says is this:

> But we do not yet hear, we whose hearing and seeing are perishing through radio and film under the rule of technology.

Just what does he mean? How is our seeing 'perishing'? And how could that be related to the technology of vision – to the camera and filming, to photography and television? In her book *On Photography*, Susan Sontag contends that,

> Despite the illusion of giving understanding, what seeing through photographs really invites is an acquisitive relation to the world that nourishes [a merely] aesthetic awareness and promotes emotional detachment.[188]

According to Sontag,

> the presence and proliferation of photographs contributes to the erosion of the very notion of meaning, to that parcelling out of the truth into relative truths which is taken for granted by the modern liberal consciousness.[189]

'Reality', she says, 'is summed up in an array of casual fragments – an endlessly alluring, poignantly reductive way of dealing with the world'.[190] And she argues that,

> Photography reinforces a nominalist view of social reality as consisting of small units of an apparently indefinite number. . . . Through photography, the world becomes a series of unrelated, free-standing particles; and history, past and present, a set of anecdotes and *faits divers*. The camera makes reality atomic, manageable, and opaque.[191]

Thus she points out that,

> [as a] way of certifying experience, taking photographs is also a way of refusing it – by limiting

> experience to a search for the photogenic, by converting experience into an image, a souvenir.[192]

There is, she insists, a 'predatory' tendency inherent in the technology of photographic vision.[193] I want to consider this 'predatory' tendency in relation to the world of suffering it 'brings forth into appearance'. Photography, as a *technē*, is a 'bringing forth' and a 'bringing to light': *apophainesthai*. If it has indeed become 'predatory', what it brings forth, into the light, what it makes *visible*, will be suffering. But it will not merely serve to *call our attention* to suffering; rather, it will also *take part* in its production: *making* it within the visible.

If we look into its etymology, our word, *photography*, means a *technē* which registers the writing of light. And if we visualize Being, the lighting of Being, presencing as, and in, a visible texture, a texture which is also a text, then we see that, ontologically considered, photography is a recording, a reading and a writing, a transcription of the lighting of Being as it inscribes itself into the flesh of the world and brings forth a field for our *technē*, a field of ontological difference, divided into the visible and the invisible. (Our visualization is also a visualization of truth: truth as an experience in the field of our vision, truth as what happens, *was ereignet*, in a 'moment of vision', truth as *alētheia*, as unconcealment in the field of visibility. We shall return to the question of vision and truth in the final chapter.) Photography is a record of the primordial writing of light. But the metaphysics at work in photography is a predatory metaphysics of presence: a metaphysics of permanent substance, *ousia*, things which stand still, things which are fixed, ready-to-hand. Photography stops the play of the lighting; it materializes the elemental illumination in things which are made to stand still *in front of* the camera.

Television, the technology of long-distance vision, vision detached and removed from its 'object', unquestionably makes it possible for people to call attention to suffering and move its distant viewers to a sympathetic response. When we think of the starving children in Ethiopia and the homeless people of Mexico City after the 1985 earthquake, we must acknowledge the beneficent function of television. There can be no doubt that, despite its detached mode of operation, television is a vision capable of touching and moving its viewers. Nevertheless, I would argue that its beneficence is exceptional, and that the tendency which prevails is its more predatory character.[194] In addition to being a vision detached and removed from its object, television is an extension of the photographic technology; and, as we have already noted, photography is very deeply rooted in the metaphysics of reified presence: its impassive gaze projects and reinforces an aggressively 'frontal' ontology. Television is certainly not an unmitigated evil; but I think we should be concerned about the future. A time could come to pass when the predatory character of the medium is tuned by technology into a state of terror, for the history of Being which Heidegger's interpretation of the history of metaphysics shows us might be taken to suggest that the predatory character of television will continue to extend and secure its domination over our visionary being and its field of light – unless there is somehow a reversal, a *Kehre*, and the beginning of a different history.

The signs of danger – of a vision totally possessed by its nihilistic will to power – are everywhere, and they should not be overlooked. It is easier for us to shut our eyes. The television image can make people suffer: the violence is inherent in the vision and the image; the process of re-presentation is itself a factor in the production of that suffering which it brings to light.

On March 10, 1983, the *New York Times* News Service reported a narrative of self-immolation. *The Chicago Tribune* carried the story under this headline: 'TV cameramen film, stand by as man sets himself on fire.'[195] Here is the account:

> *Jacksonville, Ala* – Cecil Andrews called the local television station four times last Friday threatening to set himself on fire in the town square to protect unemployment in the nation. At 11:18 p.m., he did it.
>
> Apparently drunk, the 37-year-old unemployed roofer stood near the edge of the square and fumbled with a matchbook. The first match went out.
>
> Andrews staggered to a container of charcoal lighter fluid on the ground and liberally doused his worn blue jeans and cowboy boots. Cupping his hands over a second match, he held it to his knee. A small flame sprang up.
>
> He sat down on the lawn and watched the flame,

> fanning it as it crept up his leg and then, suddenly,
> in a single burst, engulfed his body.

The 'episode' lasted 'only' 82 seconds – nearly a minute and a half – because the two television cameramen who were filming it finally decided to stop the shooting and 'try' to extinguish the flames. The cameramen later 'expressed deep regret over the incident but said they did not feel responsible for what happened'. (We need to consider, here, the connection between feeling responsible and vision without feeling or responsiveness.) Police Chief Paul Locke disagreed. 'I don't know whether Andrews would have done this had they not given him a stage,' he said. They delayed only 82 seconds; but even that was much too long. By the time the man was brought into the hospital, he was very close to death.

Now, it may be argued that there were many people in the community who felt and expressed their moral outrage; that, in fact, for every similar story about human insensitivity and cruelty, there are many more stories to be told about responsibility, kindness, compassion and caring; and, finally, that we must avoid extravagant generalization. But the point I want to make is not, I think, incompatible with the truth of that argument. Because what I want to say is that what took place manifests a social reality whose construction in the realm of our vision is distinctive of our present historical existence; and that, although it may still be an exceptional event, considered from a statistical point of view, it nonetheless is a disclosure of truth by reference to which we could perhaps understand more deeply, and more compellingly, the essential character of our time: its devaluation of all values, its nullification of meaning, its alienation, its depersonalization and dehumanization, its numbness of feeling, its narcissistic, pathological subjectivity, its drivenness, its will to power, its confusion, its violence and brutality, its predatory look and stare – everything, in short, which singled out, for Nietzsche and Heidegger, the advent of nihilism in our epoch and now compels us to feel that this singular event can only be the symptom of deepening spiritual crisis. Somehow, it lets us see more clearly into the concealed essence of nihilism. Somehow, it is an event which opens our eyes to the abyss.

I call attention to this 'episode', seemingly an event of no major historical significance, because I feel that it alone, and not the more prevalent images of kindness and benevolence, enables us to see, in a way we had not seen before, and perhaps could not have seen before, the deepest, most hidden danger confronting our civilization; and because I feel that it could move us to see more courageously into ourselves – and into its historical significance as an event (*Ereignis*) of vision in the long history of Being.

It seems to me that it is not simply the technology which makes this event inconceivable before the modern epoch. After Descartes, however, it seems virtually inevitable. But only now, I think, is the visionary character of that event an actual historical possibility for us. I give to the episode the kind of paradigmatic historical significance I give to Descartes's experience with vision – vision disembodied and essentially detached from feeling – as he gazed out his window and observed, instead of people, the movement of hats and coats. How *could* those television cameramen set up their equipment and calmly film an act of self-destruction? I try to visualize the details in slow motion. I try to put myself into their mode of presence, looking through their eyes. I feel as excluded from the 'scene' as Kierkegaard, when he tried to visualize the journey of Abraham and Isaac to the summit of Mount Moriah. Their normality, their professionalism, their very rationality – that is what makes such behaviour truly frightening and monstrous. Had they been 'certified' psychopaths or madmen, it would have been easy to understand them.

Our modern way of looking at the world made television possible. Is the technology of television now taking possession of our vision, remaking it in its own image – anonymous, mechanical, obsessed with its own images, predatory, detached from all sense of reality? Is our normal, everyday mode of looking, already ontologically forgetful, undergoing a further process of disintegration? Is it the destiny of our gaze, our vision, that it lose touch with reality and incarnate the opticality of its invention? Is it the destiny of human vision to *become* a tele-vision? Is this destruction of our visionary being as subjectivity the final phase in the *inflation* of subjectivity as will to power? Is it that moment when the will to power turns against itself and a mechanical objectivity takes possession of the visionary subject which produced it and invested it with total authority?

A sociologist discovered that, among the families of the American hostages held captive in Iran, the only ones actually shown on television were those in which some member of the family broke down and cried. Was this a conscious effort to arouse the compassion and anger of the American people? Or was it predatory, exploitative, and cruel? Was the camera simply an observer, or did it take part in the production, the re-enactment, of the pain? Near the end of his life, Edward, Duke of Windsor, old, frail and looking very vulnerable, was interviewed on camera. And he was asked whether, as he looked back over the course of his life, he felt that his marriage to a commoner had been worth his abdication of the throne of England. After the question, the television camera closed in on him and held him within the rigid geometry of its frame. This had the effect of making the viewer participate in a cold, unmoving stare, while Windsor wept uncontrollably behind defenseless hands. Observation has become cruelty. Obsessed with the picture, the image, we take part, whether willingly or unwillingly, in the production of suffering. This is what it is for us to be visionary beings in the age of the world picture.

What is to be seen in the fact that most people see nothing monstrous and intolerable in a succession of images on the screen which repeatedly surrounds pictures of war, earthquake victims, marches for civil rights and marches against nuclear power with pictures of Calvin Klein asses, toothpaste and Coca Cola? The images of reality are systematically derealized, as the images which serve the narcissism of our present subjectivity take on the halo of truth and an overpowering reality. What has happened to our vision, if we can now sit back in a comfortable chair, drink a beer or sip a cocktail, and watch – consent to watch – the televised disembowelment of a child accused of being a guerilla spy? And to what extent is our vision, our way of looking and the character of our gaze, to be held responsible for the coming-to-pass of such scenes? To what extent are these scenes reflections of our vision? What gaze do they mirror? What is the character of our responsiveness as visionary beings?

When I recall the weeping of Odysseus listening to the song of the minstrel in the Phaeacian king's palace, and when I consider how the other guests who were present there *saw him, and yet also let him be 'concealed'*,[196] I cannot doubt that we are seeing an historically new mode of vision, in which perhaps even the difference in Being between the visible and the invisible could be in danger of obliteration.

The body politic shapes our vision, structures our gaze. Correspondingly, our way of looking produces a world – a world and a political economy. It is possible to see the character of this economy, and of the vision which brought it forth, reflected back from the eyes of its victims, the subjects it has destroyed – if only we dare to look, with the simple presence of honesty, into the vacant stares of our prisoners, into the glassy stares of workers after their shift on the assembly-line, and into the dead stares of the elderly, abandoned by the institutions of our 'humanity' and waiting for the call of death. Visionary being, caught up in the will to power of this age, takes part in the production of suffering – and itself suffers in turn. Nietzsche's word for this historical event was 'nihilism'.

In 'The Age of the World Picture', Heidegger turns his thoughts concerning the history of metaphysics as a history of Being in the direction of a particular historical problematic: the rise to power of ego-logical subjectivity and the pervasive determination of Being as an object for re-presentation. If the history of Being is a story of its interpretation as force or will, will to power, then the rise to power of the ego-logical subject must constitute an essential moment in the history of that narrative. But so, also, must the dominance of re-presentation, for re-presentation is an exercize of power, and the metaphysics enacted through it is a metaphysics of the *reified* presence: presence mastered and controlled by the will of the ego-subject; presence, as the invisible depth and reserve of things, reduced to a superficial image. Today, we cannot let the sheer presence of beings simply *be*: this gift, the *Es gibt*, is always rejected. Presence must be totalized, possessed through its re-presentation. Re-presentation is the way subjectivity dominates its world, the way subjectivity imposes its will on all that is, all that presences. Re-presentation is therefore aggressive; but it is also a defense, and more specifically, an ego-logical defense, a defense of the ego, because it protects the ego's prejudices, stereotypes, and delusions: it protects the ego against the need to be more open; it helps the ego to avoid authentic encounter; it blocks perception of otherness and difference. Just as the prefix in the German word, *Vorstellung*, should call our attention to the frontal ontology that process imposes, so the prefix to its English equivalent, *re-presentation*, should call our attention to the fact that the ego-subject forever delays or postpones authentic encounter, demanding that whatever is experienced must first give itself in another way: in a form which brings it into 'total' presence *before* the willful subject and according to the subject's terms of full possession.

Since the subject of this study is the development of a capacity for vision, a capacity which fulfills our ontological potential as human beings, I would now like to begin, as promised, an analysis of the breakdown of this capacity and the avoidance of its fulfillment. This means that we consider the relation between image and suffering which constitutes the essence of narcissism – narcissistic character disorders – as a pattern of psychopathology not only unique to, but also characteristic of, our present historical situation. Narcissistic disorders are prevalent today in epidemic proportions because they manifest the suffering of an ego caught in the ontology of its own images and stuck in the dependency of its need for them. They make visible the suffering which comes when we construct a world of our re-presentations in order to avoid the struggle with reality. Thus, the existence of narcissistic disorders in the epoch of nihilism tells us a painful truth about what was forgotten and lost in the rise to power of subjectivity, while the history of the 'withdrawal' of Being in metaphysics is mirrored by another history: the history of withdrawals in psychiatric epidemiology, where the 'triumph' of subjectivity presents itself in symptoms, in defeated lives, in pathologies of the will, in falseness and sham. In these disorders of character, false subjectivities, we are now seeing the historical forms of self-destructiveness, the historical forms of suffering, which mirror by inversion the rise to power of ego-logical subjectivity; and in a suffering that betokens the despotism of images, always the re-presentedness of all that is presencing, we can see the devastating effects of a false ontology against which Plato had repeatedly warned us. In the age of the world picture, there is a peculiar pathology of the Self related to the historical reduction of Being to picture. And this psychopathology, the narcissistic character disorders, is inseparable from a corresponding pattern in the way the world is viewed and seen. In the epoch of nihilism, there is pathology in the *character* of our vision: a pathology related to the false subjectivity of an obsession with the image.

Narcissism is a false subjectivity. It is an *illness* of subjectivity, an illness whose pathology is related to the dominance of the image. As such, it is an illness whose character cannot be understood apart from its epidemiology, its historical situation: its appearance *in* our time as a pathology *of* our time. It is not accidental that narcissistic character disorders began to make their appearance as a psychiatric epidemic only in the conditions of our contemporary world. Such disorders bespeak the suffering which accompanies the 'triumph' of subjectivity as will to power; they bear witness to the emptiness we all experience in an age which requires the reduction of Being to representedness and representation; they portend the spreading of nihilism. Narcissistic disorders reflect the historical character of the present age: the one for which the world is picture, image, *Bild*. As a way of being in the world, the suffering in narcissistic disorders is one which can be understood only through the image. Although it affects different people in different ways, the dominance of the image in our culture produces a widespread pathology, for it means, in effect, being cut off from the truth of 'inner' – and that means 'one's own' – experience. We are being chained to the image and alienated from ourselves. If we become totally identified with the image, we are dispossessed: we belong only to others. The image obscures our capacity for authentic existence, true subjectivity, being true to ourselves.

In the age of the world picture, it is the image which counts. Representedness determines Being – the Being of beings – because that is the way subjectivity masters and controls. In the modern epoch, subjectivity appears as will, will to power. Will to power is domination and control. Representation, the ontology of the image, is the way this will to power, subjectivity, dominates and controls the Being of beings. Thus, the rise to power of subjectivity necessarily involves the dominance of representation, image, *Bild*. It is in the age of the world picture that subjectivity rises to power. But the age of the world picture is also the epoch of nihilism, the epoch in which we experience the nothingness of Being. What is the connection between nihilism and the rule of the image? How does the nihilism in the rule of the image affect the individual? How is it borne by the individual? The answer can be seen in the self-destructive suffering of narcissistic personality disorders, in which the being of the individual becomes totally identified with the being of the image, and loses touch with itself.

For people suffering narcissistic disorders, everything is *to be seen* in terms of power: the power of the image and the image of power. Narcissism is a pathology of the will to power: a pathology in which the will to power is totally trapped inside the images it has 'projected'. True subjectivity requires self-esteem, a certain 'care of the Self'. But self-esteem is not possible in a field of power organized around the images of impotence and omnipotence. Nor is it possible when it depends on the image, on being seen, rather than on the truth which lives within our own experience.

But narcissism is not only an individual pathology. It is, in fact, an epidemic pathology generated within our social and cultural life. A culture of the image rules over Being. Being becomes image, the will to produce and master images, the will to exercize *power* through the production, circulation and control of images. Thus, at the very moment when subjectivity appears to have triumphed, countless individuals, all suffering narcissistic disorders, begin to appear; and this suggests that what we are seeing is perhaps a pervasive disorder of society and culture, related to the historical character of our time.

In a surprising seminar on 'Discourse and Truth: The Problematization of Parrhēsia', Foucault returns to the ancient Greeks in order to raise the question of truth for critical social theory in terms of a formulation that would see it in a different light.[197] Instead of thinking about truth as such, he begins to think about the *telling* of truth. *Who* is the one who speaks the truth? Who is the one who is *free* to speak the truth? And what are the specific social and political conditions required for *parrhēsia*, the frank and courageous telling and speaking of truth? In this seminar, he shifts our attention to the question of character: the character and circumstance of the one who is speaking are at least as important as the content communicated. This shift means, for Foucault, that the question of truth raises questions concerning 'practices of the Self' – practices which, since ancient times, have embodied what the Greeks once called *epimeleia heautou*, the 'care of the Self'. This shift makes it much easier for us to connect Foucault's analysis of truth as *a social, political and cultural event* with Heidegger's interpretation of truth as *alētheia*, unconcealment. It also reminds us that the *experience* of truth as *alētheia*, which Heidegger brings to light through his rereading of the Greeks, needs to be understood in terms of character. We need to think about the character, the moral and spiritual psychology, of the one who can *experience* truth in this way. We need, as it were, an Aristotelian account of the *kind of human being* we have in mind.

Since Heidegger's interpretation of *alētheia* describes the *event* of truth in terms of our *experience* with vision and lighting, it is fitting that we should consider the *character* of this experience. Who is the one who can see the truth in this light? Who is the seer, the one who can see the truth happening in this way? What is the difference between the one who sees an 'event of unconcealment' and the one who sees a 'state of correspondence'? And what are the differences between these two ways of looking at the (truthing of the) world? At stake in these questions is our character as visionary beings. This, of course, is always a matter which concerns the Self. Consequently, I would like to dedicate the rest of this chapter to the narcissistic character disorders. Since this study is a study of vision, it is important that we give particular attention to these disorders. Narcissistic disorders are deeply rooted in the culture of a subjectivity which has risen to power through the character of its visionary being. Narcissistic character disorders compel us to think about representedness and representation, and about the ontology of the image. Thus, in brief, they compel us to give thought to vision – above all, the character of our visionary being. At this time in history.

In the age of the world picture, our character – the character, for example, of our experience of truth – is shaped by influences affecting us through the visionary medium of our being. If we want to understand the essence of this age – understand its nihilism – then we must look to the character of visionary being. We need to see its sufferings. We need to see its truth. We need to see ourselves, and our world, in the light of its pathology. The narcissistic character disorders are rooted in self-destructive tendencies structuring and dominating our visionary being: dispositions which are encouraged and promoted by the prevailing ontological attunement of this present age – that is to say, by nihilism.

In his recent book, *Narcissism and Character Transformation*, Nathan Schwartz-Salant argues from extensive clinical experience that, 'The issue of narcissism is the issue of our age, because it is the focal point for a new Self image in transition.'[198] Since he envisions a new subjectivity, his analysis is worth reading at length:

> The Narcissus myth is especially significant for our present historical time, as it has been in other transitional epochs, for the archetypal world is no longer held in tension by collectively valid religious forms, and thus has begun to constellate strongly in the human soul, acting like a magnet drawing consciousness back toward the archetypal realm. Then egos may seem to be narcissistic, but actually are being drawn back into the realms of 'not-yet existence'. They may become stuck, yielding the pattern known as the narcissistic character disorder, or they may become reborn to inner and outer object relations, relationships to other people as separate and distinct, and to the Self as the transpersonal other. A new Self structure may arise. . . .[199]

When Heidegger speaks of the age of the world picture and Nietzsche of the epoch of nihilism, they are pointing to a new social formation. But, as Christopher Lasch has argued in *The Culture of Narcissism*,

> New social forms require new forms of personality, new modes of socialization, new ways of organizing experience. The concept of narcissism provides us not with a ready-made psychological determinism but with a way of understanding the psychological impact of recent social changes – assuming that we bear in mind not only its clinical origins but the continuum between pathology and normality. . . .
>
> [T]he prevailing social conditions . . . tend to bring out narcissistic traits that are present, in varying degrees, in everyone. These conditions have also transformed the family, which in turn shapes the underlying structure of personality.[200]

(A similar view of the interaction between psychological structures and social structures is presented by Jürgen Habermas in an essay called 'Moral Development and Ego Identity'.)[201] Nietzsche, in fact, has much to say about the psychology of the modern world. In particular, he begins to see the prevalence of depressions and narcissistic character disorders in relation to the advent of nihilism. In *The Will to Power*, for example, he observes: 'The most universal sign of the modern age: man has lost dignity in his own eyes to an incredible extent.'[202] Self-esteem is a central problem in narcissistic disorders.

Schwartz-Salant contends that, in the narcissistic disorders, there is at work a 'grandiose-exhibitionistic power-drive', a 'consistent, demonic urge toward power through ego inflation'.[203] The four principal points around which our interpretation will take shape are, therefore, these: (i) that the disturbances destructive to the Self which are currently called, because of the power of the psychoanalytic school, 'narcissistic character disorders', constitute a pathology of the will as will to power; (ii) that the pathology is a pathology of the ego, and bespeaks the ego's unwillingness, or inability, to become a Self; (iii) that this pathology is connected in a distinctive way with the nihilism of our epoch; and finally (iv) that narcissistic disorders are related in an essential and unique way to the 'fact' that, in this age, the world has become picture.

Narcissistic disorders invariably occur in terms of the ego's relationship to power: not only in the ego's struggle for omnipotence, for mastery, control and domination, but also in the ego's loss of control and its experience of weakness, defeat, subjugation, and impotence. Ultimately, then, the narcissistic disorders are rooted in the ego's unwillingness, or inability, to grow beyond itself – to grow out of the unhappy dialectic of power which holds sway in the ego's field. The ego's *developmental* failure to overcome itself is thus also a *historical* failure to become a Self, and it has been, I think, a decisive factor in the historical triumph of nihilism. But the relationship between nihilism and narcissistic pathology is a vicious circle, for the spread of nihilism has created within its culture a political economy which continues to multiply and intensify the ego's difficulties in relation to the question of power. As an epidemic pathology of the modern ego, narcissistic character disorders need to be considered historically in relation to what Heidegger calls 'the unconditional domination of subjectivity', a historical phenomenon which is determinative of our political economy and reflected in our metaphysics as well.[204]

'Subjectness' has ruled from ancient times, but only in the modern epoch as will to power, a 'false' subjectivity so extreme that it becomes self-destructive. And, as we shall see, this subjectivity of the will constitutes a form of suffering in which the ego is stuck in a dialectic of power that moves back and forth, without development or growth, between loss and envy, depression and mania, passivity and rage: between, on the one hand, anxieties around matters of dependency, helplessness and impotence, and, on the other, dreams and delusions of the most godlike omnipotence.

Before we can carry forward our reflections on metaphysics, history, technology and culture, we need to consider in more detail the character of the ego stuck in a narcissistic pattern of suffering. A society in the throes of nihilism creates an environment which encourages individuals to develop a kind of character extremely susceptible to narcissistic disorders. But it is equally true that our social reality is a function of the prevalence of *individuals* living out such disorders, or manifesting tendencies of character which are not entirely free from such disorders.

'Narcissistic disorders' are not disorders of the instincts or drives; rather, they are disorders of character, disorders of adulthood which originate in the ego's stuckness, whether through inability or unwillingness, and in its failure to continue growing beyond itself in a process of continual becoming, continual self-development. Freud's paper 'On Narcissism' lets us know that he was beginning to perceive cases of what he calls 'secondary narcissism' in many of his patients.[205] For him, such narcissism is unquestionably pathological, in contrast to the 'primary narcissism' of the infant who has not yet differentiated himself from within the symbiotic relationship he has been enjoying with the mothering one. Secondary narcissism is pathological, because it involves an inappropriate, unsatisfactory, and ultimately self-destructive resolution of 'disappointed object love'. And, at the heart of this pathological complex, what we find are disturbances in the dynamics of self-development: ontological anxiety; incomplete mourning in the wake of inevitable individuation, separation, and abandonment; uncontrollable rage over the necessity in this process of loss; guilt and self-accusation; envy and mistrust in relation to others; a mask of omnipotence and aggression to conceal a profound sense of despair and helplessness; a manic state of activity to mask the emptiness within.[206]

Heinz Kohut, the American counterpart to Lacan in a revolt against Freud's ego psychology, defines 'narcissistic personality disorders' as follows:

> temporary breakup, enfeeblement, or serious distortion of the self, manifested predominantly by autoplastic symptoms . . . such as hypersensitivity to slights, hypochondria, or depression.[207]

In its most extreme form, namely, as a 'psychotic' disorder, this pattern will become a permanent or protracted delusion about reality. In more 'borderline' states, this disorder will be 'covered by more or less effective defensive structures', and these defenses will often drive the ego into one or another addiction, one or another narcotization.[208]

Now, according to Kohut,

> It is the *loss of control* of the self over the self-object that leads to the fragmentation . . . and, in further development, to the ascendancy and entrenchment of chronic narcissistic rage.[209]

Since the (archetypally masculine) ego is profoundly insecure and anxious, it needs to dominate, master, and bring everything under its control. Loss of control can therefore give rise to a terrible, uncontrollable rage. What is the cultural equivalent of this narcissistic rage in the life of the disturbed individual? In the political economy of the Western ego's investment, an economy of narcissistic production and consumption, this rage becomes the institutionalized violence of a 'rational' program to control the world with its mastery of technological power. But the nihilism concealed within this will to power is producing an apparatus which increasingly seems to grow out of our control and turn in 'revenge' against us.

Kohut argues that,

> The essential psychopathology in the narcissistic personality disorders is defined by the fact that the self has not been solidly established, that its cohesion and firmness depend on the presence of a self-object . . . and that it responds to the loss of the self-object with simple enfeeblement, various regressions [e.g., uncontrollable rage] and fragmentation.[210]

The ego suffering narcissistic 'disorder' will often feel empty and insubstantial: useless, ineffective, helpless; and it may find itself swept away by rage when it recognizes in itself an ontological dependency which it can neither accept nor avoid. In such a condition, abandoned before it could stand with dignity in the independence of an authentic individuality, the ego acts out its 'sense of inner uncertainty and purposelessness'.[211] In a panic of existential insecurity, it lives out fantasies of omnipotence, the sham of power, while arming and equipping itself in a perpetual struggle to achieve mastery and control.

In relationships with others, the narcissistic character tends to be emotionally unreliable, and is therefore experienced by others as impenetrable or emotionally inaccessible. Interpersonal relationships are defensive – there is a sham of depth, but in reality, they are shallow, superficial, and pragmatic. Because the narcissistic ego has a very poor sense of its own identity, its own 'centre of life', it has a correspondingly limited capacity for feeling, for love, for empathy, for recognizing the otherness of others and being touched and moved by their needs and concerns. Although extremely dependent, for the consolidation of its sense of identity, on the opinion of others, and on the image their reflective presence either confirms or casts in doubt, the narcissistic ego hates all forms of dependency, and is filled with rage and vindictiveness, with loathing, both for himself and for those on whose approval and admiration his identity seems to depend. The narcissistically disturbed ego is restless, moved by insatiable need for gratification, for self-validation, for an admiration and approval he cannot find within himself. Filled with envy, he seeks to be the one who is envied, envied rather than respected, and will manipulate and exploit others for the sake of their envy, their positive mirroring, or their subordination to his show of power.

Man, said Sartre, is 'a useless passion'. There is a painful absurdity in the ego's futile, repeatedly self-defeating efforts to achieve some measure of real satisfaction; it seems that the more the ego pursues its hedonistic calculus, the more it feels empty and hungers for an impossible fulfillment. The satisfactions ego pursues, being sham, being 'false', do not really satisfy. For example, the narcissistic character needs and desires self-esteem; but he pursues its mere reflection in the esteem of those whose authority derives from their show of strength. And though he desires the centredness of Selfhood, he falls into a life of self-alienation, centred around the recognition of others. An inwardly impoverished life of missed opportunities is thus lived out in the glitter of outward signs of acquired wealth, or vicariously, through the achievements of others. Even when the ego has achieved some recognized success, it fails to find sufficient meaning: the work and achievement ethic, the modern form of Freud's *reality principle,* is no longer very compelling as a source of existential meaningfulness. At bottom, the narcissistic character is deeply absorbed in himself, but this absorption is its self-defeating way of avoiding growth through self-awareness and self-understanding. and because the absorption is a withdrawal, narcissistic character disorders are manifestations of a deep, but also deeply concealed, experience of diffuse depression.

According to psychoanalysis, 'secondary narcissism' is pathological because it involves the ego's incorporation of the 'lost self-object'. Loss, then, is at the very heart of the pathology. But this can only mean that the narcissistic disorder is a phase in the evolution of depression. Clinical evidence strongly supports this inner connection between the two pathologies. What could account for the epidemic proportions, the striking historical character, of narcissistic pathology? Could this pathology be related, through depression, to the abandonment of Being, the oblivion and nihilism which hold sway with particularly devastating effect in the modern epoch? Perhaps we are nearing the beginning of an interpretation conceptually prepared to account for the epidemic incidence and prevalence; but it is essential to note that we also have the outlines of an analysis which would strongly suggest that the narcissistic disorder is not an affliction peculiar to a minority; that it is, in fact, a form of suffering rooted in specific historical conditions which today affect us all. The only difference among us, perhaps, is in fact that some people are able to resist the more extreme temptations of the dialectic of power, which draws the weakly developed adult ego into its false subjective economy of narcissistic disorders.

The myth of Narcissus is of course ancient. But there is a story concerning narcissism which belongs to modernity, and it begins with the meditations and reflections of Descartes. In Western metaphysics, there is a story which lucidly mirrors the strange pathology in narcissism.

What I am suggesting is that a pathological narcissism is the driving force in circulation throughout our metaphysics, for metaphysics is but a mirror for the narcissistic ego that rules in our time. Since the discourse of metaphysics is a reflective medium which reflects our cultural self-understanding, we should expect that it would manifest, indeed with singular clarity, our historical experience of the human condition – and in particular, our distinctive psychopathology.

Now let us consider Descartes, whose revolutionary thought stands at the very beginning of the modern epoch. I think we can see very clearly how his metaphysics both mirrors and struggles against one of the most characteristic pathologies of modern times. In his *Meditations on First Philosophy*, we can trace the degenerative stages of a very severe narcissistic disorder. The narrative of disorder begins with a grandiose vision, in which Descartes sees himself building, on a radically new foundation, a 'firm and permanent structure in the sciences'. He desires nothing less than absolute certainty, a knowledge of permanent beings which guarantees its own truth. This insolent ambition is followed, however, by a stage of doubt and uncertainty, and his contemplation of the loss of epistemological faith suddenly turns the earlier self-confidence into a mood very close to despair. Disillusioned, nostalgic, insecure, the ego now attempts to withdraw from the world: it assumes what is basically a depressive position. Narcissism soon appears, though, in the defensive double movement of simultaneous self-absorption and self-aggrandizement: the depression is avoided when the proof of God's existence is 'found' within the ego's own clear and distinct consciousness. As is characteristic of narcissistic disorder, this gesture at proof is exceedingly ambiguous, for at the very same time that it acknowledges our dependency on the existence and benevolence of God, it also surreptitiously inflates our sense of power: being able to *prove* God's existence, the Cartesian ego seems to participate in His omnipotence. The ego which felt impotent now suddenly claims a kind of omnipotence. Its dream of power, of powerful knowledge, may now come true.

But, before this triumph over depression, the Cartesian ego is severely tested, and it must pass through a madness more disturbing than narcissism. There is more than a hint of schizophrenia in the paranoid questioning of reality which begins with the assumption of a malevolent Demon; in the terrible anxiety he feels concerning waking and sleeping, and, later on, concerning the momentary deaths – the 'little deaths' – of consciousness; in the splitting apart of mind and body, spirit and flesh; and in a spectatorial detachment which borders on solipsism. 'I am, I exist, that is certain', he notes, 'But how often? Just when I think'. Again: 'I am, I exist, is necessarily true each time I pronounce it, or that I mentally conceive it'. But what happens when I am not thinking? What happens in the intervals *between* my thoughts? Do I die? Even when Descartes seems to get hold of a state he can call 'certainty', he does not immediately recognize himself. Having lost the world, he is on the verge of losing himself. 'Who am I?' he asks. Characteristically enough, the thinker's relationships with other people are likewise represented as seriously disturbed. Only a madman would mean these words: 'When looking from a window and saying I see men who pass in the street, I really do not see them, but infer that what I see is men, just as I say that I see wax. And yet, what do I see from the window but hats and coats which may cover automatic machines?'

To insist that Descartes was only feigning this madness is to miss the point entirely. What calls for thought is the striking fact that the narrative in his *Meditations* attempts to draw us into a succession of mental states which, whether they be feigned or not, nevertheless very accurately reproduce, in clinically familiar outlines, the very same configuration of psychopathology we are now seeing very commonly in the populations of our civilization. The *Meditations* reflects, like a mirror, the future character of subjectivity: egoity, power, anxiety, certainty, security, possession, domination, narcissism. It would be very difficult to deny that the postures and positions which Descartes assumes are not isomorphically, homologously correlated with narcissistic disorders. Even the depressive withdrawal is depicted. The homology suggests that there could be a concealed interaction between the history of metaphysics and the aetiology of the narcissistic disorders characteristic of the modern world.

In 'The Age of the World Picture', Heidegger asserts that,

> With Decartes begins the completion and consummation of Western metaphysics. . . . With the interpretation of man as *subiectum*, Descartes creates the metaphysical presupposition for future

> anthropology of every kind and tendency. In the rise of old anthropologies, Descartes celebrates his greatest triumph. . . . Descartes can be overcome only through the overcoming of that which he himself founded, only through the overcoming of modern, and that means at the same time Western, metaphysics.[212]

Narcissistic pathology takes over in metaphysics with the triumph of Cartesian subjectivism; and it is related to the glorification of the rationality in 'objectivity', which turns against the subject and threatens to destroy it.

Observing that, in narcissistic disorders, 'there is a grandiose power drive', Schwartz-Salant emphasizes that, 'Under its control, certainty must reign and chance be suspended.'[213] The narcissistic character, he adds, 'insists upon total determinacy of events, so that spontaneity and chance are defended against.'[214] Is it any wonder, then, that the modern phase of metaphysics, dominated by our collective ambition to enjoy certainty in all things, should coincide with the appearance of narcissistic disorders? For such a character type, self-consciousness tends to become a 'new form of order within himself which [assumes that it] could survive in isolation from the environment.'[215] This inflated posture of the ego, assuming self-sufficiency and self-mastery, is, however, just a sham or a defense: the sad truth is that the narcissistic character is profoundly insecure, haunted by groundless anxieties, and troubled by self-doubts, low self-esteem (Nietzsche spoke of self-loathing) and weak self-integration, a weak sense of his own identity, his being in its wholeness. The 'false' subjectivity of the narcissistic ego requires, and is constantly dependent upon, the mirroring of others – or perhaps a god – to confirm its most basic sense of reality. Desperately trying to avoid its inner emptiness, the narcissistic ego gets caught in a fantasy of 'fullness', 'completeness' and 'self-sufficiency'. It is hurt again and again by its own defenses – just as subjectivity is attacked in the name of an objectivity it itself established and suffers its reduction to the structure of subject and object. And the isolation the ego proclaims for the sake of self-mastery, self-knowledge, and certainty becomes a source of anxiety: the isolation is really a separation, and separation always means loss, the experience of loss and lostness – the kind of experience which undermines the very sense of one's being.

Since, for Descartes, the senses are nothing but a source of deception and the body is nothing but perishable matter – that is to say, they are challenges, in both cases, to the power of the *ego cogitans*, the ego must 'abandon' them; the Cartesian ego is a *cogito* which has dissociated, split off, from its embodiment and taken *itself* as the object of its 'love'. In order to possess absolute certainty and security, Descartes undergoes a process of separation and withdrawal, methodically abandoning all the 'objects' of the body's desires and taking himself, as purely thinking substance, for 'object'. This is narcissistic process, homologous to the process clinically recognized as the defensive comportment of severe depression. In the isolation of human beings from each other and the separation of human beings from Being, there is indeed cause for deep depression. With astonishing prescience, Nietzsche could already see the depression and interpret it as a signifier of nihilism. Both Nietzsche and Heidegger speak of apathy, indifference, intoxification, narcotization: this is what they see.

Because of its destructive narcissism, the Cartesian ego set itself up for the violent contemporary attack on it which is now taking place within Western metaphysics. Thus, for example, Merleau-Ponty has argued, against the ego's narcissistic claims, that,

> we find in our experience a movement towards what could not in any event be present to us in the original and whose irremediable absence would thus count among our originating experiences.[216]

He is reminding us that the ego which appears in the texts of metaphysics must begin to cope with impermanence, with the absence of knowledge, with incompleteness, uncertainty and insecurity: it must begin to accept the impossibility of an absolute grounding in authoritative truth. But today, as the Cartesian *cogito* is violently dispossessed, as its epistemological possession of itself is irrevocably taken away from it, the ego finds itself increasingly 'possessed' by the nihilism raging within its discourse of power. Predictably, the Cartesian ego's more worldly relative is now bent on absolute control and security by investing its will to power in the political economy of an advanced – but perhaps fatally uncontrollable – technology.

The metaphysical ego, narcissistically inflated, has long assumed the possibility of total presence, absolute plenitude. It is now being disillusioned and dispossessed. In the same way, the modern Self is being compelled to face the most extreme form of absence: an absence which, following Nietzsche, we might call 'the death of God'. Nietzsche's words confront metaphysics with the truth of a terrible absence. But the empty and groundless subjectivity of the metaphysical ego is a reflection of our historical life. Narcissistic disorders tell us something about how, as a culture, we are *living* with this absence – the death of God, the nullification of all meaning. Schwartz-Salant maintains that,

> Narcissistic problems . . . must . . . be understood as purposive, the symptoms of a new Self image attempting to incarnate, either in the individual or in the collective, or both. They may represent nothing less than the psyche's response to the call that 'God is dead', raised by Nietzsche many years ago, and painfully brought home to us in the fragmentation our modern society suffers and creates.[217]

I submit that the 'death' of God is connected in an essential way with the rise to power, and the suffering, of the modern ego. The modern ego established its power by putting God to death and making *itself* the new god. Thus egological inflation of subjectivity spells the 'death' of God. But the prevalence of narcissistic disorders indicates that this event involves a loss which causes much suffering.

For Descartes, the ego is clearly dependent on God; but this dependency should not obscure the fact that, in the process of demonstrating the existence of God, the ego has fatefully elevated itself to a position that can only be called godlike. Thus, in the technological revolution following Descartes, this ego has really begun to dream of, and increasingly thought itself able to demonstrate, a godlike omnipotence and omniscience. And yet, despite this historical rise to power, the ego continues to be absorbed in the phantoms fed by its own weaknesses: it is not at all ready to assume *higher* responsibility for itself. It nullifies the 'God' of our fathers, but cannot live, and may not survive, without the dimension of 'transcendence'.

The 'death' of God is a loss, an absence, an abandonment: an event from which we have not been able, despite appearances, to recover. The 'death' of God corresponds, in fact, to the historical appearance of a deep and pervasive cultural depression. In narcissistic disorders, depression is inseparable from inflation. Severe depressions are prevalent in our contemporary world. I see them as related, through many mediations of culture, to what Heidegger calls the 'abandonment of Being'. I see them as related, also, to a nihilism that spreads through the modern inflation of the ego.

In *The Will to Power*, Nietzsche asserts that,

> The nihilistic question 'for what?' is rooted in the old habit of supposing that the goal must be put up, given, demanded from outside – by some superhuman authority. Having unlearned faith in that, one still follows the old habit and seeks another authority that can speak unconditionally and command goals and tasks.[218]

We are throwing off the chains of external authority – imposed meaning – but we have failed, as a culture, and not only as individuals, to develop our capacity for true subjectivity: a subjectivity, that is, in which we live a life true to ourselves, making contact with the goals and tasks our ownmost body of experience can suggest, and being guided by what it gives us. Both depression and narcissism are rooted in false subjectivity: if we look into the heart of depression, what we will see is a deep despair, despair born of the conviction that existential meaning – goals and tasks – cannot possibly be generated from within the body of experience; in narcissism, the ego defends itself against depression, but the defense it constructs revolves around a self-image which is not its own, and through which, therefore, it is further alienated from itself.

At the very heart of the ego's culture of narcissism, we will always find a painful relationship to power: a self-destructive relationship to matters of strength, energy, effectiveness, potency, vitality, authority, independence, stature, status, esteem, recognition, achievement, performance, mastery, control, domination, glory, success, demonstration. As Theodor Adorno observes,

> Today self-consciousness no longer means anything but reflection on the ego as embarrassment, as realization of impotence: knowing that one is nothing.[219]

Adorno explicitly connects the psychology of nihilism with the ego's experience of impotence, a narcissistically disturbed relationship to power. Adorno is also saying that this connection is particularly intense in the present historical situation.

For further elaboration, let us turn to Jules Henry, who asserts, in *Pathways to Madness*, that,

> Life in our culture is a flight from nothingness. This fact has been evaded by recent psychology through the invention of the notion of 'effectiveness' ('mastery', *mutatis mutandis*). Psychology tells us that what we really want is to feel effective. . . . The silent anguish of many of us, however, warns that many who are frightfully effective feel like nothing nevertheless.[220]

This diagnosis takes on even greater significance when we consider it in the light of Heidegger's interpretation of the history of metaphysics. According to Heidegger, 'Being presences [in the modern world] as effecting.'[221] Consequently:

> Assurance of himself and of his effectiveness determines the reality of man. . . . Man builds upon and builds up for himself what is real as what has an effect upon him and as what he effects. What is real becomes what can be effected. . . .[222]

Tracing the history of what metaphysics calls *existentia*, existence, reality, Heidegger contends that

> *Energeia* is reinterpreted to mean *actualitas* of the *actus*. *Agere* as *facere*, *creare*. . . . Accomplishment as effecting what is effected, not allowing to presence in unconcealment, characterizes the *actus*.[223]

Thus, in time, 'Existence as *actualitas*, reality, effectedness and effectingness, becomes the objectivity of experience. . . .'[224] Depressions and narcissistic disorders tell us, through their suffering, how this history of Being affects our lives.

Jules Henry continues:

> At any rate, acting on something, being effective, is somethingness, and ineffectiveness is an abyss out of which we try to climb by acting on something. The trouble with being merely effective is that not only does it often fail to destroy the feeling of nothingness, but it also wrecks the peace of others. . . . Hence the truth of the matter is that action stemming from the feeling of nothingness, from despair, only makes the object of the action more despairing.[225]

A vicious cycle, a wheel of suffering, is immediately set in motion, when the ego's investment in the will to power assumes a character that is bound to fail. If, for the modern ego, to be is to be powerful, impotence or ineffectiveness can only spell nothingness. Since the narcissistic ego sets very high stakes, very high demands, the inevitable limitation on human power can only confront the ego with deep frustration, devastating feelings of ineffectiveness and helplessness. When Being as such is reduced to will, i.e., effecting, the being of subjectivity is also reduced. But the reduction of subjectivity means that we are either effective, i.e., powerful, or else we are nothing. Action stemming from the ego's experience of nothingness is likely, however, to be violent and destructive, moved by hatred, envy, fear and self-loathing:

> Effectiveness that is merely flight from nothingness gives rise to a darting destructiveness in which he who feels he is nothing hurls himself upon others in order to escape his illness.[226]

Jules Henry is of course referring, here, to narcissistic psychopathology as it manifests through the individual; but what he says should be understood in connection with our interpretation of cultural history. According to this interpretation, our historical present falls within, and consummates, the epoch of nihilism: since Being is reduced to power, will, will to power, any loss of power means loss of being, nihilism. Thus I want to suggest that narcissistic character disorders are to be expected in our present historical situation. Although such disorders can be seen to *contribute* to the historical unfolding of nihilism, they are also to be seen as *generated* from within the self-destructive dialectic of power that prevails during the epoch of nihilism. There is a very real basis, in the conditions of social reality, for our wide-spread feelings of impotence, ineffectiveness, and helplessness. Stated paradoxically: since real powerlessness is a normal experience of the modern Self, it could be said that we are being driven into narcissistic disorders in order to maintain some degree of 'sanity'.

Schwartz-Salant writes as follows:

> Incessant *doing* is a chronic condition of the narcissistic character. His basic belief that no center exists within [and that there is] no source of rest, results in seemingly endless activity, whether of an internal fantasy nature or an external rush to more and more achievements and tasks.[227]

Today, our technological economy is organized in 'attunement' with the restless activity of the narcissistic character. And it is in this regard that Heidegger argues that the human being,

> precisely as the one so threatened, exalts himself to the posture of lord of the earth. In this way the impression comes to prevail that everything man encounters exists only insofar as it is his construct. This illusion gives rise in turn to one final delusion: it seems as though man everywhere and always encounters only himself. . . . In truth, however, precisely nowhere does man today any longer encounter himself, i.e., his essential being.[228]

Is this not the archetypal delusion of the youth, Narcissus?

Since narcissistic disorders are associated with self-destructive defenses against insecurity and anxiety, what Heidegger has to say about the will to power in a technological economy begins to take on a heightened significance:

> The preservation of the level of power belonging to

> the will reached at any given time consists in the will's surrounding itself with an encircling sphere of that which it can reliably grasp at, each time, . . . in order . . . to contend for its own security.[229]

To be sure, as he points out, the system of defense may institute a measure of control, allowing a glimpse of power over the whole of Being:

> That encircling sphere bound off the constant reserve of what presences . . . that is immediately at the disposal of the will.[230]

However, since defenses can never be totally secure and under control, anxiety and paranoia are only bound to intensify – with the inevitable result that feelings of insecurity and powerlessness also increase. A more 'powerful' will does not necessarily lead to more, or better control. This self-destructive pattern can be discerned not only in the character of the narcissistic individual, but also in our technological drive to control the 'resources' of nature and in our national postures of military defense and nuclear armament. Furthermore, this pattern is reflected in the history of our metaphysics, where the abandonment of Being is associated with an egoity essentially involved in a grandiose attempt to secure and make certain, and where this attempt to gain control over Being and *be* in control gradually degenerates into the drive of a will to power that cannot tolerate the irreducibility of Being and makes way for the discourse of its rage: the discourse of nihilism.

The Western world is a world totally organized, today, around the activities of producing and consuming. How are these activities related to narcissism and the experience of nihilism? People suffering narcissistic disorders are haunted, and moved, by an experience of inner emptiness: lacking self-esteem, what they feel is the absence of a real centre, the absence of a Self. I think that Heidegger's interpretation of the history of metaphysics opens up a dimension of experience which can help us to understand the suffering characteristic of narcissistic disorders:

> This emptiness has to be filled up. But since the emptiness of Being can never be filled up by the fullness of beings, especially when this emptiness

> can never be experienced as such, the only way to escape it is incessantly to arrange beings in the constant possibility of being ordered as the form of guaranteeing aimless activity.[231]

And he argues for a historical correlation between this 'aimless activity' and the emergence of modern technology:

> Viewed in this way, technology is the organization of a lack, since it is related to the emptiness of Being. . . . Everywhere where there are not enough beings – and it is increasingly everywhere and always not enough for the will to will escalating itself – technology has to jump in, create a substitute, and consume the raw materials.[232]

Thus Heidegger sees a 'circularity', a wheel of suffering, in 'consumption for the sake of consumption': the 'unconditional consumption of beings' to defend against the suffering of the Self in 'the vacuum of the abandonment of Being'.

Heidegger actually pushes the diagnosis even further, for he suggests, in the very same essay – an essay, we should note, which concerns the history of metaphysics – that there is an essential historical connection between (i) our self-destructive strategies to dominate in the race for 'defensive' superiority and (ii) our equally self-destructive technological economy, which, as we shall see, is bound up very deeply with the historical prevalence of narcissistic disorders. According to him, the 'ultimate abandonment of Being' is

> hitched into the armament mechanism of the plan. The plan itself is determined by the vacuum of the abandonment of Being within which the consumption of beings for the manufacturing of technology, to which culture also belongs, is the only way out for man who is still engrossed with subjectivity in superhumanity. . . .
>
> The consumption of beings is as such and in its course determined by armament in the metaphysical sense, through which man makes himself the 'master' of what is 'elemental'. The consumption includes the ordered use of beings, which become the opportunity and the material for feats and their escalation.[233]

For the narcissistic character, the dialectic of power becomes a process of increasing self-alienation – what Joel Kovel, a radical psychiatrist, calls our 'neurosis of consumption', because the logic of capitalism in its late phase requires the constant production and commodification of new 'needs' and 'desires'.[234] The individual is gradually subdued by the power of this dialectic. Once the ego has been 'hollowed out', it can be inflated, filled with the commodities that its artificial needs pursue.[235] Producing and consuming belong, in fact, to a psychological system of defense. This defense is reflected in the metaphysical armament, the metaphysical defense, of which we have just heard Heidegger speak.

Both Horkheimer and Adorno emphasize the irony in the fact that, because of its obsession with the illusions of power, the modern ego, individualism gone wild, now finds itself controlled for the benefit of an economy which can only survive through the uncontrollable growth of production and consumption.[236] The will to control eventually produces a political economy that is *out* of our control. As narcissistic dispositions begin to take root in the traditional culture of character, they produce an environment which enormously intensifies their latent potential for nihilistic pathology. 'Self-esteem', the Self's experience of itself in its worthiness-of-being, used to be measured by productivity and effectiveness – by deeds and works. Now, in a later phase, it is increasingly turned into material worth of being, measured by the hedonistic calculus of conspicuous consumption. The 'emptied' ego, driven by consequent rage and envy, no longer possessing, but now, rather, itself a being possessed, and alienated from itself even through its available forms of resistance, is thus set up for the pain of even greater self-destruction in a cycle of historical changes from which no-one can entirely escape. We must accordingly attempt to see beyond individual psychopathology, beyond the more severe cases of 'abnormality' which attract our attention, in order to recognize, within the prevailing conditions of our civilization, the existence of massive pressures and strains, all of them significant in the appearance of typical narcissistic pathology.

In the age of images, *esse est percipi*. This reduction of human being, and of Being as such, to its representedness, its being perceived, its being seen, is what links narcissistic epidemiology to the spread of nihilism. The narcissistic character lives out, in a mostly invisible suffering, the historical condition of being *nothing* but an image, *nothing* but what can be seen, *nothing* but what is visible.

According to Freud, secondary or pathological narcissism involves the 'incorporation' of grandiose object-images as a defensive response to the insecurity, anxiety and guilt which accompanied the ego's earliest infantile movements of differentiation and individuation. Severe pathology will later manifest, therefore, when the adult ego becomes totally entangled in these images, these mere phantoms of lost objects. For Heidegger, too, there is a serious pathology when thinking gets entirely caught up in a metaphysics of representation. According to him, this entanglement spells our historical closure to a more 'healing' experience of Being as it presences, i.e., as a whole.

Jules Henry can help us understand the image as a fateful historical link between narcissism and nihilism. He writes:

> And hence the sham. For when man is nothing, he lives only by impacts from the outer world; he is a creature external to himself, a surface of fear moved by the winds of circumstance. . . .[237]

We *become* nothing when we are possessed by the image, and when the image is the measure of our being. In 'The Age of the World Picture', Heidegger asserts that, 'Where anything that is has become the object of representing, it first incurs in a certain manner a loss of Being.'[238] If, in our epoch, Being as a whole goes into the self-concealment of 'oblivion' while images or representations complete their surveillance and domination over us, the ego which is not seen, seen uninterruptedly, has no defense against the dread of its nothingness.[239] And yet, it is also true that the *visibility* of the ego is a situation of great danger, as modern technologies of surveillance should be teaching us. To the metaphysics codified in the position that *esse est percipi*, there corresponds, at the limit, a totalitarian politics of surveillance and supervision. Because this politics imposes on our being its own forms of representedness, it calls forth into being a character compelled to regard itself in terms of an alien image; in other words, it calls forth a narcissistic disorder of character.

To escape an entanglement with images which could be extremely self-alienating and self-destructive, deeply pathologizing our being, requires, as Schwartz-Salant has argued, 'a capacity to observe images [and learn from them] without identifying [and fusing] with them.[240] In narcissistic disorders, this capacity is painfully stuck in a state of confusion, for the ego swings back and forth between fascination and dread. What dominates is what glitters, what shines: the image, the illusion, the sham, the surface. So narcissistic disorders are precisely what we should expect to see in a society which, as Bryan Turner, an English sociologist, has put it, 'reality becomes entirely representational'.[241]

In our present epoch, the epoch Heidegger calls 'the age of the world picture', to be is to be perceived: seen, noticed, brought forth into the visibility of the spectacle. The Self, the ego aware of its limits, feels itself today to be nothing but, nothing if not, a fetishized commodity, an image, a collection of masks, something to be produced for, and consumed in, the spectacle of life. In *Human Nature and the Social Order*, Charles H. Cooley calls the current ego a 'looking glass self'.[242] To live in an age dominated by representation is to be constantly pressured into self-alienation, for we are compelled to become dependent, for a sense of our identity, on the power of the re-presentation – our otherness, our coercion and subjugation. Narcissistic disorders point to the destructiveness in the power of the image; they signify some of the ways in which the domination of this ontology can damage, and even annihilate, the healthy formation of character.

Narcissism and nihilism must be interpreted together, and in their essential interconnectedness, because narcissistic pathology is basically character – the being of the Self – under the spell of the image, social and cultural representations; and nihilism is the pervasive negation of Being in an epoch when quantitative re-presentations dominate all experience and cut us off from the deeper dimensions in which our well-being is (to be) rooted. Narcissism is a fateful form of suffering, for the destruction of character it makes visible constitutes an urgent epidemiological warning against the nihilism of our time.

The ontology of the image produces human suffering. And it deepens this suffering by framing it in an image. We need to see beyond the image into the depth of the suffering.

Epilogue
The Gaze and its Vanishing Point

TEXTS AND COMMENTS

Text (1). 'Another problem about which I am still not clear is the perception of the sunset and the Copernican revolution. The question is whether the sunset is a necessary representation, or whether a seeing is possible for which the sun does not set.' Heidegger, *Herakleitos Seminar 1966/67* (Seminar 8, p. 87)
Comment. Here I want to point out that the setting of the sun is the power of nature, and that a seeing for which 'the sun does not set' could be a nightmare, signifying the final conflagration, or a future time in which we would inhabit an environment whose lighting was totally controlled by our technology. Thus, the vision for which 'the sun does not set', could also appear in a story of political tyranny: the Panopticon, total surveillance, total visibility, total control of vision. Heidegger, however, wants to know whether the sunset belongs to a vision of nature that will persist, or whether our theoretical knowledge about the relationship between sun and earth will ever be so deeply embodied, so deeply inscribed in the flesh of our visionary being, that we finally see it *only* that way. Could history make that kind of impact on the visionary body? Foucault would doubtless argue that it could. I do not agree. But I would point out that Heidegger's words are indeterminate, open to many interpretations. We need perhaps to consider the possibility that there may be more than these two standpoints and viewpoints. The 'not setting' could be read symbolically – in which case, it suggests many stories of vision, many visions of Being. I am thinking, for example, of a vision that encompasses, with the imagination, the journey of the sun through the underworld: the passage of the sun, after its setting, through an invisible world. For such a vision, the sun would both set and not set. This is not a vision based on scientific knowledge or reasoning, but rather a vision produced by the imagination. The sun's never setting is not seen as a consequence of the laws of physics and astronomy, the laws of conservation and substantial identity, but rather as the mystery of unconcealment. In the scientific

'vision', the 'never setting' is a function of the totalizing homogeneity of the space-time grid into which all things are set. For the mythopoetic vision, the 'never setting' expresses the reach of the seer's imagination – a felt embrace of the sun's absence and of the darkness it leaves behind.

Text (2). 'I am in complete agreement with this [the assertion that "I possess a way of thinking which shows me that the sun of the astronomer is superior to the sun of the peasant"] and for two reasons. Recall the famous phrase from Hegel: "The earth is not the physical center of the world, but it is the metaphysical center." The originality of man in the world is manifested by the fact that he has acquired a more exact knowledge of the world of science. It is strictly necessary that we teach everybody about the world and the sun of the astronomer. There is no question of discrediting science. Philosophical awareness is possible only on the basis of science. It is only when one has conceived the world of the natural sciences in all their rigor that one can see appear, by contrast, man in his freedom. What is more, having passed a certain point in its development, science itself ceases to hypostatize itself; it leads us back to the structures of the perceived world and somehow recovers them.' Merleau-Ponty, Address to the Société française de philosophie, 1946 (*The Primacy of Perception*, 36-37)
Comment. Does Merleau-Ponty's amplification settle the point where he is standing? Does he demonstrate any superiority? We shall soon be looking more closely into that 'certain point in its development' where science imposed its geometry on the vision of the painter, the one who represents the world as a picture.

Text (3). Hegel, *The Phenomenology of Mind* (Baillie translation, Humanities, 1949): Hegel speaks of 'the objectivity of the light of the rising sun' (718) and the 'pure all-containing, all suffusing Light of the sunrise' (700), the 'Being of the Rising Sun' (726), and 'the daylight of the present' (425).
Comment. The mythology of the never-setting sun belongs to the metaphysics of presence and the patriarchal culture: the culture which rules over vision with its science and technology.

Text (4). *'Realism*: The word is used . . . to denote the pictorial attitude which depends upon a view of the world structured according to the laws of geometry, specifically linear perspective.' Samuel Y. Edgerton, *The Renaissance Rediscovery of Linear Perspective* (198-9)

Text (5). '. . . the metaphysic of Aristotelian space fell into disfavor in the late thirteenth century, [and] artists both north and south of the Alps began to accept . . . the kind of space we see empirically, [i.e., phenomenologically,] without theological preconceptions or mathematical structure. . . . Finally, in the fifteenth century, there emerged mathematically ordered "systematic space", infinite, homogeneous and isotropic, making possible the advent of linear perspective.' Edgerton (161)
Comment. I wonder, though, whether we ever, even now, see and live in a space without traces of earlier theological preconceptions.

Text (6). 'So far as we have been able to ascertain, no civilization in the history of the world prior to the time of the ancient Greeks and Romans ever made pictures according to this procedure. . . . But whatever the ancients may have known of perspective constructions vanished into oblivion during the Middle Ages in Western Europe. Nor did any of the other great cultures of the world . . . seem to have been interested in this geometric-optical way of picture-making'. Edgerton (5)
Comment. Picture-making is one thing, seeing is another. We may suppose that the most primitive peoples see, like us, with a vision more or less unconsciously informed and moved by a sense of linear perspective. Otherwise, how could they survive? And yet, Colin Turnbull, an anthropologist who lived among the Forest People of the Congo, discovered that, because of the dense vegetation and closeness of their living space, the Forest People do *not* have any ready-to-hand sense of perspective and size-distance relationships when outside their forest environment. Nevertheless, they do learn it easily and quickly. See Colin M. Turnbull, *The Forest People: A Study of the Pygmies of the Congo* (New York: Simon and Schuster, 1961). In any event, the introduction of linear perspective into the picturing of the world and later, the unquestioned domination of perspective principles, clearly require interpretation by reference to

social, economic, historical and cultural conditions. Moreover, even if it be true that the populations of Western civilization have always inhabited a visible world unconsciously structured by the principles rediscovered in the Renaissance, we still need to consider how the explicit consciousness of these principles, and the discursive hegemony they achieved through the pictures of the Renaissance schools, could have produced historically significant changes, not only in subsequent forms of picturing, but also in subsequent ways of seeing and living.

Text (7). 'We cannot help wondering why it was the artists of the Italian Renaissance who, unlike any of their fellow craftsmen outside Western Christendom, discovered or invented – once more – the geometric laws of pictorial representation. The distinction between the words "discover" and "invent" here is intentionally provocative. "Discover" implies that linear perspective is an absolute scientific truth, universal to all men regardless of cultural background or historical period. "Invent", on the other hand, suggests that linear perspective is only a convention, the understanding or adoption of which is relative to the particular anthropological and psychological needs of a given culture. The issue is moot, and the question as to whether perspective is revealed or imposed continues to stir debate. . . .' Edgerton (6)
Comment. Perhaps the debate is generated by a false dilemma, a false opposition. I submit that, with a hermeneutical understanding of the question, the debate would move to a different level.

Text(8). 'In the *Map with a Chain*, we can see clearly what happened to human pictorial orientation after the advent of linear perspective. This new Quattrocento mode of representation was based on the assumption that visual space is ordered a priori by an abstract, uniform system of linear coordinates'. Edgerton (7)

Text (9). 'The painter of the earlier picture [the fresco in the Loggia del Bigallo, painted in 1350, one hundred thirty years *before* the *Map with a Chain*] did not conceive of his subject in terms of spatial homogeneity. Rather, he believed that he could render what he saw before his eyes convincingly by representing what it felt like to walk about, experiencing

structures, almost tactilely, from many different sides, rather than from a single, overall advantage. In the *Map with a Chain*, however, the fixed viewpoint is elevated and distant, completely out of plastic or sensory reach of the depicted city'. Edgerton (9)
Comment. The earlier painting evoked a phenomenological reality within which and for which the presence of the human being as a bodily being was essential. In the later, the body and its lived experience have been suppressed.

Text (10). '*Extromission*: The ancient theory of vision which held that visual rays issue forth from the eyes'. Edgerton (196)

Text (11). '*Intromission*: The ancient theory of vision which held that rays, which prompted sight, issued from all things in nature toward the eye, converging on it in the form of a cone. This theory exerted the greatest influence in medieval and Renaissance Europe following the optics of the Arab Alhazen'. Edgerton (197)

Text (12). '*Visual ray*: The entity believed by classical and medieval optical scientists either to project from the eye or into it from the seen object. Since Euclid, always understood as traveling in straight lines, hence subject to geometric laws'. Edgerton (200)
Comment. This vision in straight lines, the shortest distance, is a well-disciplined vision. Such disciplined vision is necessary, of course, for reading; but it is a vision accustomed to linear conformity, prepared for regimentation. What would a vision moved by a more hermeneutical attunement be like?

Text (13). '*Visual axis* (*axis visualis* and *axis perpendicularis* in medieval optics; also the *centric visual ray* which Alberti dubbed the "prince of rays"): The single visual ray understood by both extromission and intromission theorists in ancient and medieval optics which conveys the image most clearly and distinctly to the brain.' Edgerton (200)
Comment. This single ray has dominated the history of metaphysics as well as post-Renaissance painting. This hegemony is more serious in metaphysics, since picturing is (in part, at least) a matter of conventions, whereas metaphysics presumes to speak the truth.

Text (14). '*Centric ray*: The single visual ray, according to the medieval science of optics, which travels from the center of the viewer's visual field to the exact centre of his eye, to the exact centre of the sensitive seat of vision, the *crystallinus*, and to the exact center of the optic nerve. Since this ray arrives perpendicular to the cornea, the *crystallinus*, and the face of the optic nerve, it remains unrefracted. Being unrefracted and also the shortest ray between object seen and optic nerve, it was thought to carry its portion of the object most distinctly to the brain.' Edgerton (195)
Comment. The paradigmatic use of the single ray in our thinking about vision belongs together with the paradigmatic assumption of a single, fixed centre and origin. Thus we see the involvement of a subject-centred metaphysics: a metaphysics which sees the subject as absolute centre and origin.

Text (15). 'We spoke metaphorically of a "mental glance" or "glancing ray" of the pure Ego, of its turnings towards and away. We brought the phenomena belonging to this context under a unity and into completely distinct relief'. Husserl, *Ideas I: General Introduction to Pure Phenomenology* (New York: Macmillan, 1931), p. 246.

Text (16). 'In every wakeful *cogito* a "glancing" ray from the pure Ego is directed upon the "object" of the correlate of consciousness for the time being, . . . and enjoys the typically varied consciousness *of* it.' Husserl (223)

Text (17). 'Just as noetically a ray of love proceeding from the Ego splits up into a bundle of rays, each of which is directed towards a single object, so too there are distributed over the collective object of affection as such as many noemetic characters of love as there are objects collected at that time. . . .' Husserl (314)
Comment. Starting with the assumption of a single, fixed centre of subjectivity and of 'consciousness' as inherently 'raylike', Husserl's phenomenology seems incapable, here, of illuminating the experience in question. The account he offers provokes thought; but first, I think, it provokes laughter.

Text (18). 'Nowhere in the Manetti account does it say what the visual angle was at Brunelleschi's eyepoint inside

the doorway of the Duomo. The biographer only mentions that the artist painted what could be seen *a uno sguardo* – "at a glance" from where he stood. In some of the old optics treatises, it was stated that the human eye could take in as much as ninety degrees; but in *Della prospettiva,* which may better reflect Florentine opinions about the science during Brunelleschi's own lifetime, the visual angle was always specifically regarded as "always acute and never as an obtuse or right angle" '. Edgerton (141)
Comment. Perhaps we should consider what would happen if such a glance, rather than the frontally confrontational look, were to become the prevailing paradigm for vision and its access to truth.

Text (19). '*Vanishing point*: The Term in English since the eighteenth century to express the phenomenon of convergence of parallel lines in distant vision. Referred to similarly as the *punto di fuga* (point of flight) in Italian. . . . Originally designated only as the *punto centrico* by Alberti, referring of course to the centric vanishing point in frontal perspective. . . . Since the science explosion of the seventeenth century, the notion of the vanishing point as a mathematical as well as artistic function has been more and more appreciated. Hence we are led to believe today that this phenomenon is an established fact of vision. However, the medical science of optics taught just the reverse – that the so-called illusion of perspective convergence (i.e., the vanishing point) was a mere *Fata Morgana* and could be disproved by simple geometry. Brunelleschi and Alberti, therefore, had their work cut out for them. They had to show that this strange illusion could actually be recapitulated and predicted by applied geometry. In the beginning they were only concerned that the 'centric point' worked with enough uniformity to make their pictures seem convincing in light of the contemporary belief that anything that functioned by mathematical law must be inherently in harmony with God's masterplan'. Edgerton (200)
Comment. The vanishing point is the point which vanishes, flees, eludes visual control. Perspective paintings give the impression that this point is completely under control within the structure of representation. We must not forget that this point points into that into which it vanishes: the point structures the surface of the painting and then withdraws into the invisible – into a space which Euclidean geometry

can neither master nor regulate. This (truth) is hard to see, so long as our vision comes under the sway of the correspondence theory of truth.

Text (20). 'While optics was a science, it was also a branch of the Faith. The initial impetus to study it in medieval Europe was that its geometric concept of light-filled space provided some kind of rationalization of how God's grace pervaded the universe. Meister Eckhart, the early fourteenth-century German mystic, noted that the soul "sees" God just as the eye itself beholds a picture on the wall. . . . Even St. Antonine, the ascetic archbishop of Florence (1446–1459), adapted ideas from optics for the purposes of moral allegory throughout his *Summa theologica*. In one whole chapter, entitled "Concerning These Things Which are Required for Good Seeing Both Corporeally and Spiritually", . . . he drew liberally from the science in order to make earnest correlations between spiritual "seeing" and visual pathology.' Edgerton (60-61)
Comment. Although optics originally demonstrated the rationalization of space as a way of glorifying God, the progressive rationalization of vision and its space gradually threatened the existence of God. The total rationalization of space rigorously excludes the presence of God. The optical eye cannot enjoy a vision of God. Meister Eckhart's analogy is worth noting in light of the transformation of the world into a 'world picture'. Even Eckhart unwittingly served the purposes of this transformation! What St. Antonine sought to make visible, namely, correlations between *spiritual seeing* and *visual pathology*, is very much the task of this book.

Text (21). In the *Opus Majus*, 'Bacon included a section on optics, whose geometric laws – he wished to show – reflected God's manner of spreading His grace throughout the universe'. Edgerton (16)
Comment. Such indeed is the 'cunning of history', that a vision of the world promoted in order to celebrate God and His Creation should turn into the instrument which contributed to His Death.

Text (22). 'Yet the Renaissance "rediscovery" of perspective was not immediately heralded as a victory of objective reality over medieval mysticism. To the contrary, the early users of the new art-science thought of it as a tool which might help restore the moral authority of the Church in a world becoming progressively materialistic'. Edgerton (7)

Text (23). 'The picture, as constructed according to the laws of perspective, was to set an example for moral order and human perfection.' Edgerton (24)
Comment. The picture, thus constructed, was to serve as a model for our *vision*. Implicit in this painterly instruction, we may detect the conviction that, if our *vision* would only conform to these optical laws, it would serve the moral order and human perfection. We can now see the effects of this viewpoint.

Text (24). 'For all its errors, medieval optics was a completely rational and mechanistic science, relatively uncluttered by theological mystery. It cemented together the best scholarship and talents of three great civilizations. . . . No wonder, then, especially as the science reached a wider population during the fourteenth and fifteenth centuries, that people not only took for granted these explanations, but also tended more and more to "see" the natural world according to these Euclidean or geometric rules.' Edgerton (65)
Comment. Since vision is not just a biological function; since vision has changed in relation to science and culture, we have to correlate the visionary 'psychopathology' considered in this book with the need to change society and culture. Thus I find Edgerton's use of quotation marks – he writes "see" – provocative. Do these marks turn the seeing in question into metaphor or embellishment? Or are they to indicate a hermeneutical movement? What if vision is inherently hermeneutical? Edgerton's title speaks of the Renaissance event as a 'rediscovery' of linear perspective. This word suggests the uncovering of something hidden *through* culture, and perhaps an original discovery, *for* culture, of something hidden within *nature*: in any event, a hermeneutical situation: perspective is neither found nor created, but rather . . . I prefer to say 'retrieved' instead of 'discovered'. More to the point, I want to think of the historical event in terms of *phainesthai*, *phōs*, *physis*, and *alētheia*. The Renaissance event was historically distinctive, even though *knowledge* of perspective can be traced back through Roger Bacon, Galen, Ptolomy, Vitruvius, Euclid, Aristotle, Plato, and farther back, perhaps to Pythagoras.

For it was not until the Renaissance that an ancient heritage of texts, practices and experiments, reports and letters – in short, discourses – burst into painting, burst on the scene, burst into the world of painting itself. Only in the Renaissance did *this* vision come into being as and in a *picture*. The Renaissance world re-presented the happening of vision happening to itself. Through this picture of the world, vision saw itself *in* perspective *as* perspective for the first time in the history of (Western) civilization.

In *The Order of Things: An Archaeology of the Human Sciences* (New York: Random House, 1973), pp. 308-311, Foucault discusses 'Las Meninas', a painting by Velasquez that illustrates the historical developments Heidegger, Jean Paris, and Edgerton are concerned with. Foucault points out that, in the classical picture, there is never any representation of the act of representing – nor any of the subject who is engaged in representing: in other words, there is no attempt at self-representation. But, without this self-consciousness, there is no sovereign subject, for sovereignty requires self-mastery and self-control, a precondition for the sovereign domination of the visible world by means of a panoramic vision. 'Las Meninas' is a moment in the history of painting which correlates with the Cartesian and Kantian revolutions in the history of philosophy.

The Renaissance retrieval brought perspectivity forth in a visually striking and beautiful way; it made it, or rendered it visible *to* vision as the inborn law of its own methods and movements. The fact that knowledge of perspective was *applied* to painting for the first time during the Renaissance period, the fact that theory suddenly became ready-to-hand knowledge, is of much less significance than the fact that this shift from theory to practice took place *in painting*, that is to say, in vision's self-representations. Because the event took place *in the field of vision*, it happened as a mirroring, and therefore reflexive process. The picture showed vision to itself; and it was this mirroring of vision in front of itself which recast history.

Historians can trace knowledge of perspective very far back – farther back than Plato and the pre-Sokratics. But this knowledge was encoded in esoteric texts, reports, prescriptions: treatises on optics, in theology, on medicine. Apparently, it did not move from speculation and theory into the practice of the painter until Alberti, Brunelleschi and Ghiberti made it the foundation of their art and let that moment of vision *be seen* through the window of the picture. They let the picture show vision to itself; they let the laws of perspective, to which they understandingly submitted themselves, show themselves to vision as its own laws, and furthermore, as laws it accepts with deep aesthetic rapture. This event of unconcealment, *alētheia*, brought forth a new truth about vision, a new vision, and even a new vision of truth. Vision's self-representation gave it something to think about. The picturing of perspective raised the rational order of the visible to a higher level of (cultural) awareness, and raised this kind of order to a powerful secular vision of the world: a vision that brought into history the modern world of science and technology.

It is possible to get a useful sense of the modern history of vision by considering how Grosseteste (1168-1253) understood his work in 'optics'. According to Edgerton (76), 'Grosseteste neatly synthesized the neo-Platonic doctrine of light, with all its deep implications for Christian theology, with the basic geometry of Greek optics and the Aristotelian notion of a physically rational universe'. If the difference between the ancients and the moderns shows up here, it is not too difficult to imagine the earliest, most archaic visions, theoretical visions and amplifications, in which perspective, size-and-distance relationships, and the interactions of light and vision might have been examined very carefully, thought and enjoyed very carefully.

Text (25). '. . . Certain artists [e.g., Jan Van Eyck], interested in the reflections of such curved mirrors, were inspired to represent appropriately distorted perspective effects on their depicted faces in pictures'. Edgerton (135)
Comment. Of course, by 'their depicted faces', Edgerton undoubtedly means the (sur)faces of the *mirrors* depicted in paintings. But suppose that our reading is careless and our glance sees another reference in this phrase. Suppose 'their depicted faces' should refer to the faces of the *observers*, in this case, the *painters*, as they are to be depicted in the mirrors – on the surfaces of the mirrors. The picture would accordingly show faces responding to the demonstration of 'perspective effects' with 'appropriately distorted perspective effects on their faces'. What would we see through this supposition? What effects would we see registered, mirrored, on their faces? Amazement and perplexity, perhaps? And a sense of enchantment? Perhaps. But I can

also visualize anxiety, skepticism, or an experimental squint.

The Renaissance 'rediscovery' of linear perspective was indeed an event of mirroring. For the very first time, vision was seeing itself depicted, and therefore seeing itself, seeing itself seeing itself, seeing itself as the seen and depicted, seeing itself as the one seeing and depicting, seeing itself reflected in a mirror and as the one reflecting on this reflection. In his *Phenomenology of Perception*, Merleau-Ponty speaks of this mirroring, which in his later thinking showed him the simultaneous *co-emergence* of subject and object, self and other, seer and seen, showed their *relativity*, and showed their *intertwining*:

> To see is to enter a universe of beings which display themselves, and they would not do this if they could not be hidden behind each other or behind me. In other words . . . they remain abodes open to my gaze, and, being potentially lodged in them, I already perceive from various angles *the central object* of my present vision. Thus every object is the *mirror* of all others. (68)

But in the Renaissance, what was seen through this mirroring did not undermine, as it does today, the standpoint and viewpoint of subjectivity; did not call into question the structure of subject and object, picturing the reversibility of seer and seen. On the contrary, it strengthened it, because it made visible to this vision the new historical possibilities it had been granted to make use of the laws of perspective and mirroring, the rational geometry of the visible world; and it moved vision to contemplate theoretically a rational mastery of the whole visible world.

Text (26). 'The notion of geometrical space, indifferent to its contents, that of pure movement which does not by itself affect the properties of the object, provided phenomena with a setting of inert existence in which each event could be related to physical conditions responsible for the changes occurring, and therefore contributed to this freezing of being which appeared to be the task of physics'. Merleau-Ponty, *Phenomenology of Perception* (54)

Comment. In the Renaissance, the rationality of vision was made visible to vision for the very first time, because the painting – the representing, the picturing – of perspective and of the mirroring of objects, shapes, lines, and dimensions, showed, made visible to vision, the rationality of the visible field. Thus, this picturing both showed and instanced, self-reflexively, the power of vision's rationality: its power to turn this vision of rationality into a mastery and domination of the visible world according to rational principles. A science and technology of power thus came to light in that historical moment.

We see in the Renaissance a new historical vision coming to be in a burst of aesthetic pleasure. What was 'new' in the Renaissance was the fact that an ancient *theoretical* knowledge was finally put to use, put into *practice* in painting, and that this representation showed vision to itself, and mirrored its mirroring. We see in the aesthetic beauty of Renaissance paintings – and not only in the pictures of the south – the triumphant, rapturous, almost delirious awakening of sight to the vision of a geometrical, completely rational world.

Perspectival vision as such we may suppose was not new; the ancients, even very primitive peoples, saw the world perspectivally, i.e., in accordance with these perspective principles. Nor can it be said that the theoretical knowledge of perspective put to work in the Renaissance constituted a new body of knowledge. Again it must be noted that such knowledge already existed in ancient civilizations – Greece, Rome, and the Arabic world. What was genuinely new was vision representing itself to itself, confronting itself in challenging new ways.

Text (27). Brunelleschi used a mirror to demonstrate his experiment with perspectival painting. Edgerton says: 'In what better way could he prove that his perspective picture reproduced the very process of vision as it was understood in his day? . . . Thus the observer could imagine himself, as he was led through this demonstration, to be looking into an actual eye through its pupil, to be observing the image being formed in the way that the optical scientists adduced.' (152)
Comment. Thus, here, for a brief historical moment, the eye of vision looked into itself and saw with amazement, pleasure, and some humility, the beauty of its perfection, its rationality – its power.

Today, we see in this mirroring a process of decentering and displacement, a questioning of standpoints and viewpoints, a problematizing of subjectivity and its projections, a deconstruction of source and origin. We see in this mirroring an intertwining of subject and object, self and other: an interdependent co-emergence. And, for us, this experience awakens our vision to its future, a time when it is no longer *caught* in the metaphysics of vision which has held sway since the beginning of modernity.

We need to see the Renaissance event as a decisive moment in the history of vision – and, too, in the history of a metaphysics dominated by the visual paradigm. Erwin Panofsky is right to see Renaissance perspective as a 'symbolic form' and to point out that, in phenomenological reality, straight lines are *curved*, that the visual field is *not* flat, and therefore, that the 'rediscovery' of linear perspective is *not* the 'rediscovery' or 'recollection' of an ultimate (metaphysical) truth. (Edgerton, 153-163)

More specifically, we need to see in the Renaissance event a mirror for us to see *ourselves*, and see ourselves – our present historical vision as visionary beings – in relation to them. What we see, as Panofsky and his teacher, Ernst Cassirer, also saw in the mirroring of this great event, is that each historical period and epoch is governed by its own historical perspective, which prevails in a way that influences individual vision and the collective realization of the potential, the capacity, which is implicate in our visionary nature.

I would like to remark, finally, a curious turn of events. In the Renaissance, indeed as far back at least as Roger Bacon, perspectivism was in the service of what postmodernist writers are calling 'logocentrism'. Now it seems that perspectivism has become unruly, even anarchic, and confronts 'logocentrism' with great animosity. Somehow, it has changed signs: from positive to negative. I will take sides with perspectivism as a critical or negative dialectic; but my quarrel with logocentrism and its effects on subordinate fields (the single ray, the frontal position, the fixed standpoint, the straight line, the central axis, the 'fons et origo', the foveal domination) is ultimately for the sake of a deeper gathering by the *Logos* and a deeper centering.

For Galen, the crystalline lens was the 'seat of visual power'. (See Edgerton, 72) We could *sit* with this thought, these words, and let our body's *feelings* resonate to the experience Galen is pointing out. We could even, in fact, let the feelings these words awaken and bring forth suggest their own image, their own vision: a vision, perhaps, that would change the historical course of our lives and our civilization no less dramatically than Galen's changed the course of things in *his* time. Perhaps we could achieve, through our visionary body, a recollection of the field of Being, the *Logos* of visionary beings, as a deeper centre, the deeper, more invisible 'seat' of our visual 'power'.

CHAPTER 2

Crying for a Vision

'Waters with tears of ancient sorrow. . . .'

Emerson, *Threnody*

(1) [Vision] is that gift of nature which Mind was called upon to make use of beyond all hope, to which it was to give a fundamentally new meaning, yet which was needed, not only to be incarnate, but in order to be at all.

Merleau-Ponty, *Phenomenology of Perception*[1]

(2) . . . not aware of what is going on there, not aware of what must be thought in the true thinking of Being as a whole, namely, that such thinking is a cry of distress, arising from a calamity.

Heidegger, *Nietzsche*, II[2]

(3) O God, and the God of our fathers, . . . the eyes of thy people wait in hope on thee.

Service for *Yom Kippur*, Day of Atonement[3]

(4) [P]erhaps the most important reason for 'lamenting' [*hanblecheyapi*, i.e., 'crying for a vision'] is that it helps us to realize our oneness with all things, to know that all things are our relatives. . . .

Black Elk, *The Sacred Pipe*[4]

(5) When we go to the center of the hoop we shall all cry, for we should know that anything born into this world which you see about you must suffer and bear difficulties. We are now going to suffer at the center of the sacred hoop; and by doing this, we may take upon ourselves much of the suffering of our people.

Black Elk[5]

Part I
The Breaking Open of the Vessels

(1) Chaos, *khaos, khainō,* means 'to yawn'; it signifies something that opens wide or gapes. We conceive of *khaos* in most intimate connection with an original interpretation of the essence of *alētheia* as the self-opening abyss (cf. Hesiod, *Theogony*).

Heidegger, *Nietzsche,* II[6]

(2) But the central plan of Creation originates in the lights which shine in strange refraction from the eyes of *Adam Kadmon*. For the vessels which, themselves consisting of lower mixtures of light, were designed to receive this mighty light . . . from his eyes and so to serve as the vessels and instruments of Creation, shattered under its impact. This is the decisive crisis of all divine and created being, the 'breaking of the vessels', . . . [also expressed in the image of] the 'dying of the primordial kings'. . . . [T]he death of the kings, from lack of harmony between the masculine and the feminine elements, described in the *Zohar*, is . . . the 'breaking of the vessels' [which is] also a crisis of the powers of judgment, the most unassimilable of which are projected downward in this cataclysm, to lead an existence of their own as demonic powers. . . . [A]fter the crisis nothing remains as it was. . . . Nothing remains in its proper place. . . . The breaking of the vessels continues into all the further stages of emanation and Creation; everything is in some way broken, everything has a flaw, everything is unfinished.

Gershom Scholem, *On the Kabbalah and Its Symbolism*[7]

(3) [The 'God of revelation in traditional Judaism' is understood by the Kabbalah as] the masculine principle, which through the breaking of the vessels has departed from its original unity with the feminine and must now be restored on a new plane and under new aspects.

Scholem, *On the Kabbalah and Its Symbolism*[8]

(4) We are like an empty vessel. We have neither prophecy nor vision. We walk and feel our way like the blind.

Service for *Yom Kippur*[9]

(5) The shell must be broken and what is contained in it must come out; for if you want the kernel, you must break the shell.

Meister Eckhart, *Sermons*[10]

(6) In healing ceremonies, light, water, love, release of emotions, energy flow, circulation, harmony and crystal clear water are all descriptions of curative experiences. The water is a description of free flowing energy which cleans the body by unlocking egotism and its resulting cramps. . . . Since the water is a healing agent, the diseases it cures must be characterized by rigidity. But rigidity has many aspects. . . . A water experience is holistic and unifies the entire personality so that ego, Self, dreams, body, inner and outer come together in one human being. The more rigid the ego . . . , the more threatening the flow of the body or the psyche appears. A rigid and frightened personality becomes terrified, split off from nature, and cannot believe that a Self or a body consciousness exists that can organize behavior once ego rulership is given up.

Arnold Mindell, *Dreambody*[11]

(7) The critical ontology of ourselves has to be conceived as an attitude, an ethos, a philosophical life, in which the critique of what we are is at one and the same time the historical analysis of the limits that are imposed on us and an experiment with the possibility of going beyond them.

Michel Foucault, 'Nietzsche, Genealogy, History'[12]

This work on vision began, not with a vision, but with an experience of crying. Crying for the earth, the earth itself, whose devastation I see all around me. Crying over the plundering of the land. Crying from the depths of my ancestral body for the victims of the Holocaust. Crying for the Indians massacred in my country, for the last of the dying guardians of the earth. Crying *with* them. Crying in kinship with all suffering beings. Crying in participation, in sacrifice; but also in thankfulness for their spirit of sharing.

With the crying, I began to see, briefly, and with pain. Only with the crying, only then, does vision begin. . . .

Our eyes are not only articulate organs of sight; they are also the emotionally expressive organs of crying. The site where vision takes place is sometimes a site where a very different kind of process takes place. We will now give some thought to the character of this process. What is crying? Is it merely an accidental or contingent fact that the eyes are capable of crying as well as seeing? Or is crying in the most intimate, most closely touching relationship to seeing? Is crying essential for vision? What is the ontological significance of crying as a mode of visionary being?

Only human beings cry. Animals are beings endowed with sight; but only we are capable of crying. What does this show about us? What does this show *to* us? Is it this capacity for crying, then, which ennobles our vision, makes it human? And is it not the *absence* of this capacity which marks off the inhuman? By the 'inhuman' I mean the monstrous and the inwardly dead: the Nazi commandant, for example, and his victim, the Jew, locked into a dance of death, neither one, curiously, able to shed a tear: for different reasons, their eyes are dry, empty, hollow. What we have seen, we who are alive today, of human cruelty and evil demands that we give thought to this capacity for crying and examine, looking into ourselves, the nature – or character – of its relationship to vision. What does this capacity make visible? What is its truth? What is the truth it sees? What does it know as a 'speech' of our nature? How does it guide our vision?

Crying is not something we 'do'. Crying is the speech of powerlessness, helplessness. In an epoch when power is nihilism, our visionary being is engaged in a historical struggle, and the stakes are high. As a response to what history has made visible, crying calls for vision, for thought, for understanding; we need to *see* what *it* makes *visible*.

What is the significance of the fact that philosophical discourse has not seen any need to consider the cultivation of this capacity in its vision of the truth? Why only now, here and now, has the voice for this capacity interrupted the philosophical discourse, tearing into its tissue of projections?

The self and society, the other, are interdependently co-emergent, constituted in and through social practices. But these practices emerge from a matrix of nature. Crying is a 'speech' of nature, an experience which constitutes a strong connection with this matrix: a strong and immediate connection not totally mediated by existing social practices. The self can undergo important processes of individuation by developing this inherent connection and bringing it forth, into the world of its vision. The self always exists within social practices. But social practices do not totally determine the being of the self, for the self belongs to nature and the realms of the spirit – to sleep, to dreaming, to the underworld, to the transhistorical body.

The self, a body-self, exists within society, within social practices, as a centre of interiority, of awareness, brought into being through the social cultivation of symbolic, reflective, visionary processes. The self is capable, however, of generating and developing, from out of this interiority, practices of its own. But the self will not experience deep satisfaction until these practices are brought in a meaningful way into the social and cultural world.

Crying, of course, is involuntary. But the experience of crying, with which we are all familiar, can be taken up by the self, taken to heart, and turned, through the gift of our thought, into a *practice* of the self. The practice is concerned with the cultivation of our capacity for care, *Sorge*, feeling: our care-taking capacity, that is, as visionary beings. What I want to think, here, is, then, the significance of crying as an experience of the care-taking self, an experience taken up into its guardian awareness, its thoughtfulness, and channelled into social practices, practices of visionary being. Crying becomes a critical social practice of the self when the vision it brings forth makes a difference in the world, gathering other people into the wisdom of its attunement.

We are wont to see crying and seeing only in a pattern of absolute difference and opposition. But what Herbert Guenther observes in regard to the dualism of heart and mind may be pertinent here, too:

> In splitting these, we have made them quite inadequate as a means of dealing with our self-imposed problems.[13]

Thus, in this chapter of our work, we will attempt to

reintegrate the perceptivity of crying into the larger process of vision, letting it show itself as a moment of extremely important learning.

I would like to think more, here, about the crying which comes from our individual suffering of history: the crying which comes from what we have seen; the crying which expresses the need we need to see; the crying which expresses a need for vision; the crying which gives our vision its historical task.

Some words of Heidegger are so entirely fitting, here, that I would like to let them speak for crying:

> Instead of furnishing [ready-made] representations and concepts, [it] experiences and tries itself as a transformation of its relatedness to Being.[14]

The experience of crying I am attempting to think makes known to us our historical need for a vision 'of the whole': a vision belonging to the whole; a vision serving and preserving the 'wholeness' of things. We need a vision which *comes* from a sense of wholeness – a vision very different from the one which totalizes and enframes.

We could think of our eyes as capable of three kinds of mood: (i) the ontical moodedness of everyday seeing, which can differentiate and articulate what it beholds only in a more or less dualistic, objectifying, re-presentational manner; (ii) the transitional moodedness of a seeing which cries for vision, immersed in painful seeing, immersed in the processes of its subjectivity; and (iii) the moodedness of a more joyful, more fulfilled seeing, clear and bright and articulate, and capable of being deeply touched and moved, even at a distance, by what it is given to see. Since the crying in question here is a *historically motivated form of suffering* which comes not only from what has been seen but also from what has not, the moodedness characterized in (iii) is essentially related to the unfolding of our visionary capacity as an 'effective' practice of the worldly self. In this regard, I should therefore like to note John Rawls's assumption of what he calls, in *A Theory of Justice*, an 'Aristotelian principle':

> The [Aristotelian] principle states that, other things being equal, human beings enjoy the exercise of their realized capacities (their innate or trained

> abilities), and that this enjoyment increases the more the capacity is realized, or the greater its complexity. . . .[15]

I agree, though I suppose that Rawls and I would hold different interpretations of the concept of 'realized capacities'. For me, it is the 'Question of Being' which takes the true measure of our capacities.

Crying is the rooting of vision in the ground of our needs: the need for openness; the need for contact; the need for wholeness. These needs are the needs of the mature and sagacious inner self, not of the ego. According to Heidegger, 'the comportment [*Verhalten*] of the human being is pervasively tuned by the openness of Being as the whole'.[16] Inasmuch as vision is a mode of comportment and embodies the mood of our relationship to this openness of Being, it should follow that, as he says in *Gelassenheit* (translated as *Discourse on Thinking*), 'the field of vision is something [intrinsically] open. . . .'[17] Merleau-Ponty puts it this way: 'To say that I have a visual field is to say that I have access to, and an opening upon, a system of beings, visible beings. . . .'[18] Ontologically speaking, what they both are saying is true. And yet, it cannot be denied, as we contemplate the history of our civilization, that the visionary capacity with which we have been endowed also bears within itself a certain disposition to become restricted, closed, fixed, power-mad and violent.

Vision *needs* real closeness, contact. 'My eye', writes Merleau-Ponty, 'is a certain power of making contact with things.'[19] But, as we have seen, this 'power' has been sacrificed in the drama of history, yielding to the more powerful movement constitutive of 'theoretical' vision: detachment, separation, abstraction, rationality without heart.

When crying is taken up by the self as a practice of care, it can be an opening of our eyes to their transhistorical rootedness in the primordial field of vision. When crying opens our vision to this rootedness, it also connects us very deeply to other living beings, for the roots of all visible beings are inseparably intertwined in the invisible. This intertwining is something that can be seen – but only with eyes that have become the organs of this invisibility. It is this process of becoming as an experience of visionary self-development which concerns us.

That in crying which perhaps most deeply calls for thought is the phenomenological fact that it can help us to realize how desperately our vision needs to feel its belonging to the matrix of the visible as a whole. This need is a need, we might say, for healing. A visionary *sense* of the whole of Being is 'healing'. Xenophanes reminds us of this primordial need when he states: 'It is the whole that sees, the whole that thinks, the whole that hears.'[20] Crying can root our vision in this (sense of the) whole. Crying may thus be described, using Heidegger's words, as an experience which 'reveals beings as a whole'.[21] If we could reach the well-spring of suffering; if we could be flooded with cleansing tears, tears that 'sing the healing whole', perhaps we might be moved beyond the pain, beyond the vale/veil of tears, beyond the *alētheia* of suffering, into the dawn of another historical vision.[22]

Heidegger, shadowing Xenophanes, observes that the enlightened seer is: 'one who makes present and belongs in an exceptional sense to the wholeness of what is present'.[23] The seer is 'the one who has already seen the wholeness of what is present in its presencing.'[24] The seer, he says, is 'the one who . . .': a significant phrase most readings will overlook. The seer is an individual; indeed, the seer is a striking example of individuality. And yet, the seer's unity, coherence, identity and 'oneness' do not constitute the ordinary 'one', but the deeper, more invisible 'one': the oneness of a seer whose seeing is rooted in the intertwining, where it is coupled, as in the archetypal *coniunctio*, with that which is seen. The seer's 'oneness', as 'the one who . . .', manifests a rootedness in the indivisible oneness and wholeness of the visionary field.

'Already seen' also calls for thought. The sense in which the wholeness can be described as 'already seen' is a sense which requires that vision be aware of, and accordingly governed by, its rootedness in the field as a whole – and by an abiding sense of the indivisible oneness of all visible beings. What is 'already seen' is the presencing of the background, the field: that which gives (*Es gibt*) what is there to be seen. Heidegger therefore immediately adds a word of caution, because the 'wholeness' he has in mind is different from the 'totality' sought in metaphysics: '[This] does not mean that what is present is nothing but an object wholly dependent upon the seer's subjectivity.' I suggest that 'totality' should refer to the wholeness metaphysics turns into an object, and that 'wholeness' should refer to what cannot be totalized in this way, though it *can* be experienced through cultivated feeling, sensibility, *aisthēsis*. The difference is fundamental. Calculative reason only understands 'totality', because it is grasping, instrumental, quantitative, and disembodied. The heart understands a wholeness which cannot be totalized – a wholeness which is also open. Because it is in touch with feeling, because it comes from the body of feeling, crying can help us to experience the indivisible wholeness of Being in its invisible presencing as the field of our vision. Crying *is* visionary feeling, and feeling is inherently closer to a sense of wholeness than the disembodied intellect.

This sense of wholeness is a *need* of our visionary being. It is also a need *of our time*. But most of us are, most of the time, deeply separated from an awareness of this need. The need is split off; and that dividedness in our being – which our metaphysics reflects – is related to the *character* of our present age. We must somehow begin to see the nothingness of Being, our present historical situation, as it manifests in relation to our visionary condition, our 'own' visionary being. As individuals, we need to see and understand that there is an essential *connection* between our sufferings, our pathologies, as visionary beings, and the 'history of Being' which is now unfolding in our historical world. Our sufferings cannot be understood, nor can our needs be realized, until we can *see* this connection.

Pathology speaks; it *is* a speech: a *logos* of the *pathē*. If Being is thought as *Logos*, we may ask ourselves how our visionary suffering – that, for example, which is deep enough and intense enough to move us to tears – could be the speech, the *logos*, of Being, of the *Logos*: how, that is to say, our suffering is a manifestation of nihilism, the history of Being in our present epoch.

Crying is a breakdown of vision. The breakdown is twofold. It expresses the inadequacy of vision as a guide for action, as a mode of *praxis*, as an organ of response. But it also expresses the vulnerability of a false ego-logical composure. If only for a moment, the egobody's control over vision – over our visionary being – is overwhelmed and overcome. This experience is potentially significant, not only as an experience of individual maturity, of further development and individuation, but also as an experience in which the historically prevailing mode of visionary being, egoity, is temporarily suspended. Like our experience with the breakdown of a tool, crying is an experience which compels us to feel our 'exposure to the disclosedness of beings'.[25] What was taken for granted – our openness to beings and their accessibility to our projects – is put into question. Our visionary power, our sense of power, of agency, as visionary beings, is put into question. The capacity of vision, vision as clarity, as light, as ego-logical, is put into question. Crying can expose the practices of vision, the social practices, we take for granted; it can suspend them in a very radical ontological critique. Like the *epochē* in skepticism, it can suspend the 'natural attitude', the ontical, ontologically forgetful attitude of anyone-and-everyone, of *das Man*, which 'normally' regulates our visionary comportment towards beings, and towards the field of Being as a whole. The human being 'clings', as Heidegger says, 'to what is readily available and controllable'.[26]

But crying, which is always beyond our control, can deny us the ego's ultimate delusion of visionary mastery and control. The breakdown of vision brings with it a temporary breaking up of the ego's power. This moment, however, can be continued and deepened, further integrated into daily living: continued, that is, as a practice of the Self, a practice inseparable from our participation in the social practices constitutive of our belonging to a historical world. Only then, I want to say, do we begin to see: see, that is, with eyes which have *learned* something from that moment of breakdown. Visionary self-development – individuation – is inseparable from, and accordingly dependent upon, corresponding changes in the historical conditions. The egoity characteristic of our present historical culture is embodied in the individual, and it affects, even regulates, the possibilities and opportunities for the unfolding of our visionary capacities, the capacities we received as a gift from nature.

Crying *becomes* a practice of the Self when it constellates a task: a task for us as individuals, and equally so, as beings of history.

In *The Tantric View of Life*, Herbert Guenther contends that,

> in the difference that is set up between subject and object, all the torment that usually accompanies the rift in Being asserts itself.[27]

There is pain, there is pathology, there is the speech of suffering, there is a visible speech of the *Logos*, when the *difference* between subject and object sets up a pattern of *indifference* to the ontological difference (between beings and Being as such). 'Pain rends', Heidegger says. But he goes on to say that it is also 'the joining of the rift'.[28] Perhaps.

What *have* we seen? What have we seen in our lifetime? What did *other* generations see? To what did they testify, what witness? What horrors, what violence, what hatreds and injustices? What sufferings? What we have seen, including what we have seen only indirectly, through the eyes of others, calls for a response from the depth of our visionary being. But before a response can emerge, it may be necessary for vision to cry, to weep, to feel the inadequacy of its visionary power, to sense the proportions of its responsibility for what *is* to be seen; to make contact with 'reality' through a body of feeling.

The visibility of evil is always a mirror. To see the 'evidence' of nihilism is to see into our capacity for destructiveness as visionary beings. It is to see destructiveness in relation to the *collective* vision. It is to see vision – the pathology of vision – in history, and in relation to the history of Being.

When crying becomes conscious, conscious of itself, and seeks to understand its pain, its affliction, it becomes a practice of the historical Self, crying for vision. Although crying is an individual experience and we shall be considering it as an individuating process or practice, it is also essential to understand, here, that the 'causes' of this pain are not necessarily internal, private, or individual. What is calling for such a practice, our crying need, is always in some way profoundly connected, and profoundly responsive, to events in the historical life-world. Our attention to crying as a practice of the Self must not cause us to forget these 'external' circumstances. Indeed, I would argue that crying, as a practice, *needs* to be thought in essential relation to the conditions which brought it forth as a mode of responsiveness to what was seen, and in relation to the possibility of a more adequate, more appropriate response. As a practice of the Self, crying is an experience in which deep contact with the reality of the situation is made and felt. Such contact is not regressive; it is rather, I suggest, an essential first step in realizing the meaning of the situation. It is the 'strength' of our openness to seeing, the 'strength' of our being affected by what we see, and is the 'strength' of a first understanding. The possibility of a deeper, more adequate and more appropriate understanding essentially depends on this initial realization.

Crying, as practice, means working with oneself, with one's experience, in order to see *with understanding*. But seeing with understanding means nothing unless it is engaged by the world and issues forth in social practices. It must not be supposed that crying is merely a private, personal experience without relation to social and political conditions. I would trace this supposition to our metaphysics, which turns dialectical polarities into dualisms, and to an ideology the domination of which *subjectivizes* the Self by encapsulating it within the egobody's skin. Crying is *not* merely 'inner'. *Feeling* is not merely 'inner'. To suppose that they are reinforces an oppressive dualism and supports the prevailing social conditions.

In thinking about visionary being in relation to history, we must of course consider the advent of nihilism. But nothing in the historical experience of our civilization can be compared to the nihilism manifest in the Holocaust. Thus I want to begin this chapter on crying and the movement beyond with some recognition of this event as an event in the realm of vision. I do so principally as a gesture of remembrance, gathering our vision into a struggle with its meaning. What the eyes of its victims saw; what their eyes, their vision suffered, can hardly be imagined, and even less, comprehended. Primo Levi, a survivor of Auschwitz, tells us, in his story, that '. . . we saw our women and our children leave towards nothingness'.[29] What kind of vision is possible after this?

Heidegger never once mentioned the Holocaust. But, in responding to the thought of Anaximander, he asked himself, and us, too, this question:

> Do we stand in the twilight [*Vorabend*] of the most monstrous transformation our planet has ever undergone, the twilight of that epoch in which earth itself hangs suspended? Do we confront the evening of a night which heralds another dawn? . . . Is the land of evening [*Land des Abends*] only now emerging? Will this land of evening [*Abendland*] overwhelm Occident and Orient alike, transcending

> whatever is merely European to become the location [*Ortschaft*] of a new but primordially fated history?[30]

I do not believe that these questions can be answered. I think that they should remain with us; that we should *carry* them with us; that they are needed, as questions, to summon our vision to its most extreme vigilance.

Another survivor of this nightmare, Elie Wiesel, has also given us his testament, a story simply called *Night*, helping us to see what he saw.[31] Because I share his conviction that what he saw must not be recalled abstractly, but remembered with a concreteness that will continue to challenge our collective historical vision, I want to inscribe into this text on crying, right here, where our eyes are now passing, some traces, some imprints, some shadows: something to compel the eyes, if only for an instant, to recognize, to see, the horror and the terror of that night of the soul. We cannot grasp 'the meaning' of the Holocaust. I will *not* attempt to illuminate it. It compels me to leave it in the darkness of pain. I cannot *see* the vision which could emerge from this night without enduring pain. But there is one thing I do feel I know: that no vision must *forget* this historical cataclysm, and that we cannot go on without passage through this vale of tears, an abyss of pain and grief.

One of the striking things about Wiesel's chronicle is the frequency with which he describes his ordeal by reference to visionary experience, imagery and symbolism. The testament is a painful work of vision, an arduous task, a difficult exercise in remembering. But it is also an *appeal* to vision. The speech of suffering is addressed, in this work, to vision. The suffering needs to be seen; only then is it remembered. ('Remembering' is a task for the body of vision as a collective historical being. But interpretation is already under way in this remembrance.)

The author speaks of what the eyes have seen and suffered; he speaks of *how* they have suffered, how they were violated and assaulted; he speaks of weeping and crying, and of the ultimate impossibility of tears, of pain so deeply burned into the flesh, so searing, that the eyes are left dry; and he speaks of the virtually total destruction of the human capacity for vision, for insight and understanding. The nihilism rages in precisely that destruction. Its 'historical essence' is in there, too, and I would like to see whether we can let it show itself, make itself darkly visible. I am going to set before our reading eyes, on a paper appropriately adorned by figures shrouded in black, without idle comment, as a gesture of respect, some passages from *Night* through which our eyes, our capacities for vision, need to pass:

> Never shall I forget that night, the first night in camp, which has turned my life into one long night, seven times cursed and seven times sealed. Never shall I forget that smoke. Never shall I forget the little faces of the children, whose bodies I saw turned into wreaths of smoke beneath a silent blue sky. . . . Never shall I forget those flames which consumed my faith forever. . . . Never shall I forget that nocturnal silence which deprives me, for all eternity, of the desire to live. Never shall I forget those moments which murdered my God and my soul and turned my dreams to dust. Never shall I forget these things, even if I am condemned to live as long as God Himself. Never. (43-44)
> This day I had ceased to plead. I was no longer capable of lamentation. On the contrary, I felt very strong. I was the accuser, God the accused. My eyes were open and I was alone – terribly alone in a world without God and without man. Without love or mercy. I had ceased to be anything but ashes. . . . (73-74)
> [My father's] last word was my name. A summons, to which I did not respond. I did not weep, and it pained me that I could not weep. But I had no more tears. . . . (112-113)
> One day I was able to get up, after gathering all my strength. I wanted to see myself in the mirror. . . . I had not seen myself since the ghetto. From the depths of the mirror, a corpse gazed back at me. The look in his eyes, as they stared into mine, has never left me. (116)
> . . . eyes . . . become blank, nothing but two open wounds, two pits of terror.[31]

In *Beyond Good and Evil*, Nietzsche wrote that, 'when you look long into an abyss, the abyss also looks into you.'[32] It was not only God who died in the Holocaust. A vision of humanity – the old vision of humanism – died too. We need to see, to *let* ourselves see, the nihilism in the Holocaust. The testimony of its survivors is necessary for this seeing. Only through their eyes of remembrance, eyes which have gazed into the abyss of meaning, will we, as a global humanity, be able to envision a world in which the repetition of this nihilism could never be seen again. But Heidegger's words, in 'The Anaximander Fragment', are unsettling: 'What mortal', he asks, 'can fathom the abyss of this confusion? He may try to shut his eyes before the abyss. He may entertain one delusion after another. But the abyss does not vanish.'[33]

'Man,' Nietzsche says, 'would sooner have the void for his purpose than be void of purpose. . . .'[34] I think that the passages I have taken from Wiesel's story confirm this point: 'two pits of terror' is still, in Nietzsche's sense, the 'creating' of meaning. Wiesel survived. To have been totally void of purpose would have been his death. 'Suicidal nihilism' is not to will at all. Wiesel came extremely close to this, but in the struggle for truth he succeeded – if that is the word – in giving to his experience the value of an 'effective' interpretation: an interpretation, in particular, without delusion.

Nihilism appears, as Nietzsche writes in *The Will to Power*,

> because one has come to mistrust any 'meaning' in suffering, indeed in existence. One interpretation has collapsed; but because it was considered *the* interpretation, it now seems as if there were no meaning at all in existence, as if everything were in vain.[35]

The meaning Wiesel constitutes (I prefer this word to Nietzsche's word, *creates*) is therefore 'superior' to the meaningfulness which was brought forth in ancient times by the Jewish slaves. These slaves were genuinely 'creative', not 'passive': their nihilism was an 'active nihilism', and not at all suicidal. It enabled them to survive. However, the interpretation they gave to their experience was, in a sense, a form of delusion: they looked for meaningful truth *outside* the world of daily experience. Their ontotheological interpretation certainly gave meaning to their experience of powerlessness; but it also gave their suffering a legitimation which diverted energy from worldly action. According to Nietzsche, this otherworldly legitimation of suffering became the very essence of Christianity, whose self-defeating interpretation of life later prevailed.

Nihilism is at work in history whenever there is a deep contradiction between the collective experience of life and the culturally available interpretation. However, this nihilism is only intensified when the contradiction is resolved, not by a new interpretation emerging from experience and enabling that experience to thrive, but rather by a retreat into otherworldly delusion or a renunciation of the actually lived experience – standing by the prevailing interpretation, rather than taking a stand on the ground of real historical experience. 'The most universal sign of the modern age', says Nietzsche, is that 'man has lost dignity in his own eyes'.[36] What this means can be gathered from another note, where Nietzsche makes his point more specific: it means, in part, 'not to esteem what we know [*erkennen*]'.[37] We need to learn how to esteem, value, and trust our own experience. By not trusting it, and not trusting ourselves, we can only make our experience *less* reliable. The body of experience can, as Gendlin puts it, 'talk back' to history.[38] The individual body of experience can never be totally reduced to history – except by its death.

Although crying often expresses – particularly in children – only the ego's narcissism, it can also gather people together and deepen the roots of collective vision in the ground of feeling, enabling new social practices to emerge. It could even open up the possibility of an ontological movement: as a cry from the depths, a motion of the eyes, as organs of thought, in response to the visible 'loss of Being'. Heidegger speaks about this in 'The Age of the World Picture'.[39] It would be a mistake to think that the 'loss of Being' is only a textual phenomenon, perhaps 'only' a metaphor, in the interpretive discourse of post-metaphysical thinking. Heidegger's words should refer us to our actual historical situation. What can we *see* through these words? What kind of vision is called for? And how is vision *changed* by its awareness of this 'loss'?

Loss, in our life-world, means pain, suffering, grief, mourning, depression. Clinicians say that they are seeing a very high incidence of depression today. Depressions are epidemic today: an epidemic pathology, of the self, of the soul, of the spirit. If Heidegger's ontological interpretation be true, then we should not be surprised by these depressions. Our entire culture, host of a raging nihilism, has now become host to a multitude of depressions. There are *many* 'dead souls'.[40]

Loss of Being constitutes an environment, a climate, an entire social and ecological system: conditions conducive, in ways that are mediated by many factors in unique confluence – including the genetic, the constitutional, and position within the political economy – to the formation of depressive pathology.[41] Nietzsche himself experienced and reflected this form of suffering, though he did not explicitly associate its appearance in the Western world – the historically distinctive characteristics of its emergence and formation – to the advent of nihilism.[42] But if nihilism involves 'loss of Being', we should be looking into the ontological significance of widespread depressive pathology.[43] The speech of depression is the speech of loss. In terms of our visionary being, 'loss of Being' means a crying need for vision: the kind of vision which comes from the very depths of depression, our collective experience of this loss. The thinker, here, is someone who is moved, moved even to tears, by the loss of Being which is visible in the sustaining light our world still admits.

To be or not to be: to see, or not to see, this loss of Being. That is the ontological question which confronts us today as visionary beings. Insofar as the spirit of depression is collective, what is needed, what is called for, is a collective vision. But the roots of such a vision must first draw sustenance from the waters of life, the natural elements, the tears of the soul.

In his *Comparative Psychology of Mental Development*, Heinz Werner can see only regression and degeneration in perception with a 'high degree of unity between subject and object'.[44] But his interpretation is valid, if at all, only as a diagnosis of 'physiognomic perception'. In what he calls 'primordial perception', there certainly is a 'lack of differentiation', and, looking backwards, a 'decline in the polarity of subject and object'. Crying, seen in this way, is unquestionably regressive: regressive vision. And yet, when we consider that the dualism of subject and object is constitutive of the foundation of Western metaphysics, and that the historical task of problematizing and overcoming our metaphysical legacy must become a task for our individual and collective vision, we can see how the 'regressiveness' of crying, crying, that is, when it is crying out of a need for vision, could perhaps be decisive in motivating an essential ontological movement. Hermeneutical movements can sometimes *appear* as regressive, because they always involve a phase of 'return', a 'reversal' of the 'fate' in linearity.

Our *first* experience of loss, and consequently, of depression, of depression as a condition of being, can of course be traced back to the infant's separation from its symbiotic attachment to the mother's breast. But Norman O. Brown amplifies this moment as an interpersonal, hence socializing experience of archetypal significance:

> At the mother's breast, [to put it] in Freudian terms, the child experiences that primal condition, forever after idealized, 'in which object-libido and ego-libido cannot be distinguished'. . . . The primal childhood experience, according to Freud, is idealised [and remembered with nostalgia] because it is free from all dualisms. . . . Psychoanalysis suggests the eschatological proposition that mankind will not put aside its sickness and its discontent until it is able to abolish every dualism.[45]

How does this dualism, which I understand to be a reification of a more fluid process of polarizations, cut us off from an integrating experience of Being in its wholeness?

The weaning of the infant is a potentially traumatic process of socialization. It begins the adaptive, self-individuating formation of the social ego. But Freud argued, in *The Ego and the Id*, that 'the character of the ego is a precipitate of abandoned object-cathexes and that it contains [and continues] the history of those object-choices.'[46] The ego begins with, or as, an experience of withdrawal, abandonment, separation and loss. Its 'original position' is therefore depressive. Is there no connection, no parallel, between this early developmental process in the individual, this experience of loss with the emergence of the ego, and our present historical experience of the 'abandonment' and 'withdrawal' of Being?[47] Let us note, first, that, in his work on *Inhibition, Symptom, and Anxiety*, Freud characterized the ego as the 'seat' of anxiety. And let us recall that 'egoity' is very much in question in 'Overcoming Metaphysics', where Heidegger articulates a connection between 'egoity' and nihilism. How is 'the abandonment of Being' felt and seen today? How does it show itself? How are we affected by it as visionary beings? Could our historical passage *beyond* the present social ego and its (body of) vision depend on our capacity to 'retrieve' or 'recollect' an *earlier* experience of Being?

In *Beyond Good and Evil*, Nietzsche asks:

> Wanderer, who are you? I see you walking on your way without scorn, without love, with unfathomable eyes; moist and sad like a sounding lead that has returned to the light, unsated, from every depth – What did it seek down there? –[48]

What has the sage seen? Why does he weep? Who is this character, a man who cries like a child? The word 'like' should warn us against seeing a regression here. In this same text, Nietzsche also writes that,

> A man's maturity – consists in having found again the seriousness one had as a child, at play.[49]

I take Nietzsche to be commenting, here, on some words from Herakleitos, words to which Heidegger also refers his thought, many years later: 'Time is a child playing at draughts; the kingship is in the hands of a child.'[50] Heidegger even reflects on an anecdote which sees Herakleitos playing dice with children in the precincts of a temple.[51] In any event, Nietzsche, to whom I want to return, asserts that,

> With the strength of his spiritual eye and insight grows distance and, as it were, the pace around man: his world becomes more profound; ever new stars, ever new riddles and images become visible for him. Perhaps everything on which the spirit's eye has exercised its acuteness and thoughtfulness was nothing but an occasion for this exercise, something for children and those who are childish. Perhaps the day will come when the most solemn concepts which have caused the most fights and suffering, the concepts of 'God' and 'sin', will seem no more important to us than a child's toy and a child's pain to an old man – and perhaps the 'old man' will then be in need of another toy and another pain – still child enough, an eternal child![52]

Sartre has argued, in *The Transcendence of the Ego*, that 'the essential role of the ego is to mask from consciousness its very spontaneity.'[53] And Heidegger has argued that *Dasein*, human being, is 'openness-for-Being'. Perhaps the thinker needs to cultivate his capacity for vision by returning to this 'eternal child'. Or by struggling with the pain, the sense of a historical impossibility, that can *oppose* the return. Heidegger writes of a 'truth' of Being, which

> will be given over to man when he has overcome himself as subject [i.e., as master and lord of the earth], and that means when he no longer represents that which is as object.[54]

In the vision and crying of the infant, there is a symbiotic (con)fusion of subject and object; likewise, in the primitive cultures of the world, there is a perceptual structure noticeably different from the one which has prevailed, with increasing rage, since the beginning of the modern age. The child's evolution continues until its 'closure' at the normal stage of social ego development. The development is powerfully influenced by the existing social conditions, which it then perpetuates, through its own ego-logical dynamics, in its mastery and domination of the earth. If crying is the rooting of vision in the (con)fusion of subject and object, perhaps the thinker's psychological and historical return to these roots is a necessary step on the way to a vision which could carry history *beyond* the social ego of the modern world.

The historical issue focuses on power and powerlessness, pathologies of the will in the epoch of nihilism. The modern social ego is deeply split between power as mastery and domination, power as active nihilism, and power as defeat, despair, hopelessness, a passive nihilism so deeply destructive of the self that we succumb, as Nietzsche, Heidegger, and Robert Jay Lifton have noted, to a process of psychic numbing.[55] But there is a third possibility in between. What if we were to begin, as Jeremy Rifkin suggests in his *Declaration of a Heretic*, by 'renouncing the use of power as a means of obtaining security'?[56] This gesture would certainly be 'active', but in renouncing power as domination and affirming a care-taking task, caring for the self, caring for others, caring for the earth, it would not be a continuation of modern nihilism.

Heidegger maintains that,

> Objectification blocks us off against the Open. The more venturesome daring does not produce a defense.[57]

Although philosophers might be inclined to read into these lines only their critical function in the discourse of metaphysics, their exceptional merit, I should think, pertains to the fact that they can readily be understood to propose a truth which makes good sense in the discourses of psychology and politics. Many schools of thought in psychology – Freudian, Jungian, Reichian, Gestalt, transpersonal, Lacanian – are based on this truth. The politics of deep ecology is based on this truth.

In *Gelassenheit*, Heidegger argues for the Nietzschean point that the ego-object relationship is

> only an historical variation [hence, only *one* interpretation] of the relation of man to the thing, so far as things can become objects.[58]

And he completes his argument with the claim that, 'The same is true of the corresponding historical change of the human being to an ego.' This needs to be understood, however, in relation to Nietzsche's perception that 'the entire history of higher culture is permeated by the ever-increasing spiritualization and "deification" of cruelty.'[59] Are cruelty and pain inherent in the historical structure of ego and object? Can we *see* this? Or *could* we? What if historical change *depends* on our seeing this? Seeing the variety of historical forms assumed by the structure of subject and object does not have to be interpreted as a basis for extreme relativism or a cause for despair. Their history *could* be a basis, instead, for the hope that our understanding and awareness would make a difference, turning 'our fate' into an emancipatory opportunity.

In his analysis of time, Merleau-Ponty notes that phenomenological reflection 'discloses subject and object as two abstract "moments" of a unique structure, which is presence.'[60] An experience with crying can be an intensification of such presence, and in that sense, it can function as a rudimentary form of what he calls 'radical reflection':

> The task of a radical reflection, the kind that aims at self-comprehension, consists, paradoxically enough, in recovering the unreflective experience of the world. . . .[61]

This passage makes the hermeneutical paradox explicit. The paradox he points to here is *the same* as the paradox which appears in our interpretation of the task of vision, namely, that crying for a vision, an apparent regression, constitutes an essential phase in its challenge.

In *Dreambody: The Body's Role in Revealing the Self,* Arnold Mindell, a Jungian analyst, notes his observation that, 'In becoming cultural beings, we humans lose contact with the vegetative nervous system.'[62] In the symbolic, mythopoetic discourse of our culture, crying is associated, not only with water, but with the vegetation dependent on water. And the vegetative realm has traditionally been under the care-taking of the feminine consciousness.[63]

We briefly reflected, earlier, on the nature of crying as the root, or rooting, of vision; and it now seems fitting that we return to this matter. Why do I propose that we think of crying as the elemental root of seeing? Why not say, for example, speaking in more Kantian terms, that crying is a transcendentally necessary condition for the possibility of 'seeing'? Of course: the Kantian formulation *approximates* the ontological truth of the matter. But what needs to be said, what needs to be seen, is meant to move us even closer to its disclosure. First of all, what we are trying to understand, namely, the connection between sight and crying, is not an *a priori* principle of pure understanding, but rather, in words Merleau-Ponty might have used here, an 'organismic a priori' of the human body as we live and feel it. Second, what we are concerned with is much more primordial than a condition born in, and borne by, the faculty of pure thought. In keeping with its very essence, crying is not, and cannot be, like vision – 'clearly articulate'. Thus it is essentially different from Kant's *a priori* principles. Third, it is important to bring out the special nature of the connection: specifically, this means that it is more fruitful to understand the connection as an integrative process of sensuous rooting, rather than as an abstract process of cognitive synthesizing in accordance with certain rules. Fourth, the *historical need* which tears at our eyes until they cry is a need which stems from the fact that the figures our vision 'articulates' are uprooted, in the prevailing culture of vision, from their elemental ground. Many things can move us human beings to tears. But this uprootedness is a distinctive structurational disposition of modern visionary practices, and it is something which, as a question of ontology, calls for thought.

Crying begins the process of reversal when thinking takes it up into its care. The thinker will perhaps begin to see these figures rooted in their 'nurturing' ground. (The ground is that which gives, *es gibt*, what is to be seen. The seen comes forth, comes out. 'Nurturing' is a way of speaking about this process, this *apophainesthai*, in a symbolic discourse belonging to our cultural history. The symbolic, the mythopoetic, are constitutive in many ways, some very subtle, of our experience. We need to *see* the figures at the centre of our visual focus as gifts from the ground, the field.)

Crying is the rooting of vision in the world, and in the whole of Being. But the rooting of vision, as a practice of the Self, gives insight, and a very radical ground for new vision, new understanding. The vision which emerges from this rooting could perhaps proffer some guidance for new social practices of a visionary character.

In the conversation in *Gelassenheit*, the Scientist avers that, as he puts it, 'The transition from willing into releasement [*Gelassenheit*] is what seems difficult for me.'[64] So it is – not, however, for the Scientist alone, but for each and every one of us. In the realm of vision, crying is an involuntary expression of this difficulty: it is *a will-breaking process* of letting go: letting go of our fixed ways of seeing things, our metaphysical habits, our cultural typifications, our obsessions, our defenses. In the realm of vision, therefore, crying must be a crucial phase in the transition from willing to 'letting be', a new practice of vision.

In the water of tears which well up spontaneously and flow uncontrollably in the care-taking task, the dualisms of willful vision could perhaps begin, finally, to dissolve. Water is a precious image in our visionary life, and its flowing character makes it a moving symbol of *Gelassenheit* for our visionary being. Even though it comes from pain, crying can certainly be, for us, a very beautiful experience. As Saraha, one of the Buddha's eighty-four great disciples (mahasiddhas), once wrote: 'spontaneity's actuality is there and [is always] beautiful.'[65] The breaking down of the social ego's traditional defenses, its systems of anxiety, of attraction and aversion, is therefore an opportunity to 'break open', to abolish a painful 'difference' between subject and object, to see, in *some* way, beyond its historical shadow.

Sartre speaks, in *The Transcendence of the Ego,* of a challenging existential need of our time: 'to effect the liberation of the Transcendental Field, and at the same time, its purification.'[66] According to my reading of Sartre, it should be possible, through phenomenological method, to envision a visionary process in which 'the Transcendental Field, purified of all ego-logical structure, recovers its primary transparency.'[67] We are of course reading Sartre in a new light, seeing, through his conception of the 'transcendental field', a way of shifting his point to designate, in the context of our discourse, the visionary field of Being as a whole. 'Transparency' thus means visibility – in the sense of unconcealment. The field becomes more transparent, more deeply unconcealed, but also more deeply invisible, when visionary being is 'purified' of ego. As a practice of the self, crying can *become* this gesture of purification, dissolving the power of old experiential structures at work in the social body of vision.

Thinking needs to come from our experience in living. But our experience in living has needs which need to be thought in the context of our cultural discourse: a discourse with a long history of symbolic, mythopoetic stories and interpretations. Thinking (*Denken*) is different from poetizing (*Dichten*). But thinking must work, must be prepared to work, with those stories and interpretations: if it disregards them, it will lose touch with the experience which once, once upon a time, came to speech through their poetizing. To think of 'tears', for example, as either 'real' or 'metaphorical', either 'physical' or non-existent, is to perpetuate the prevailing metaphysical categories, rather than to attend to the task, our 'deconstructive' project, which is: to *see* through them, and see *through* them. Our discourse here, in this chapter, exploits the play, the give, in our words. The ambiguity of 'tears', in this text, is what makes it radical, subversive, fruitful. In my use of the word, in my thinking with the gift of this word, I have flowed – I have *let* myself flow – back and forth, *between* taking it to mean 'real' crying and taking it to mean a certain process of awareness. Both ways could help us to develop (our understanding of) this awareness in visionary being, if we draw upon the resources of our cultural heritage. Sometimes, what is called for is an experience with the imagination, vision's image-making capacity. 'Tears' *could* be taken up by thought in *that* way: as a projection of our metaphorical vision; as an image to help us develop that in our capacity for vision which understands the gift of this experience.

Poeticizing 'tears', we can give to thinking, give into its care, and give to make it more caring, the body of feeling, the body of vision, awakened by vision's image. Let us imagine, then, for a moment, as a way of bringing this discussion to a conclusion, that there is a visible world, beyond the veil/vale of our tears, whose beauty and goodness we could celebrate with tears of joy; and let us imagine that our crying could become the eyes' libation, their sacrifice, giving thanks for their vision. After the Holocaust, how can we? Better to see the nothingness, the nihilism, the horror, than not to see at all. Even out of the void, a different vision could someday emerge.

Part II

Beyond the Veil of Tears

Opening Conversation

(1) Our minds are continually active, fabricating an anxious, usually self-preoccupied veil which partially conceals the world.

Iris Murdoch, *The Sovereignty of Good*[68]

(2) If enframing is a destining of the coming to presence of Being itself, then we may venture to suppose that enframing, as one among Being's modes of coming to presence, changes.

Heidegger, 'The Turning'[69]

(3) . . . another destining, yet veiled, is waiting. . . .

Ibid

(4) Vision is an action, not, that is, an eternal operation . . . but an operation which fulfills more than it promises, which constantly outruns its premises and is inwardly prepared only by my primordial opening

upon a field of transcendence, that is, once again, by an *ekstase*.

Merleau-Ponty, *Phenomenology of Perception*[70]

(5) . . . a just and loving gaze. . . . I believe this to be the characteristic and proper mark of the active moral agent.

Iris Murdoch, *The Sovereignty of Good*[71]

(6) My body is wherever there is something to be done.

Merleau-Ponty, *Phenomenology of Perception*[72]

(7) It is the whole that sees, the whole that thinks, the whole that hears.

Xenophanes[73]

(8) I would like ever and again to know that only the constellations dwell above me, which from their distance see everything at once, as a whole, and so bind nothing, rather leave all things free in every way. . . .

Rilke, in a letter to his wife[74]

1 The Field of Vision as Subject and Object

(1) How does metaphysics belong to man's nature? Metaphysically represented, man is constituted with faculties as a being among others. His essence constituted in such a way, his nature, the what and how of his being, are in themselves metaphysical: *animal* (sensuousness) and *rationale* (non-sensuous). Thus confined to what is metaphysical, man is caught in the difference of beings and Being, which he never experiences. The manner of human representation which is metaphysically characterized finds everywhere only the metaphysically constructed world. . . . Who is man himself within this natural metaphysics? Is he only an ego which first thoroughly fixates itself in its egoity through appealing to a thou in the I-thou relationship?

Heidegger, 'Overcoming Metaphysics'[75]

(2) I am all that I see, . . . an intersubjective field.
Merleau-Ponty, *The Visible and the Invisible*[76]

(3) . . . we are a field of Being . . .
Merleau-Ponty, *The Visible and the Invisible*[77]

(4) There is . . . no problem of the *alter ego* because it is not I who sees, not he who sees; because an anonymous visibility inhabits both of us, a vision in general, in virtue of that primordial property that belongs to flesh, of being here and now and of radiating everywhere and forever.
Ibid.[78]

(5) In my organic consciousness I am as much identified with the not-self as with the self. . . . Philosophers [are accustomed to] disregarding our organic consciousness, which does not discriminate between subject and object, and thanks to which we are already virtually free [of dualism].
Hubert Benoit, *The Supreme Doctrine*[79]

(6) Out of what is in itself an indistinguishable, swarming continuum, devoid of distinction or emphasis, our senses make for us, by attending to this motion and ignoring that [i.e., by going along with our attractions and aversions], a world full of contrasts, of sharp accents, of abrupt changes, of picturesque [i.e., ego-entertaining] light and shade.
William James, *The Principles of Psychology*[80]

(7) . . . the paper seen and the seeing of it are only two names for one indivisible fact which, properly named, is the datum, the phenomenon, or the experience.
Ibid.[81]

(8) When Subject looks at itself, it no longer sees anything, for there cannot be anything to see, since Subject, not being an object as subject, cannot be seen. . . . That is the 'mirror-like void' – the *absence* of anything seen, of anything seeable, which subject is.
Wei Wu Wei, *All Else Is Bondage*[82]

(9) Thou couldst not see the (true) seer of sight, thou couldst not hear the (true) hearer of hearing, nor perceive the perceiver of perception, nor know the knower of knowledge.

Brihadaranyaka Upanishad[83]

(10) . . . a universe comes into being when a space is severed or taken apart. The skin of a living organism cuts off an outside from an inside. So does the circumference of a circle in a plane. By tracing the way we represent such a severance, we can begin to reconstruct . . . the basic forms underlying linguistic, mathematical, physical, and biological science, and can begin to see why the familiar laws of our own experience follow inexorably from the original act of severance.

G. Spencer Brown, *Laws of Form*[84]

(11) . . . this separation [*écart*] . . . is a natural negativity, a first institution, always already there. . . .

Merleau-Ponty, *The Visible and the Invisible*[85]

(12) . . . only the experience of our own confused psyche has led us to begin our mythology with a chaos out of which, as we assert in defiance of all probability, order developed. This again is only a projection of our incomplete experience of the genesis of our order-giving consciousness. Even today our consciousness, in its striving to comprehend itself, has for the most part failed to see that the development of this order and light of consciousness is contingent on a pre-established order and a primordial light.

Erich Neumann, 'Creative Man and Transformation'[86]

(13) We saw with reference to the perceivedness of the perceived that on the one hand it is a determination of the perceived entity but on the other hand it belongs to the perceiving – it is in a certain way objective and in a certain way subjective. But the separation of subject and object is inadequate; it does not make possible any access to the unity of the

phenomenon [in an 'ecstatic-horizonal temporality'].

Heidegger, *The Basic Problems of Phenomenology*[87]

(14) [There is a poetizing, mythopoetic vision in which] the unitary reality is apprehended – a deeper, more primordial, and at the same time more complete reality, that we are fundamentally unable to grasp with our differentiated conscious functions, because their development is oriented towards a sharper [ego-logical] perception of . . . polarized reality.

Neumann, 'Creative Man and Transformation'[88]

(15) There is nothing mystical about the symbolical unitary reality and it is not beyond our experience; it is the world that is always experienced where the polarization of inside and outside, resulting from the separation of the psychic systems, has not yet been effected [e.g., in the pre-personal, pre-ontological dimension of visionary begin] or is no longer in force [e.g., in the transpersonal and ontological dimensions]. *Ibid.*

(16) . . . even the severing [*Trennen*] is still a binding and connecting.

Heidegger, *Basic Problems of Phenomenology*[89]

(17) When we have an experience, for example, of seeing a tree, all that takes place at the time is the perceiving of *something*. We do not know whether this perception belongs to us, nor do we recognize the object which is perceived to be outside ourselves. The cognition of an 'external object' already presupposes the distinction between inside and outside, subject and object, the perceiving and the perceived. When this separation takes place . . . the primary nature of the experience is forgotten, and from this an endless series of entanglements, intellectual and emotional, takes its rise. [This primary nature, or process] refers to the time prior to the separation of mind and world, when there is yet no mind standing against an external world and receiving its impressions through various sense channels. Not only a mind, but a world, has not yet

> come into existence. This we can say is a state of perfect emptiness. . . . [But then] there arises a thought in the midst of emptiness: this is . . . the separation of unconsciousness and consciousness, or, logically stated, the rise of the fundamental dialectical antithesis.
>
> Suzuki Daisetz, *The Zen Doctrine of No Mind*[90]

(18) We must keep in mind that when Lacan speaks of the 'reality principle' he means the ego's distorting, negating, and oppositional manner of adjusting things to suit its own rigid style, especially its resistance to the growth of the subject. . . . The emphasis seems to be on the ego's aggressive resistance to growth and change, blocking the lifelong coming-into-being of the subject.

William Richardson and Thomas Muller, *Lacan and Language*[91]

(19) . . . the ego is structured exactly like a symptom. . . It is the human symptom *par excellence*; it is the mental malady of man.

Jacques Lacan, *Les Écrits techniques de Freud*[92]

(20) If Perceptive Organs close, their Objects seem to close also.

William Blake, *Jerusalem*[93]

In 'Eye and Mind', Merleau-Ponty calls upon us to look, with him, into 'the immemorial depth of vision'.[94] Thus, we are gathered into a process of 'recollection': the task, and the gift, of thought. We will bring to light, in that depth, a guardian awareness, and we will redeem our preliminary vision of an emerging body of understanding.

Merleau-Ponty writes that,

> We must rediscover, as anterior to the ideas of subject and object, the fact of my subjectivity and the nascent object, that primordial layer at which both things and ideas come into being.[95]

'Rediscover' is a word which points to a problem, but the problem persists despite that word: the word *itself* is problematic. What Merleau-Ponty is trying to conceptualize is a hermeneutical process: a process, therefore, which is not accurately described by calling it either 'discovery' or 'invention'. The implicit is always already 'there' where we 'find' it. And yet, the process of making 'it' explicit is a process which changes 'it'. We know what 'it' is only in the context of this transformational process: we do not enjoy any privileged metaphysical access *outside* of this process. Since we do not encounter the implicit as it is 'in itself', we cannot say that we 'find' or 'discover' it. But, since the process is constrained, nonetheless, by factors beyond our present grasp, we also cannot say that 'the implicit' is only our invention. The process of enquiry reflects and repeats the correlativity it makes explicit as it looks, beyond subject and object, into their immemorial depth.

As we have seen, Merleau-Ponty's meditation on temporality breaks through the habits and routines of our most 'natural' attitude and: 'discloses subject and object as two abstract "moments" of a unique structure which is presence.'[96] To speak of disclosure is, however, to acknowledge, or point to, a concealment. Why has this 'unique structure' been in concealment? What is the historical significance of its coming-to-disclosure in the work of this passage? Merleau-Ponty anticipates these questions himself, and formulates his answer in a reflection on reflection, a reflection, that is, on the metaphysical presuppositions controlling the history of our methodological practice:

> The fact that this may not have been realized earlier is explained by the fact that any coming to awareness of the perceptual world was hampered by the prejudices arising from objective thinking. The function of the latter is to reduce all phenomena which bear witness to the union of subject and world, putting in their place the clear idea of the object as *in itself* and of the subject as *pure consciousness*. It therefore severs the links which unite the thing and the embodied subject, leaving only sensible qualities to make up our world. . . . But in reality, all things are concretions of a setting, and any explicit perception of a thing survives in virtue of a previous communication with a certain atmosphere.[97]

Later in the text, he adds this:

> For it is reflection [i.e., as 'representational thinking'] which objectifies points of view or perspectives, whereas when I perceive, I belong, through my point of view, to the world as a whole.[98]

Our vision needs to retrieve that sense of 'the world as a whole', because the culture of representational thinking has not encouraged us to stay in touch with the primary experience – our belongingness.

Heidegger touches on the historical question when he asks, in his commentary on Herakleitos, Fragment B16:

> Why is it that we stubbornly resist considering even once whether the belonging-together of subject and object does not arise from something that first imparts their nature to both the object and its objectivity, and the subject and its subjectivity, and hence is prior to the realm of their reciprocity?[99]

What Is Called Thinking? seems to identify this prior realm of integration with the Being of beings, and makes the same point in another way: 'The Being of beings', he writes, 'is the most [intensively and vividly] apparent; and yet, we normally do not see it – and if we do, only with difficulty.'[100]

In the 1966–1967 Seminar on Herakleitos, presided over by Eugen Fink, Heidegger even invokes the concept of 'suppression' to describe what we might call the 'psychological' aspect of this occlusion of Being, the field of our vision.[101] And he does not hesitate to communicate his conviction that this situation is a ground of deep distress and that we need, accordingly, to give the most persistent thought to this field of Being: that which, in its unifying integrity, holds together, and in this way grounds, the subject-and-object structure.

As *Logos*, as the primordial *Legein*, Being is that which *gathers* subject and object into their structure, *laying down* for them, at the same time, their field and ground. For Heidegger, attending to the Being of beings means, in the context of vision, that we keep our eyes open to receive, in the manner that is most appropriate, the lighting of Being, anterior to every gaze, every object. Can our visual perception (*Wahrnehmung*) take up (*nehmen*) and maintain (*bewahren*) today, as a guardian awareness, the primordial contact of subject and object, and their inherence, as a structure, in the field of Being as a whole? (Note that the English word 'awareness' is related to the German words for preserving, protecting, and truth.)

Western psychology, uncritically reflecting the narrowed vision of ontically forgetful everydayness, continues to obscure our experience of this 'immemorial depth of vision'. Even Carl Jung, who struggled – against Freud – to conceptualize a theory of 'depth psychology', found it virtually impossible to see clearly beyond the realm of ego's constitution, where subject and object appear together in a structure of extreme polarization. Nowhere in his published writing is this difficulty more in evidence than in his 'Psychological Commentary' on *The Book of Great Liberation*, an ancient Tibetan text on visionary experience as a path to enlightenment. Here Jung confesses his bewilderment, confronted with a depth psychology that recognizes a body of non-dual awareness which is *not at all unconscious, yet also not centred around, or constituted by, the limited sensibility of the socialized ego*. Jung, mystified, writes:

> The Ego may be depotentiated, . . . but so long as there is awareness of something, there must be somebody who is aware.[102]

Of course there must be somebody, but why must this 'somebody' be reductively identified with ego? According to Jung, then, there cannot be intentionality – an 'awareness of something' – without an ego. This is precisely, however, what the *Abhidharma* psychology, and its theoretical refinement in the *Madhyamika*, would vehemently deny.[103] Nagarjuna's most famous treatise, *The Mulamadhyamika-karika*, is a radical deconstruction of ego psychology and its metaphysics. And it reads very much like Nietzsche's critique of metaphysical concepts, published in *The Will to Power*. The ego is a *social* ego: socialized being. As such, it is a historical product. Can we not schematize, in the imagination, a centering of awareness without the domination of this ego? Jung can only imagine this process of centering as unconscious; he does not consider the possibility that this unconscious process could be taken up into our awareness; nor does he realize that this possibility could become the ground of a practice – a visionary practice. In another paper, 'Consciousness, Unconscious, and Individuation', he affirms that,

> Historically as well as individually, our consciousness has developed out of the darkness and somnolence of the primordial unconscious. There were psychic processes and functions long before an ego-consciousness existed. 'Thinking' existed long before man was able to say: 'I am conscious of thinking'.[104]

But Jung does not consider the retrieval, the hermeneutical recollection, of this primordial 'thinking'. While accepting the Freudian theory that the ego is established in our adaptation to the 'reality principle' of the prevailing social consensus, Jung still follows Freud in not understanding the consequences. If the ego is largely the 'effect' of our prevailing social practices, then, since we can change them, we can change our ego-logical identity. Jung can see what precedes the ego, but he cannot see, here, any development beyond it. In particular, he does not see that the very awareness he articulates – that 'psychic processes and functions' existed 'long before' the emergence of ego-logical consciousness – already *opens up* the barriers of ego structure. But Jung earnestly struggled with the problematic:

> no consciousness can exist without a subject, that is, an ego. . . . We know of no other kind of consciousness, nor can we imagine a consciousness without an ego.[105]

But what about the 'thinking' which Jung says 'existed long before man was able to say: "I am conscious of thinking."'? In his *Phenomenology of Perception*, Merleau-Ponty noted that, 'When I turn toward perception, [I] find at work in my organs of perception a thought older than myself, of which those organs are merely a trace.'[106] What happens when this trace is 'found'? What difference does it make?

Despite his unnoticed assumption that the 'subject' *must* be an 'ego', Jung concedes that,

> we are practically compelled to believe that the unconscious cannot be an entirely chaotic accumulation of instincts and images. There must be

> something to hold it together and give expression to the whole. Its centre cannot possibly be the ego, since the ego was born out of it into consciousness and turns its back on the unconscious, seeking to shut it out as much as possible.[107]

And then, concluding the drift of his reasoning, he asks: 'Or can it be that the unconscious loses its centre with the birth of the ego?' Both Heidegger and Lacan have questioned the historical 'inevitability' of the ego-centre. Since the birth of the ego is its emerging from submergence in the elemental field of Being, there where it exists only as merged with its object, the assumption that the ego represents the end of this process seems altogether arbitrary. Moreover, in view of the fact that the ego represents the reality of social consensus, the assumption constitutes an ideological position.

According to Merleau-Ponty, radical phenomenological reflection discloses a 'pre-objective realm' of being inhabited by an 'anonymous', 'pre-personal', and of course 'pre-reflective' body-subject. When our ego-logical consciousness *subsides* into this embodiment, there is a possibility that we can experience, by grace of the body, a 'subsidiary' awareness of the field as a whole. By grace of the body, there is a rudimentary but panoramic intentionality gathering and centering our ekstatic existence. Jung cannot understand a subjective centre which is not ego-logical because he forgets the structuring synergies of the body. When we remember the body, there are not, as Jung supposed, only two kinds of process: consciousness and the unconscious. There is a consciousness carried by the body: a gathering and centering through bodily felt awareness. In *Phenomenology of Perception*, Merleau-Ponty asserts that,

> To say that I have a visual field is to say that by reason of my position I have access to, and an opening upon, a system of beings, visible beings, that these are at the disposal of my gaze in virtue of a kind of primordial contact and through a gift of nature, with no effort made on my part; and from which it follows that vision is prepersonal.[108]

Let us take leave of our comfortable home in the realm of the ego, and go boldly into the darkness of the prepersonal.

There is a wisdom to be retrieved here, for an ego caught up, more often than not, in the pathology of the structure it built and inhabits.

But what takes place in the field of vision, the field of Being, when radical reflection opens our visionary being to channels of awareness not controlled by the social ego? Surely, Merleau-Ponty would not want us to believe that the disclosure of a prepersonal vision leaves our more personal modality of vision unmoved and unchanged. Unfortunately, Merleau-Ponty has nothing to say about this, although his interpretation of the phenomenological method of reflection would seem to require a dialectic in the relationship between reflection and experience. But experience can be profoundly changed in response to what reflection discloses. Thus I submit that the ego's vision cannot remain totally unchanged in the light of such a radical interpretive disclosure. The text of *Phenomenology of Perception* does not return to the process of experiencing after the reflective practice has formulated its interpretation.[109] What happens, what changes are wrought in our experience, when we deeply experience the truth Merleau-Ponty brings to language with the words 'primordial contact' and 'gift of nature'? For the next developmental step – a thinking which would speak from out of its changed experience, its experience of the difference interpretation makes – we must turn to some of the late manuscripts published in *The Visible and the Invisible*.

Returning to the anonymous, prepersonal vision of our bodily felt awareness, we retrieve an experience of vision that could change us profoundly. Prepersonal vision is more panoramic, global, physiognomic, more affected by our situational moodedness, more immediately in contact with the field as a whole. It is a *more diffuse* presence, more immediately attuned by, and to, the primordial 'clearing' and 'lighting' of Being. When we retrieve this vision, making it explicit, bringing it to light, the consensually legitimated vision of the social ego is radically called into question: and this questioning comes from the Question of Being. If we are moved into the openness and expansiveness of the clearing, it is likely that ego-logical habits and routines will be disturbed by the subsidiary guardian awareness which arises to question them.

Dasein, for Heidegger, is ekstatic. By the same token, vision is an *écart*: opening, dispersing, separating. The formation of perceptual structures – the process of *Gestaltung* – involves the ekstatic *écart*. In *The Visible and Invisible*, Merleau-Ponty reflects that 'to be conscious=to have a figure on a ground – one cannot go back any further'.[110] The figure-ground structure constitutes a primordial level or phase of perceptual articulation – *legein* – to which Being lends itself: 'it is that separation [*écart*] first of all that is [the opening up of] perceptual meaning.[111] It is a primordial moment in the organizing of perception according to the principle (*archē, nomos, legein*) of the ontological difference, the difference between beings and the Being of beings.

But the ontological difference takes place in many different ways: different cultures, different historical times, different societies and political economies *make* a difference. In our present historical world, seer and seen interact through a structure which takes the form of subject and object. They stand directly opposite one another, fixed in a straight line of frontal confrontation. (The Germans speak of 'Vor-stellung'. The 'vor' says it all.) This structure permits the subject an optimum visibility and control. Thus it manifests the character of the ontological difference as it figures in the figure-ground differentiation of our visionary existence. In our present political economy, the separation of figure and ground in visual perception is subject to the efficacy of an inveterate – and sometimes pathological – tendency, whereby we become so spellbound, so caught up in the figure or object, that we forget, neglect, and lose touch with its contextual ground. This inborn ontologizing tendency, which our social conditions bring forth and reinforce, can, however, be resisted, and an awareness of the ground can be deliberately cultivated; but such attitudes require an individuating vigilance, discipline and practice, because they go against the grain of our interpersonal socialization, our modern history.

In his study on the teachings of Herakleitos concerning 'Logos', Heidegger reflects on the Greek word, *chorizō*, separate. Distressed by the deeply compelling character of our inveterate tendency – he calls it our *Gewohnheit* – he reminds us that, 'even the severing is still a binding and connecting.'[112] A binding not only in the sense of joining, but also in the sense of claiming and enjoining. The *Gestalt* is the articulation and layout (*legein*) of a claim on our visionary capacity.

The authors of *Gestalt Therapy,* Fritz Perls and Paul Goodman, contend that,

> For the gestalt to be unified and bright – a so-called 'strong gestalt' – all this varied background must become progressively empty and unattractive. The brightness and clarity of the figure is the energy of the excitement of seeing-the-square which is freely being drawn from the empty background.[113]

A 'good gestalt' is a 'unified figure against an empty ground'. But I must disagree, because they turn a true phenomenological description into an affirmation of universality and normativity. What they are *accurately* describing is 'normal' everyday perception in its ontical character as ontologically forgetful; but they are seriously mistaken when they assume that the familiar character of the structure they describe expresses a universal ideal norm, or an inevitable phase of termination in the development and fulfillment of our visionary capacity. The 'emptied background', here, is in fact an *historical production*. In the final analysis, we are compelled to acknowledge that there is a wild, unknown, unrealized potential inherent in the 'nature' of our perceptual experience; it is transhistorical and pre-civil, being a gift of nature, and we must take care not to identify this potential with any particular historical organization of the visionary dynamic.

Western metaphysics, emerging from, and therefore also reflecting, the character of a perception formed within the socialization constitutive of everyday life, sees only the separation, and not also the connecting. It reflects the ontical concerns of an historically distinctive 'rationality': a rationality of reification incarnate in our perceptual experience. As William James points out in his *Principles of Psychology,*

> If to hold fast and observe the transitive parts of thought's stream be so hard, then the great blunder to which all schools are liable must be the failure to register them, and the undue emphasizing of the more substantive parts of the stream.[114]

John Welwood, a clinical psychologist, comments on this

passage in a way that is pertinent to our foregoing analysis, and, I think, quite helpful:

> . . . focal attention, man's analyzing capacity, can easily fall into error when it fails to recognize its interdependence with the prearticulate aspect of mind that can be experienced fleetingly in the transitive moments [and felt shifts] of the stream of consciousness.[115]

Our substantival ontology – *Vorhandensein* and *Zuhandensein* – is a reflection of, and a long commentary on, the historical character of our perception, our vision. The character of our *perception* is a manifestation of our *character*. The historical forms of desire – the 'moodedness' of our being, which constitutes our attractions, aversions and indifference – are what determine the basic *character* of our gaze. In the 1966-1967 Seminar on Herakleitos, Eugen Fink points out that,

> In distinction to ancient philosophy, modern philosophy does not think appearance so much from the issue of what is in the openness of a general presencing, but rather as an object presenting itself for a subject. . . . But what is cannot be understood with the categories of attraction and repulsion.[116]

Indeed it cannot. But subjects, as egos, are precisely so constituted that their experience of 'what is' tends to be determined and patterned by these categories. Seeing things in terms of attractions and aversions is the *normal* way of seeing for ego-logical subjects. The ego's field of vision is a 'hodological space' normally structured by its desires and projections.

The present world is a world of science and technology whose political economy, both capitalist and communist, is organized around ideologies of mastery and domination: possession for control and security. This world calls forth, simultaneously, and in interdependent co-emergence, an ontology and a vision, a way of seeing beings and their Being. Whatever presences during the time of this frontal ontology is to be seen, confrontationally, by a sharply and narrowly focused, raylike gaze, looking at its object, if possible, in the perspective of a perfectly *straight* geometric line.[117] The object of vision is (to be) isolated and abstracted from its surrounding world, and, of course, is (to be) perfectly illuminated. Under optimal conditions, the object is (to be) placed (*gestellt*) directly *in front* of the eye, permitting it a maximum visibility, maximum control. The ideal gaze, the gaze most adequate to our ontology, is therefore (supposed to be) a disinterested and dispassionate gaze – the masterful, well-disciplined gaze of 'the rational Man'. Platonic rationalism, as we know, values the permanence and unchangingness of things. These values continue in the vision and ontology of Cartesian rationalism, which values certainty above all, and therefore insists on the need for 'apodeictic' clarity and distinctness. The gaze of these forms of rationalism, which later adapted itself very easily to the values of science and technology, desires total visibility, total presence in the light. This desire constitutes a distinctive economy in the history of visionary existence.

The structural differentiation of figure and ground, in which and as which the ontological difference presences, is deeply affected by the prevailing character of this economy. Just *how* deeply can perhaps be measured by the fact that Perls and Goodman, the authors of *Gestalt Therapy*, can see only the nothingness, the emptiness, of the ground of vision. They cannot see, nor can they understand, its hermeneutical presencing. And I submit that this is indicative of the widely unrecognized pressure on our vision to conform to the nihilism which rages throughout our present world. Perls and Goodman cannot see the presencing of the ground as ground because vision is accustomed to attend to the being of objects. But the ground is not an object; it cannot be experienced as it is in its presencing by a gaze whose intentionality, or attentiveness, is focused according to the metaphysical values of rationalism and instrumentalism. Their experience of the figure-ground differentiation is governed by a gaze which is not yet prepared to receive, and be responsive to, the presencing of the ground as ground. What they experience is determined, is constrained, by a perception which brings to this presencing the values of an economy which forgets the Being of beings and reduces the hermeneutical dimensionality of (the play of) the ontological difference. What Perls and Goodman see – no ground of meaningfulness – is, however, precisely what we are *all* bound to see, with a greater or lesser degree of awareness and resistance, in the epoch of nihilism.[118]

In the economy of nihilism, it is a contribution to the critical subversion of its historical vision to question the prevailing character of the structural differentiation of figure and ground, and to attempt a deconstruction of the ideology which invalidates, and consequently masks, the ontological significance of the separation, or difference, which takes place in a time and culture of subject-object dualism.

But our questioning of the historical character of our perceptual 'separations', and our challenge to the cultural interpretation of this experience in terms of our metaphysics, must not be misunderstood. I am not giving voice to a mystical call for a return to the One, the absolute unity of the beginning. I am not recommending a dissolution of the self, the individual, in an undifferentiated matrix of Being; nor am I proposing the fusion and confusion of subject and object identities. I do not want to see an infantile or psychotic regression, but rather a creative, critical appropriation of our present conditioning. For this could give way to a new historical process of structural differentiation in our visionary experience and being: a process, however, which requires a *hermeneutical* return to the 'primordial moment' of perceptual separation, so that the binding, the intertwining, the connective tissue of the matrix of vision, can also be felt.

Thus I would like to recall to mind what Merleau-Ponty says, in his *Phenomenology of Perception*:

> The unity of either the subject or the object is not a real unity, but a presumptive unity on the horizon of experience.[119]

The same thing may be said of the unity of subject and object. The 'unity' we are attempting to retrieve (*récuperer, wiederholen*) by our hermeneutical return belongs to an unconscious experience of wholeness which preceded the interdependent co-emergence of the subject and its object in a socially constructed and validated structure of perception. The 'unity' we experience in this way, in this phenomenological practice of thought, is not to be *found* any more than it is to be invented *ex nihilo*. It is, rather, our nature, our human nature in the making; and it comes into being as our awareness challenges itself and stretches itself to 'complete' the impossible return. The 'unity' *in question*, here, is the 'unity' assumed by our metaphysical tradition. The 'unity' *at stake*, here, is a 'unity' achieved through growth, through a process of individuation which prospers in differentiation, in the *play* of differences, because of a deeply felt sense of present dynamic unity in the presencing of the ground.

The eye cannot see itself looking; it cannot *objectify* its own act of seeing. (See Wittgenstein's *Tractatus Logico-Philosophicus*.) The gaze, therefore, as looking, transcends and eludes an objective, total fusion. Nor can we ever hold captive that which our gaze beholds in a panoramic vision of its absolute totality: for there is no thing-in-itself, no thing in absolutely isolated abstraction from the (socially constituted) field of light – the context which lets it be visible. Therefore, the ontological movement of thought, embodied in the task of radical reflection, can achieve only a much-needed *integration*, and never a total *fusion*. Nothing more is possible; nothing more needed. What our vision cries out for – what we need – is *not* the 'pure immediacy' perpetually espoused by Western metaphysics, but rather a meaningful experience of real connectedness, a grounding sense of interdependency, and a positive, productive, caring interaction:

> I find much more than an object: a Being of which my vision is a part, a visibility older than my operations or my acts. But this does not mean that there [ever] was [or now could be] a fusion or coinciding of me with it: on the contrary, this [process can only begin to] take place because a sort of dehiscence opens my body in two, and because, between my body looked at and my body looking, my body touched and my body touching, there is an overlapping or encroachment [i.e., at most a *partial* fusion, never complete], so that we must [more accurately] say that the things pass into us as well as we into them.[120]

In the Seminar on Herakleitos, which he gave with Heidegger, Fink asserts that 'There is a constitutive distance between seeing and what is seen', and he adds that this distance is 'in the unity of the overarching light that illuminates and makes visible'.[121] In the lighting and clearing of Being, in the field of our visionary being, seer and seen are disclosed, not in any mystical union that would *dissolve* their individual identity, but rather in the integrity of an *intertwining*, deepening their original contact and expanding the existential meaning of their reciprocal presence. This is, to be sure, a vision. But a vision is always a task, a task of promise.

This task, as I conceive it, is, and must be, a *radical* one. Now, 'radicality' is a term which, according to its etymology, designates a 'return' to the roots, the origin, the beginning. But our method of reflection, a hermeneutical phenomenology, ensures that this return will be doubly radical, because in attempting a 'return' to 'the' primordial ground, 'the' primordial unity (*archē, nomos, logos*) of subject and object, our thinking is radicalized even further, in that its phenomenological and hermeneutical character makes it unavoidably self-referential. Thus, the return turns critical, deconstructive, like Foucault's archaeology of knowledge and his genealogy of power: the return ends up undermining and subverting the necessity, the authority, the metaphysical legitimation of all perceptual structures – including the construct, or projection, of an absolute, and absolutely graspable, primordial unity. The radicality of a hermeneutical return consists in the fact that the move by which we break out of the structure of subject and object situates us in the 'unity' of a field whose very nature compels us to recognize our relative positionality and acknowledge the elusiveness of any absolute unity.

The 'unity' of 'the ground' is the unity of a dynamic field of *intertwining* presences and absences: a field of luminous presences endlessly pointing beyond themselves into the invisible. The 'unity' with which we are concerned, here, is only 'primordial' in a *relative* sense: it is a function of *our present capacity* to disentangle ourselves from an excessive and unnecessary identification with the prevailing construction of reality, the historical structure of ego and object. The goal of our return is not a state of total fusion, or the attainment of the 'original' unity, in any case inconceivable, but rather our release from virtual confinement within the prevailing epistemological and ontological structures of experience: a release, by grace of the bountiful matrix of vision, the gift of the lighting, which consists in an expanded sense of visionary possibilities, and an opportunity to pursue the luminous interconnections already laid down for the rooting of perception by grace of the intertwining dimensions of its field.

In his Seminar on Herakleitos, Fink observed that

'. . . light, with seeing, spans the eye and what is seen'.[122] It is in the *expanse* of this span that our vision, no longer bound to the structure of subject and object, will experience its 'release'. We may call this expanse the 'ground', the 'field', or (less felicitously) the 'foundation' of the structure in question – but only if we remember that the expanse is the clearing of the Open. This is not the 'ground' or 'foundation' which Western metaphysics has traditionally sought. (See my chapter on 'The Ground and Its Poetizing', in *The Body's Recollection of Being*.) The prevailing principle of the ground, the long dominant historical form of this metaphysical principle, is a logocentric, masculine principle. A new form of the principle is called for, and the form it needs to take must involve a recognition of, and an integration with, historically excluded experiences of the ground. Because the masculine principle has dominated the history of our metaphysical interpretation of the ground, we need, as a culture, to retrieve and redeem a truth which has been, for too long, served and preserved only by virtue of a traditionally feminine experience. In the later parts of this chapter, we will give further thought to this matter.

2 The Intertwining

> Things attract my look, my gaze caresses . . . things, it espouses their contours and their reliefs, between it and them we catch sight of a complicity.
>
> Merleau-Ponty, *The Visible and the Invisible*[123]

> I give myself as spouse to all beings.
>
> Padmasambhava (Tibet's most beloved bodhisattva, eighth century A.D.)[124]

In *Der Satz vom Grund* (*The Principle of the Ground*), Heidegger states that,

> The foundation, the ground of the judgement, is that which, as the unifying unity of subject and predicate, bears their connection.[125]

This, of course, is intended as a thesis belonging to the discourse of 'transcendental' logic. We shall, not surprisingly, assume its truth. But the question I want to ask, since we are concerned, here, with the 'logic', or ontology, of vision, is this: Can we translate Heidegger's thesis from its place in the discourse of transcendental grammar to a place in our discourse on perception without losing sight of its truth or removing it from a fundamental – that is to say, ontological – position? Because Heidegger's thesis is true to the depths of our experience, it is deeply resonant, melodious with a multitude of meanings. Consequently, when we listen to its utterance with ears emptied of metaphysical noise, or at least attempt to hear it without conforming to tradition and taking its discursive field for granted, it becomes possible to hear it propose, in one of its innumerable echoes, the sense of an interpretation which I would like to formulate, staying as close as we can to the formal structure of the original thesis, as follows:

> The field, the ground of the visual perception, is that which, as the unifying unity of subject and object, bears their connection.

In the context of our discourse on vision, on visionary being, this may be taken to mean that,

> The clearing and lighting, [presencing as] a field of sheer lucency, is that which, as the unifying unity of seer and seen [the one beholding, the one beheld], bears the radiant ecstasy [ek-stasis] of their embrace.

The thesis is no longer a thesis of abstract thought, a proposition in the formal discourse of transcendental grammar, but rather a thoughtful recollection of Being in a phenomenological discourse which bears very directly – that is to say, concretely – on our experience as visionary beings. In our translation, the thesis is set in motion: it refers us to our (body of) experience; dialectically, it engages our visionary self-awareness; it functions hermeneutically, reminding us to be more thoughtful. It reminds us that existence is ecstatic, that to be a human being, *Dasein*, is to stand out into the openness of Being, and that we are, in an ontological sense, centres of visionary ecstasy, and not ego-logical subjects. It reminds us that vision is a gift from the clearing and lighting of Being, a 'unifying unity' presencing elementally as the medium which makes possible, and conditions, the emergence of all visionary structures. It reminds us that the historically prevailing structure of our vision takes place in responsive beholdenness to the gift (*Es gibt*) of a unifying unity: the clearing and lighting of Being. It helps us to recollect that primordial ground and field by grace of which our vision is first made possible. By reconnecting us to the unifying unity of the primordial field, it helps us to see our bodily felt 'centre of ecstasy' and makes of our vision, our visionary 'transactions', a hermeneutical disclosing – a *homologein* – of the primordial ecstasy of Being, seen, ontologically, as presencing in the opening-up of a (self-differentiating) field of sheer lucency. Thus we can see that, as a statement in the 'logic' of vision, our translation proposes an ontological task, a project; for it recalls our vision to itself and calls our vision into question, questioning its hermeneutical capacity, its skillfulness, in serving as a non-egological centre of disclosedness, of truth, for the ground of Being.

In the 1966-1967 Seminar on Herakleitos, Heidegger looks back on *Being and Time* and makes this comment:

> In *Being and Time*, Dasein is described as follows: Dasein. The *Da* is the clearing and openness of what is, as which a human stands out. . . . In this you must not reflect, but rather see. . . .[126]

And then he asks: 'In what is objectivity, and that which is represented, grounded?' Fink, taking up this question, then replies: 'That wherein consciousness and object play'. And to this Heidegger says: 'Thus the clearing in which something is present comes to meet something else present. Being opposite to . . . presupposes the clearing in which what is present meets a human'.

Responding to our crying historical need for a glimpse, a sense, of the unifying unity of the field of our vision, radical reflection takes us into the depths of an experience of sheer lucency, where the subject of vision and the object of its beholding are intertwined, and belong together in the primordial flesh of the lighting. In the presencing of Being, sheer lucency, we may see an intertwining play of radiant energies.

Thus, in the self-abandonment of crying, we touch the very *root* vision, the intertwining of Eros which entwines the copulation of subject and object, seer and seen, and embeds them in what Merleau-Ponty calls, in *The Visible and the*

Invisible, a 'symbolic matrix'.[127] This matrix is 'polymorphic',[128] a play of radiant, ecstatic energies, and it is only with an awareness of this play of lighting that subject and object, an embracing couple, can cry in bliss.[129]

In *Civilization and Its Discontents*, Freud defends the status he assigned to the ego, arguing for its supreme position and suggesting that the only alternatives are infantile or delusive regressions.[130] Since he *assumes* that the ego-logical structure, the structure of subject and object, constitutes the only, and the most supreme achievement of self-development, any developmental changes which involve the dissolution or deconstruction of that structure can only be considered regressive, infantile, or psychotic.

But Romain Rolland once wrote a letter to Freud in which he challenged Freud's theory of the ego and expressed deep unhappiness over the fact that Freud 'had not properly appreciated the ultimate source of religious sentiments'.[131] Freud granted that, for Rolland, this source 'consists in a peculiar feeling, which never leaves him personally, [and] which he finds shared by many others. . . .' As Freud understood it,

> It is a feeling which he would like to call a sensation of 'eternity', a feeling as of something limitless, unbounded, something 'oceanic'. It is, he says, a purely subjective experience, not an article of belief. . . . One may rightly call oneself religious on the ground of this oceanic feeling alone, even though one reject all beliefs and all illusions.[132]

In his reply, Freud observed, with disarming honesty, that, 'I cannot discover this "oceanic" feeling in myself. It is not easy to deal scientifically with feelings.'

Focusing, however, on 'the ideational content which most readily associates itself with the feeling', Freud proceeded to describe it – accurately, I think – as 'a feeling of indissoluble connection, of belonging inseparably to the external world as a whole'. And while he freely acknowledged its reality, or at least its subjective existence, he declared that he 'can only wonder whether it has been correctly interpreted. . . .' Freud then said this:

> The idea that man should receive intimation of his connection with the surrounding world by a direct

> feeling which aims from the outset at serving this purpose sounds so strange and is so incongruous with the structure of our psychology that one is justified in attempting a psychoanalytic, that is, genetic explanation of such a feeling.

But the interpretation Freud proposes covertly *rescinds* the recognition he had earlier given to the 'oceanic feeling'. The psychoanalytic interpretation claims 'only' to expose the 'origin' of this feeling. But the 'origin' it brings to light is infantile and regressive. The interpretation is, therefore, hostile and disingenuous, because it degrades the experience and denies it reality. Let us note that 'Man' is the point of reference: it is 'his' ontological connectedness, through feeling, which is in question. The oceanic feeling is said to be 'incongruous' with the structure of what Freud manipulatively calls 'our' psychology: so incongruous, in fact, that 'one' is 'justified' in attempting an interpretation which would reduce such a feeling to the status of its origin. Whose psychology? And who is 'justified'?

Freud's text, however, continues:

> Normally, there is nothing we are more certain of than the feeling of our self, our own ego. It seems to us an independent unitary thing, sharply outlined against everything else.

But, he adds, taking a step which must have been difficult and courageous, but which, in the final analysis, is not daring enough:

> That this is a deceptive appearance, and that, on the contrary, the ego extends inwards, without any sharp delimitation, . . . was first discovered by psychoanalytic research. . . . But towards the outer world at any rate the ego seems to keep itself clearly and sharply outlined and delimited.

Freud does grant that there is a state of mind – but he emphasizes that there is only *one* state of mind – in which the ego 'fails' to do this:

> an unusual state, it is true, but not one that can be judged as pathological. At its height, the state of

> being in love threatens to obliterate the boundaries between ego and object.

Freud is quick to point out, though, that,

> From pathology we have come to know a large number of states in which the boundary lines between ego and outer world become uncertain, or in which they are actually incorrectly perceived. . . . So the ego's cognizance of itself is subject to disturbance, and the boundaries between it and the outer world are not immovable.

This is all, however, beside the point, for what Rolland was alluding to, by way of his reference to an 'oceanic feeling', was not pathology, but rather a spiritual *transcendence* of the subject-object structure.

Freud nevertheless struggles with the problem, for he follows that paragraph with this gesture of concession:

> Further reflection shows that the adult's sense of his own ego cannot have been the same from the beginning. It must have undergone a development, which naturally cannot be demonstrated, but which admits of reconstruction with a fair degree of probability.

It should be noted, here, that, in his reconstruction, Freud argues that the ego comes into being as we set up boundaries to defend ourselves against pain:

> The tendency arises to dissociate from the [emerging formation of] ego everything which can give rise to pain, to cast it out and create a pure pleasure-ego, in contrast to a threatening 'outside', not-self.

The tenure of the ego thus corresponds to our defensive adaptation and the triumph of what he calls the 'reality principle'. But let us note his 'reconstruction' of the 'original state' of the developmental process, which he takes to be 'consummated' with the installment of the ego:

> In this way, the ego detaches itself from the external world. It is more correct to say: Originally, the ego

> includes everything, later it detaches from itself the external world. The ego-feeling we are aware of now is thus only a shrunken vestige of a far more extensive feeling – a feeling which embraced the universe and expressed an inseparable connection of the ego with the external world. If we may suppose that this primary ego-feeling has been preserved in the minds of many people – to a greater or lesser extent – it would co-exist like a sort of counterpart with the narrower and more sharply outlined ego-feeling of maturity, and the ideational content belonging to it would be precisely the notion of limitless extension and oneness with the universe – the same feeling as that described by my friend as 'oceanic'.

Here we see that Freud acknowledges the 'reality' of an experience of the world which is not structured ego-logically; but his reconstruction of the process of ego-formation can grant it validity only as an experience of infancy prior to the emergence of an ego. He asks:

> But have we any right to assume that the original type of feeling survives alongside the later one which has developed from it?

For Freud, the various religions are nothing but different cultural attempts 'to reinstate [the] limitless narcissism' of our most infantile stage of life: in other words, since religions attempt to ground spiritual life in an 'oceanic' feeling first experienced in early childhood, the 'self-development' they encourage is really nothing but a process of regression.[133] Freud is unable to conceptualize a development beyond ego structure which would take the form of a hermeneutical movement: a regressive return to retrieve a dimension of experience left behind not only in the developmental transition of the Western individual from infancy to adulthood, but also in the cultural transition from pre-modern forms of life to forms which are distinctively modern. What I have in mind, then, is a movement which is not (so to speak) 'completed' until the 'oceanic' experience, the wisdom of interconnectedness and wholeness, has been brought back, brought into the present, and appropriately integrated into present living.

I want to argue that this 'archaic' feeling, which once, in Freud's own words, 'embraced the universe', this guardian awareness of our interconnectedness with all other beings and our grounding in the wholeness of Being, and which Freud himself, under pressure, did finally recognize, is of the most decisive importance for our present critical task. It is this 'feeling', this awareness, which preserves, through all the vicissitudes of ego-logical history, a sense of the ground, the 'unifying unity' of subject and object. Its retrieval and redemption are therefore *necessary* for the deconstruction of structures of experience reified under the influence of our prevailing metaphysics. The 'archaic' feeling does not have to remain forever archaic; it remains archaic only as long as it is split off, excluded, in some way denied.

'Love', says Freud, 'threatens to obliterate the boundaries between ego and object'. Is this challenge to the boundaries inherently, or necessarily, pathological? Why must love be understood (only) as a 'threat'? It seems to me that, if our age is indeed threatened by the continued domination of the subject-object structure, what is needed is *an historical transformation* of this structure through the retrieval and redemption of what Freud is calling 'love': the experience of a unifying unity which grounds all the structural formations of our perception, and in the 'play' of which, as Fink phrases it, the difference between 'consciousness' and its 'object' opens out into the lighting of Being.

Casting light on our subject, Merleau-Ponty concludes a discussion of the medical history of Schneider, a patient, we may recall, who is diagnosed as suffering from war-time injuries to the brain, by arguing that, in his case, we can see very clearly that 'perception has lost its erotic structure, both spatially and temporally'.[134] Now, I submit that, if we are justified in taking the specialness of this case to imply that an erotic structuring is inherent in the very nature of perception, and that, barring conditions of neurological damage and conditions of extreme deprivation, neglect, and abuse, conditions such as those involved in childhood autism, which may profoundly, and sometimes irreversibly destroy the psyche's capacity to feel, to care, to relate, what would presently fulfill the historical need in our perceptual endowment is to be sought in the further development of this erotic structuring.

We must look to, we must *erörtert*, the place, the *Ort*, where this erotic structuring, a 'symbolic matrix' under the spell of Eros, can be seen to originate. Since the analysis of Schneider's case reinforces the plausibility of our suspicion that part of what is involved in the 'dimming down' of everyday vision is its loss of an erotic global and holistic sensibility – a loss which would also be a 'loss [in the lighting] of Being' – our task, a task of thinking, must take the form of a radical attempt to disclose the mysterious place of luminous darkness, where Eros is lost, or is hiding. Our visionary project, our project as visionary beings subject to the conditions of our history, must consist in a 'redemption' of this erotic experience that radically deconstructs and transforms the prevailing metaphysical reification of perceptual structures.

This is my answer to the question with which Freud leaves us at the end of *Civilization and Its Discontents*:

> The fateful question of the human species seems to me to be whether and to what extent the cultural process developed in it will succeed in mastering the derangements of communal life caused by the human instinct of aggression and self-destruction. In this connection, perhaps the phase through which we are at this moment passing deserves special interest. Men have brought their powers of subduing the forces of nature to such a pitch that, by using them, they could now very easily exterminate one another to the last man. They know this – hence arises a great part of their current unrest, their dejection, their mood of apprehension. And now it may be expected that the other of the two 'heavenly forces', eternal Eros, will put forth his strength so as to maintain himself alongside of his equally immortal adversary.[135]

(I would like to register, here, my unwillingness to regard aggression and self-destructiveness as instinctual impulses. As I see them, they are rather the effects of social experience. The assumption that they are biological instincts deflects our attention from their social causes and discourages efforts to bring about social changes. Nor is it necessary to explain the difficulties we encounter when we attempt to modify them by postulating 'eternal' biological forces beyond the control of our civilization. These difficulties can be explained, to my satisfaction, when we consider that there are, operating in the background of individual lives, larger, and extremely complex, social structures.)

I am suggesting, then, that the retrieval of the unifying unity of the gathering and belonging-together, the *legein* of subject and object, is also, at the same time, the discovery and restoration of their binding – their *religio* – under the spell of Eros: a binding whose vital structure we have, as individuals, and as a culture, lost touch with.

Since light, present and absent, is the universal medium, the element, which constitutes the grounding field of our vision, the historical task of radical thinking – thinking radically and critically steered by the question of Being – is to retrieve for thought this dimensionality of illumination and make it visible, in its truthing, as that in which the radiant energies of the field are indeed always and already ecstatically intertwined. And I think we will find this ecstatic intertwining of which Merleau-Ponty speaks precisely where Heidegger looks to see the clearing and lighting of Being. 'Intertwining' names the bodily felt character of the ontological experience. Recollecting the clearing and lighting of Being takes place together with an experience of the intertwining of seer and seen in a matrix that embraces them both. Here, in an elemental intertwining, is where the unifying unity of the visionary situation, the field of vision structured around the dualism and antagonism of subject and object, incessantly takes place; here, rooted in the intertwining and ecstasy of radiant energies, rooted in the interplay of constellations of light, is where the most primordial contact of seer and seen, interdependently co-emergent from the same elemental, universal flesh, perpetually takes place.

In *The Notebooks of Malte Laurids Brigge*, Rilke writes:

> I am learning to see. I don't know why it is, but everything penetrates more deeply into me and does not stop at the place where, until now, it always used to finish. I have an inner self of which I was ignorant.[136]

Phenomenologically considered, this 'inner self' is a *place of rest*, an *Aufenthalt*. It is the place of deeply felt experiencing, where subject and object are gathered into the intertwining of the field, the unifying unity. It is a 'place of rest' because its wholeness provides relief from the oppositions and pressures which are at work within the structure of subject and object. I would like to quote from D. W. Winnicott, an eminent English psychoanalyst, to clarify this point. In *Playing and Reality*, he says:

> My claim is that [there is] an intermediate area of experiencing, to which inner reality and external life both contribute. It is an area that is not challenged, because no claim is made on its behalf, except that it shall exist as a resting-place for the individual engaged in the perpetual human task of keeping inner and outer reality separate yet interrelated.[137]

When our visionary being is consciously, thoughtfully grounded in an experience of the intertwining, we will realize the dynamism of the field and will feel very alive; but we will also feel that we have found a 'place of rest'. Paradoxically, it is precisely because the intertwining is dynamic, is experienced as a deep source of our aliveness, that it is a ground, a place, of rest. Dynamism, flow, aliveness, unity and wholeness are what we lose, what we miss, what we need, insofar as we are pressed into the reified structure of subject and object. As we shall see in the next section, finding this resting place is an essential achievement in the development of our capacity to see with *Gelassenheit*.

In responding as we have to Heidegger's discourse on the principle of the ground, we are following and exploring a hint to be heard in the *Phenomenology of Perception*, where Merleau-Ponty writes:

> Let us try to see how a thing or a being begins to exist for us through desire or love, and we shall thereby come to understand better how things or beings can exist in general.[138]

Underlying the *cognitive* intentionalities of the normally mature ego, we will find the *corporeal* intentionalities of desire, of bodily feeling and sensibility. As the form of our basic relatedness; as a movement towards or away from, cognitive intentionality is always *originally* manifest as a *conatus*. Attraction and aversion, the vicissitudes of ego-logical life. Jung is therefore entirely correct when he asserts that the notion of *libido* 'expands into a conception of *intentionality* in general'.[139] It is *Eros* whose 'original promiscuity' awakens and motivates perception.[140] It is Eros, already engaging us in a 'participation mystique', which *moves* us to explore our surrounding space – that elemental nature about which Merleau-Ponty asks: 'is it not of its essence to embrace every being that one can imagine?'[141]

So if 'my gaze *pairs off* with colour', as Merleau-Ponty asserts,[142] this is because the matrix of vision is also an ecstatic field of radiant energy, a field perpetually dividing itself into different historical variations of seer and seen, yet perpetually holding the two together in the unity of a *prior* intertwining:

> The thing is *inseparable* from a person perceiving it, and can never be actually *in itself*, because its articulations are those of our very existence, and because it stands at the other end of our gaze, or at the terminus of a sensory exploration which invests it with humanity. To this extent, every perception is a communication or a communion, . . . or, on the other hand, the complete expression outside ourselves of our perceptual powers, and a coition, so to speak, of our body with things.[143]

And Merleau-Ponty adds a comment which we have already had reason to note, but which, at this point, we might usefully consider again:

> The fact that this may not have been realized earlier is explained by the fact that any coming to awareness of the perceptual world was *hampered by the prejudices arising from objective thinking*. The function of the latter is *to reduce all phenomena which bear witness to the union of subject and world*, putting in their place the clear idea of the object as in itself and of the subject. . . . But in reality, all things are concretions of a setting, and any explicit perception of a thing survives in virtue of a previous communication with a certain atmosphere.[144]

This passage suggests that we need to consider to what extent objective thinking, the thinking at work in our metaphysics, is in the service of an ego which needs to suppress all traces of experience bearing witness to the ecstasy of an erotic union in the primordial matrix of visionary Being. Thus I want to ask: to what extent is the 'reduction' exercised by objective thought an effect of the social, historically conditioned *repression* which occurs in the light of our erotic, and originally polymorphic, bodily nature?[145]

Pre-ontologically, prepersonally, and pre-egologically, the body of visionary being is an erotic play of structural difference, figure-ground differentiations, deeply but unconsciously attuned by, and deeply but unconsciously attuned to, the primordial ecstasy of light. Thus, should there be an effort to make contact (again) with the play of Eros, the visionary field itself, the sheer lucency of Being, may make its ecstatic appearance. And, since Eros, and the recognition of Eros, will decisively *transform* the structure in which subject and object are related primarily through the antagonisms of socially constituted desire, the things that vision beholds, when it begins to retrieve its erotic character as an openness, as itself a clearing within the field, may likewise, in an aesthetic co-responsiveness, make visible their primordial emergence from the clearing of Being:

> If the qualities *radiate* around them a certain mode of existence, if they have the power to cast a spell and what we called just now a sacramental value, this is because the sentient subject does not posit them as objects, but enters into *a sympathetic relation* with them, makes them his own, and finds in them his momentary law.[146]

Seeing this radiance of qualities requires a vision opened up in a responsive way to the synergic presence of the lighting. It will not be visible to a gaze in bondage to the structure of subject and object. Thus I agree with the argumentative strategy of Mark Taylor, who bases his deconstruction of theology, in *Erring: A Postmodern A/theology*, on a simple truth of the body:

> The boundary between bodies is a permeable membrane; it has gaps and holes to let the inside out and the outside in. . . . When inside is only inside and outside is only outside, when [breathing,] eating, drinking, . . . [and the processes of

> elimination] . . . stop or are stopped, vital current no longer flows and the body truly dies.[147]

But Eros also appears in many other guises, other dimensions of our experience. Thus, for example, in Heidegger's 'Building Dwelling Thinking', Eros assumes this form:

> When I go toward the door of the lecture hall, I am already there, and I could not go to it at all if I were not such that I am [always already] there. I am never here only, as this encapsulated body; rather, I am there, that is, I already pervade the room, and only thus can I go through it.[148]

(Unfortunately, Heidegger does not realize that the truth of this experience contests the claim to absolute truth in the reified structure of subject and object.) Sartre formulates the matter this way:

> My body is everywhere. . . . [It] always extends across the tool which it utilizes . . . it is at the end of the cane on which I lean . . . it is at the end of the telescope which shows one the stars; it is . . . in the whole house, for it is my adoption to these tools.[149]

And here is Bergson, whose work, *The Two Sources of Morality and Religion*, perhaps had an influence on Sartre's account:

> For if our body is the matter upon which our consciousness applies itself, it is coextensive with our consciousness. It includes everything that we perceive; it extends unto the stars.[150]

In *Survival in Auschwitz*, Primo Levi helps us to see what I would consider to be another aspect of the presence of Eros. Reflecting on the survival of people deprived not only of home, family and friends, but also of clothes, shoes, hair, and finally, even a name, he writes:

> But consider what value, what meaning is enclosed even in the smallest of our daily habits, in the hundreds of possessions which even the poorest beggar owns: a handkerchief, an old letter, the

> photo of a cherished person. These things are part of us, almost like limbs of our body; nor is it conceivable that we can be deprived of them in our world, for we immediately find others to substitute for the old ones, other objects which are ours in their personification and evocation of our memories.[151]

The texts from Heidegger, Sartre and Bergson describe the erotic character of perception by pointing out its panoramic embrace, its extendedness, its gathering and unifying; whereas Primo Levi's text describes how this erotic character connects us very strongly with the particular things that are present around us in the world of our daily living. In the next passage we will consider, a passage from Giambattista Vico's *The New Science*, we can see how these erotic connections create a multidimensional universe of cultural symbols:

> . . . in all languages, the greater part of the expressions relating to inanimate things are formed by metaphor from the human body and its parts, and from the human senses and passions. Thus, head for top or beginning; the brow and shoulders of a hill; the eyes of needles and of potatoes; mouth for any opening; the lips of a cup or pitcher; the teeth of a rake, a saw, a comb; the beard of wheat; the tongue of a shoe; the gorge of a river; a neck of land, an arm of the sea; the hands of a clock; heart for center (the Latins used *umbilicus*, navel, in this sense); the belly of a sail; foot for end or bottom; the flesh of fruits; a vein of rock or mineral; the blood of grapes for wine; the bowels of the earth. Heaven or the sea smiles; the wind whistles; the waves murmur; a body groans under great weight. The farmers of Latium used to say the fields were thirsty, bore fruit, were swollen with grain; and our rustics speak of plants making love, vines going mad, resinous trees weeping. Innumerable other examples could be collected from all languages.[152]

These metaphors of language are possible because perception itself is spontaneously and irrepressibly metaphorical: it makes connections, it gathers and unifies, it joins, it creates, it reproduces and multiplies; it is, in a word, erotic. Vision, being inherently metaphorical, responds primordially to things with a gaze very different from the literal gaze of the scientific eye. For the erotic gaze, things correspondingly show themselves, as Robert Romanyshyn has pointed out,

> only through other things; and in this regard, the elusiveness of things is betrayed. . . . This elusive character of reality involves not only the changing face of things – the sun at night and in the morning – or the paradox of identity/difference – the forest of flowers and rocks; it also involves the way in which one thing is what it is, only through something else. The tree, for example, shimmers in the water which flows beneath it, and through that reflection the tree *is* the water, just as the water in visiting the tree on a hot summer day is the tree. Indeed, . . . reflections and shadows . . . demonstrate this elusiveness of things and . . . indicate how each thing establishes a network of allusions to other things. The reflection visibly shows that the tree *is* the water, and in this regard, can one ask for a more visible demonstration of the originally *metaphorical* character of things? To say that the tree *is* the water affirms only what we see, this metaphorical play of things. . . . Through reflections and shadows, things *deepen* each other. The originary metaphorical character of things is the depth of reality. It is nothing less than the way in which *material* reality *matters*.[153]

Reflections can show the tree being water; the shadow can make a double who follows us around; the contours of some hills can remind us of the legendary face of a cultural hero; and the sight of a colour can evoke within us a passing tremor of emotion: all this is possible, because our vision primordially, and perhaps only, in fact, for an instant, is an ecstatic organ of Eros. Eros is the spirit of communication, relationship and love; it is also the spirit of play, and the poetry in vision. Eros is the spirit which symbolically inhabits the intertwining.

Arnold Mindell, a Jungian analyst who helps people integrate, into their living, channels of awareness which constitute a system of experience he calls our 'dreambody', tells a story of psychotherapeutic insight and change that will contribute, I think, to our understanding of the intertwining. The story will demonstrate that it is vision channelled through the dreambody which experiences the intertwining character of the 'unifying unity', and which appreciates the presence of the field as that which grounds and embraces the structures of subject and object.

It seems that one of Mindell's students was suffering from severe backache. Here, now, is Mindell's report:

> She had a dream in which she awoke in the middle of the night with a terrible shock. We started to work on her pain. She told me that she felt as if someone had put a tremendous amount of pressure on her back. One of the other students then went behind her and started pushing on her back and amplifying her experience of the backache. That afternoon we were working in a cabin in the mountains, and while he was pushing on her back, there was a sudden bang from the road nearby. It was a car accident. One car had backed up and crashed into the front of another. . . .
>
> The student [with the pain] . . . was especially disturbed by the accident. When I asked her why it should disturb her so much and why it should have happened right then, she thought it could have been [an instance of what Jung calls] synchronicity.[154]

Mindell notes that her focus of awareness had shifted: first she was focusing on her back; then there was a 'channel switch' and she was focusing on the event of the street. He asked her what she was experiencing.

> 'Well', she answered, 'I saw that woman in the front back her car into the car behind her. It was not he who was pushing her but she who backed into him'. (*Ibid.*)

For some strange reason, the particular details of the accident seemed to touch her and move her, seemed to bear a message of personal significance for her. The accident curiously affected her, and she could sense very keenly, through the channels of this uncanny attunement, the meaningfulness of a symbolic connection. She had the feeling that the accident was connected with her problem.

She experienced the accident as a metaphor for the significance of her present life-situation. And when she listened to her body of experience, an insight, a vision, suddenly emerged. The body of experience, the body of vision, allowed to speak for itself, unequivocally advised her: 'Let life be and let things happen. Let go and relax.' Then, according to Mindell, 'she began to cry, her eyes still closed . . . and tears started running down her cheeks'. Coming out of this experience, she communicated a further insight: 'Y'know, I've lived my life wrong. I always tried to push my life and could never believe that life was right the way it was.'

It is crucial to note, here, that it was an experience of wholeness, a sense of some unifying unity, which gave rise to her insight, her vision; she could suddenly see her life as a whole; she could see it in its wholeness; and she experienced this larger outlook as grounding and healing. From out of that shift in visionary channels, involuntary, or anyway unconscious, she was able to begin to see how she might change her life.

Mindell comments:

> This student's backache, dream symbol and automobile accident were different channels, different perceptions of her dreambody. The dreambody in this case was her [inner] God telling her . . . not to push backwards against fate but go with it and live humbly. . . . Becoming yourself can be understood as knowing your dreambody, becoming whole or round, developing your full experience through awareness of each of your different channels. (*Ibid.*)

And he explains that the channel-switch 'from a body process to an occurrence in the world indicates that this woman's dreambody was in her back and in her environment as well'. This explanation should recall some texts we read earlier: the ones by Merleau-Ponty, Heidegger, Sartre, and Bergson. Mindell observes that 'Switching channels can be a pretty mysterious thing'. In a sense, this is true – because we do not understand ourselves very well. Much of our lives are lived more or less unconsciously and in daily routines that substitute 'going along' for motivation. Furthermore, the metaphysics circulating in our culture does not understand, nor even acknowledge, dreambody awareness: the kind of resonance, the kind of attunement, the kind of deeply felt interconnectedness, which, because it happened to be keen, was able to *see* in the accident a personal significance. But in another sense, this dreambody insight is not at all mysterious. I want to *resist* the tendency to mystify perfectly comprehensible, and ultimately familiar, experiential processes.

There is no need to assume or project, here, a mysterious, supernatural event. Nor are we compelled to suppose that the accident was not really an accident – that it was fated to happen, that it was in any way causally related to her presence there, or to her backache. All we need to understand is that she was bodily attuned to her environment, responsive to its network of what Baudelaire called 'correspondences', in a way that made it possible for her to see a connection – see it metaphorically – between her life and the accident. The connection she sees, through the channels of awareness which constitute her dreambody, is an interpretation, a connection she *makes* out of the experience she happened to feel by grace of her rootedness in a lived sense of the intertwining of all things. As Merleau-Ponty says, in 'The Intertwining – The Chiasm', the flesh of the body is 'Visibility sometimes wandering and sometimes reassembled.'[155]

In *The Religio Medici*, Sir Thomas Browne noted that '*All flesh is grass* is not only metaphorically, but literally, true'. In *our* time, however, what needs argument is not the 'literal', but rather the 'metaphorical'.[156] How is vision grass? The *dreambody* is not a substantial body; it is not encapsulated by my skin. It cannot be understood metaphysically. It is a synergic body whose boundaries, and therefore connections, are co-extensive with its capacity to feel, to sense, to be open to its connectedness, its field-dependent identity with other beings – and even with the matrix of all beings, the field of Being itself. The dreambody is *nothing* other than our hermeneutical awareness of this pervasive connectedness – and this connectedness itself. It bespeaks the truth of our *Befindlichkeit*, which is to be situated in a world through our moodedness, for moodedness gathers together what 'wanders' into a primordially felt unity. The dreambody is our metaphorical experiencing of the world. It is what we *make* of our capacity to see relationships, to see things hermeneutically, in a round of cross-references. Its vision is made possible by the phenomenological fact, the *Es gibt*, of an elemental intertwining, a unifying unity, a field or ground which brings forth a play of identities and differences, an ontological difference which manifests in a weave of differentations, in figure and ground. And its vision is potentiated through the careful development of channels of awareness, of feeling and perception, which can always root us more deeply and extensively in (the character of) this elemental field.

Now, what kind of concept is this – the 'intertwining'? It is, I suggest, a symbolical or metaphorical concept: a hermeneutical concept which comes from the inherent poetizing of radical phenomenological thinking. It is a concept genuinely *grounded* in our experience of the elemental, the primordial. Its 'poetry' bespeaks its groundedness in the dynamism, the inherent creativity, of experience. As I have argued elsewhere (in a paper on 'The Poetic Function of Phenomenological Discourse'), all language which is genuinely thoughtful, and therefore deeply rooted in the aliveness of our experience, and which *remains* in contact with this dynamism, will 'sound poetic', because what we call the 'poetic sound' simply *is* the polyphonic resonance of deeply felt, deeply lived experience – experience really *alive* in our thinking and saying.[157] But there is also another factor we need to consider, in order to account for the symbolical concepts Merleau-Ponty used in his later thinking, viz., that the dimension of experience in question is primordial, and can therefore be entered only by a method of thinking with a hermeneutical respect for the way in which the primordiality of the primordial discloses itself. It is only as a 'dreambody', with eyes not controlled by a merely personal and ego-logical identity, that we can *see* this intertwining.

The intertwining is already in effect in our prepersonal, pre-egological embodiment. It is constitutive of a pre-ontological understanding of our visionary being in relation to the field of Being as a whole. It is, at first, unconscious. It constellates, psychologically, the 'collective unconscious': an elemental flesh within which we are gathered as visionary beings. But it can be raised into awareness, and then our contact with a prepersonal sense of intertwining, our contact with a pre-egological sense, makes possible an *opening up* of our socially organized life, our life as personal, ego-logical, interpersonal beings. With the retrieval of the intertwining, a recollection which takes place, so to speak, through the dreambody, our prepersonal sense of belonging to an elemental visionary intertwining – what got split off during the process of our growth into the normal state of personal-interpersonal living – can develop, can unfold, into a *transpersonal* experience of intertwining. Likewise, our pre-egological sense of the intertwining, split off during the struggle for the mastery of ego-logical existence, can then develop, growing into its individual fulfillment in an experience of the intertwining which will begin to sustain a life not dominated by the ego – a life lived without selfishness, or the need to master and dominate.

In *Mothers and Amazons: The First Feminine History of Culture*, Helen Diner writes:

> Knitting, knotting, interlacing, and entwining belong to the female realm in Nature, but so does entanglement in a magic plot . . . and the unraveling of anything that is completed.[158]

Mythopoetically – and in a sense which is relative to the cultural conditions, at once emancipatory and oppressive, that have determined the appearance of the mythic archetypes in our time, the flesh, the intertwining, is the realm of the Feminine Principle:

> the Great Mother, adorned with the moon and the starry cloak of night, is the goddess of destiny, weaving life as she weaves fates.[159]

But in the interplay of identities and differences, the Great Mother is also Eros, is Hecate, is Moira, is the disturbing presence at the crossing of the roads. The Great Mother is a wisdom hidden hermeneutically in the warp and woof of the intertwining. The intertwining is the weave of our destiny, woven into the tissue, the very flesh, of our visionary being. It is also an inscription upon the flesh, and if we learn how to read it, the text of our promise, a programme, a task, an opportunity.[160]

3 GELASSENHEIT: VISION WITHOUT DUALISM

Opening Conversation

(1) . . . to let a line muse . . .

Merleau-Ponty, *Eye and Mind*[161]

(2) Animals trustfully stepped/ into his open glance, grazing,/ and captive lions/ stared in as into inconceivable freedom;/ birds flew straight through him,/ the open of soul. Flowers/ gazed back into him/ big, as into children.

Rainer Maria Rilke (in a letter to Lou Andreas-Salomé, Paris, 1914)[162]

(3) But even they [the gods] have not acquired this eye at a single stroke: seeing needs practice and preschooling, and he who is fortunate enough will also find at the proper time a teacher of pure seeing.

Nietzsche, *Daybreak: Thoughts on the Prejudices of Morality*[163]

(4) The [Buddhist] teaching of the Ayatanas [in the texts on psychology in the *Abhidharma*] suggests that we attend to the feeling-tone aspect of experience, which we often neglect through our fascination with the objective field.

Stephen Beven, 'Bases of Cognition'[164]

(5) What is unconscious generally are patterns of organismic structuring and relating which function holistically as background of focal attention.

John Welwood, 'Meditation and the Unconscious: A New Perspective'[165]

Among the Ainu, an aboriginal tribe of people living on an island now belonging to Japan, there is a sacred song, the 'Song of the Spider Goddess', which weaves into the text, at regular intervals, the following refrain:

> . . . doing nothing but needlework,
> I remained with my eyes
> focused on a single spot,

and this is the way
I continued to live
on and on
uneventfully. . . .[166]

A perplexing communication, which the anthropologist who recorded it does not explain. Since he does not, I would like to suggest an interpretation myself. For the interpretation I offer, it can at least be said, I think, that it is provocative and useful. As I interpret that strophe, it is an allusion to the goddess's visionary wisdom, a wisdom she teaches as a meditative practice, a discipline and skill of the eyes: a practice with vision which can develop a capacity for concentration and focus. (In Tibetan, *zhi-gnas*.) What is to be developed by this practice is a capacity which modifies the normal character of our visionary concentration and focus. Speaking phenomenologically, I want to suggest that the 'single spot' is a resting-place, an *Ort der Stille*, a place where the eyes can become, and remain, unperturbed, tranquil, at peace, and relaxed: a place often experienced as providing a therapeutic sense of eternity and immortality. The practice is not one of fixation, a gaze rigid, tight, and unmoved, but one, rather, of centering: a centering, moreover, which *decentres* the form of subjectivity which lives only in the narrowness of the atomized, immediate present. The 'Song of the Spider Goddess' can teach us *Gelassenheit* as a practice for the *development* of our visionary potential.

Normal perception, the ontical perception of anyone-and-everyone, is inveterately grasping, as the very word itself should remind us. It is an anxiety-driven, restless intentionality: a grasping *of* light and a grasping *in* the light. But such perception cannot see the whole of things, because wholeness, unlike totality, is not something that can be grasped. In *Being and Time*, Heidegger points out that,

> Letting something be encountered is primarily circumspective; it is not just sensing something, or staring at it. It implies circumspective concern, and has the character of becoming affected in some way. . . .[167]

'Circumspective concern' characterizes a vision, a gaze, which is deeply affected by – which allows itself to be affected by – the presencing of the visionary field as a whole: a gaze whose global character as panoramic caring lets that field be present in its wholeness. A gaze moved by 'circumspective concern' is a vision in touch with its felt sense of the visionary field, whose primordial wholeness it embraces through its inherent ontological attunement:

> Occasionally, we still have the feeling that violence has long been done to the thingly element of things and that thought has played a part in this violence, for which reason people disavow thought instead of taking pains to make it more thoughtful. But in defining the nature of the thing, what is the use of a feeling, however certain, if thought alone has the right to speak here? Perhaps, however, what we call feeling or mood . . . is more reasonable – that is, the more intelligently perceptive – because more open to Being than all that reason which, having meanwhile become *ratio*, was misinterpreted as being rational.[168]

This argument for the philosophical recognition of feeling appears in 'The Origin of the Work of Art'. It is a position which encourages me to spell out my conviction that, in an important sense, a sense we need to recognize and understand, *Gelassenheit* is an *achievement* of sensibility, a development of our ability to feel, and that it is a possibility which requires thoughtful discipline and practice. *Gelassenheit*, in fact, is a practice of the Self: a practice involving us in the ongoing development of our capacity for 'circumspective concern' – for caring. Thus I want to think, here, about what 'circumspective' means as a term for describing the *character* of our gaze in relation to its visual field. And I will argue that *non-duality* is at the very heart of *Gelassenheit*, the realization of our visionary life, and that non-duality requires ongoing practices with the emotionality, the affective and motivational character of our vision: practices which involve the development of our capacity for a different style of visual motility – relaxation, neutralization, steadiness, equilibrium, and focusing with holistic awareness.

Let us first take a look at a passage from Nietzsche's *Twilight of the Idols*:

> One has to learn to *see*, one has to learn to think, one has to learn to *speak* and *write*: the end in all three is a noble culture. – Learning to *see* – habituating the eye to repose, to patience, to letting things come to it; learning to defer judgment, to investigate and comprehend the individual case in all its aspects. This is the *first* preliminary schooling in spirituality: *not* to react immediately to a stimulus, but to have the restraining, stock-taking instincts in one's control. Learning to *see*, as I understand it, is almost what is called in unphilosophical language 'strong will-power': the essence of it is precisely not to will, the *ability* to defer decision.[169]

Not to will is precisely the attitude of *Gelassenheit*, achieved – if you like, paradoxically – only by the greatest of ongoing exertions, practices of the Self involving the gaze in the neutralization of its inveterate tendency to grasp, secure, master, and dominate; involving the gaze in exercises which relax the eyes and diminish their deep-seated ontological anxieties, since these are the cause of restless and aggressive motility; involving the gaze in exercises to transform the incessantly grasping, desire-driven character that is typical of the ego-logical vision which prevails in modern times; involving the gaze in a gentler, more restful centering and focusing; involving the gaze in a different way of structuring the gift of the lighting. *Gelassenheit* is *beyond* the nihilism of the will, beyond the attitude of control; but its actualization is an ongoing process which needs our self-control and self-discipline: it does not happen 'all by itself'. It is a task, a project, and sometimes, an achievement, of the Self. It is our self-critical, self-aware vigilance, as visionary beings; and this vigilance is our commitment to response-ability: at the very deepest level, our commitment to the actualization of a care-taking, thought-taking function, the on-going development of our visionary circumspection.

I call attention to some 'working notes' written by Merleau-Ponty:

> understand perception as differentiation, forgetting as undifferentiation . . . disarticulation which makes there be no longer a separation (*écart*), a relief. This is the night of forgetting. Understand that the 'to be conscious'=to have a figure on a ground, and that it

> disappears by disarticulation – the figure-ground distinction introduces a third term between 'subject' and the 'object'. It is that separation (*écart*) first of all that is the perceptual meaning.[170]

Later, he proposes that we characterize the experience of the field not totally determined by an identification with the subject-object polarity, in terms of polymorphism:

> Show that, since the *Gestalt* arises from polymorphism, this situates us entirely outside of the philosophy of the subject and the object.[171]

It is the vision of *Gelassenheit* which goes *outside* our historical metaphysics of subject and object. But, if we are to see this, we must now consider what Heidegger says about *Gelassenheit*. In the Memorial Address he gave in Messkirch, Heidegger said:

> Releasement [*Gelassenheit*] toward things and openness to the mystery belong together. They grant us the possibility of dwelling in the world in a profoundly different way. They promise us a new ground and foundation upon which we can stand and endure in the world of technology without being imperiled [and brutalized] by it.[172]

Our vision is in crying need, now: in need of a vision of *Gelassenheit*. But what kind of vision is this? That is our critical question.

Heidegger seems to be suggesting that, instead of what he calls 'the madness of vision', we might look forward to the possibility of a vision no longer agonized by the socially constituted polarization of subject and object.[173] What I take Heidegger to be telling us about this other vision, this vision of *Gelassenheit*, is that it would involve a gaze which 'lets what is present as such become present in unconcealment'.[174] In other words: it 'lets [the] entities which are accessible to it be encountered unconcealedly in themselves'.[175]

Now in fact, this mood, this attitude, is already implicit in the German word for perception: *Wahr-nehmen*. This word reminds us that perception is always, potentially, a care-taking gesture: a taking or receiving which maintains and preserves beings in their wholeness-of-being: it reminds us that, in order to perceive beings in their truth, we must let them show themselves as they are. And this means: not on *our* terms, but rather, in accordance with their own necessities. Perception is *true* when it becomes a 'thoughtful maintenance of Being's presence'.[176] But what, more concretely, does all this mean? Heidegger's concept of *Gelassenheit* is suggestive; but it needs to be fleshed out. Until it has been spelled out in terms that refer us phenomenologically to our life-world of experience, to our gaze, for example, its historical significance cannot be determined – nor can it bring forth, bring about, any significant changes in our world.

In her *Padma bKa'i Thang, The Life and Liberation of Padmasambhava*, a text composed in Tibet during the ninth century, Yeshé Tsogyal observed that,

> the evidence of the Body of Metamorphosis is tranquillity of the senses, gentleness, a peaceful heart, quietness, and the holy perfections.[177]

This description is, I think, a useful starting-point for our attempt, here, to give more experiential determinacy to the attitude of *Gelassenheit*. Its usefulness lies in the fact that it refers us to our embodiment. It enables us to articulate the character, the meaning of *Gelassenheit* for our visionary life.

The ideal of *Gelassenheit* calls for a gaze which is relaxed, playful, gentle, caring; a gaze which moves freely, and with good feeling; a gaze which is alive with awareness; a gaze at peace with itself, not moved, at the deepest level of its motivation, by anxiety, phobia, defensiveness and aggression; a gaze which resists falling into patterns of seeing that are rigid, dogmatic, prejudiced, and stereotyping; a gaze which moves into the world bringing with it peace and respect, because it is rooted in, and issues from, a place of integrity and deep self-respect.

Tranquillity depends on steadiness, on relaxation – a distinctive modalization of the tensions inherent in intentionality. This modalization depends, in turn, on 'circumspection', a different relationship to the ground, the encompassing field. In the achievement of *Gelassenheit*, the gaze would move with a motility grounded in a deep sense of serenity. It would not be moved by a need for violence. As organs of *Gelassenheit*, the eyes would move smoothly, gracefully, as if inscribing a melody in the medium of light. Visual *Gelassenheit* is therefore possible only when our emotional attachments, the energies of desire that invariably polarize our visual situations, have been appropriately neutralized, tempered, balanced. In the present conditions of everyday life, our vision normally gets caught up, almost at once, in relationships of intense attraction and avoidance. Being representational, our vision structures the situations it inhabits along lines of opposition. Whether rejecting or desiring, the gaze is normally involved in forms of aggression. It is obvious how rejection involved aggression. But even attraction leads to aggression. For, whenever the gaze is attracted and moved by desire, it seeks to master and possess. (Sartre understood the pathologies of vision very well; but he projected these pathologies into the very *essence* of vision, thereby cutting off all possibilities of transcendence.) Equanimity will be possible only when we have been successful in neutralizing these intense emotional energies. Only then may we see with *Gelassenheit*.

The gaze of enlightenment moves with stillness; it is rooted in a place of deep inner stillness. Even when it moves, as it must, of course, it moves with a graceful stillness: firm, steady, even, beyond perturbations.[178] Borrowing some words from Foucault, I would like to suggest that *Gelassenheit* is a vision which moves 'towards the truth, towards steadiness, towards the ground'.[179] Only such a developed gaze is capable of living the great vision of *Gelassenheit* and bringing it into society.

In *Painting and Linguistics*, Jean Paris comments on J. F. Lyotard's philosophical appropriation of the analysis of vision proposed by two scientists, Barre and Flocon, in a book called *La perspective curviligne*. According to the summary by Jean Paris, it seems that Barre and Flocon want to argue that,

> 'Education and habit stimulate our desire to match reality with our concepts'. . . . But the weakening of the peripheral vision to the benefit of the foveal one deprives us of a 'vast lateral fringe' that 'the focalized attention represses'; and the ocular mobility, by structuring only the 'knowable', . . . [succeeds] . . . in making us reduce the 'true' space, which is curved, to our three arbitrary dimensions.[180]

Therefore, according to Paris's representation of the authors in question, what we need to do is practice a different kind of gaze:

> One good advice: 'set your eyes on a [single, fixed] point for a long time, and let laterally come to you, without turning towards it, all that is eliminated by the prehensile, secondary, organized vision' – and soon, 'instead of the rectangular, stabilized, constant, central space of the foveal vision', you will discover 'this curved twilight, evanescent, lateral space', the true space of the truth.[181]

I want to point out, here, that this practice is strikingly similar to the practice of visionary wisdom transmitted in the 'Song of the Spider Goddess'. In both cases, the character of the focusing is *different* from the normal. The concentration, the one-pointedness, allows the gaze to settle down, to rest. Focused by self-control, focused in a gentle, but disciplined way, the gaze is not fixated; on the contrary it becomes relaxed and restful. This relaxing is crucial, because it is only when the gaze is thus modified that it can also be diffused, spacious, open, alive with awareness and receptive to the presencing of the field as a whole. This holistic awareness of the field, however, is itself, in its turn, a source of further emotional rest, a ground of tranquillity, equanimity, and healing. A sense of wholeness is always healing, always beneficial. When the restless gaze is gently disciplined, brought to rest, to tarry and linger for a while in one-pointed concentration, there is an aesthetic diffusion of emotional energies throughout the visual field. This diffusion *neutralizes* the inveterate tendency to reify the field in a polarized structure. In this way, then, the practice of one-pointed focusing *deconstructs* the structures of subject and object, and actually *heightens* our capacity to see what is there to be seen – what is *given* to us for our beholding.[182]

As we know, the question of the ground has been a central question in the history of Western metaphysics. Thus it seems reasonable to suggest that, if we want to understand how our vision sustains it; and if we want to understand the impact of our metaphysics on the historical character of our vision, then we must give thought to the being and presencing of the ground in our vision of figure and ground. What *is* metaphysics as a way of structuring the visual field? Following Heidegger, I would say that our metaphysics is a reification and totalization of (luminous) 'presence'. Our metaphysics requires three things: (i) constancy of presence, (ii) permanence of presence, and (iii) fixity or rigidity of presence. Thus I submit that it is not at all an historical accident that Gestalt psychology and Gestalt therapy should have discovered, as a result of their extensive experimental and clinical work, precisely these 'motives' at work in gestalt processes, and that it should see in them the character of much contemporary psychopathology.[183]

Tarthang Tulku, a Tibetan scholar and teacher of Buddhism, has suggested, with such pathology in mind, the following practice with vision:

> If you first close and then open your eyes very carefully, subtly preserving your connection with the more gentle, embracing light, then the light you discover will be very different from the ordinary sort. . . .[184]

This, I would argue, is a practice of great importance for the development of *Gelassenheit* as a visionary style of historical being.

Let us continue exploring the relationship between *Gelassenheit* and the 'metaphysics of presence', taking up this question as a question for our experience with vision. If the vision of *Gelassenheit* is a vision no longer stuck within our historical 'metaphysics of presence', we may suppose, *a fortiori*, that it engages a vision which no longer finds itself compelled to experience the presencing (*Anwesenheit*) of Being with a need to require, and impose upon it, the metaphysical conditions of constancy, permanence, and fixity. *Gelassenheit* must be a vision which is open to a different historical relationship, a different historical response-ability, to the presencing of Being.

In order to focus our reflections on the question of the meaning of 'presence', I want briefly to consider a recent position-paper by John Sallis. Accurately rendering Heidegger, Sallis writes that,

> In *Being and Time*, 'presence' means predominantly, though not exclusively, *Vorhandenheit* (presence-at-hand, in the usual translation). This is to be

> understood in its correlation with pure seeing, with *noein*, with intuition (*Anschauung*): When something gives itself to one's sheer gaze, when it is simply there for one's looking, displaying itself before and for apprehension, then it has the character of being present-at-hand. Such a character is to be contrasted with that of things with which one deals concernfully, when one manipulates things and puts them to use.[185]

Now, it is to be noted, here, that, for Sallis as for Heidegger (the Heidegger, that is, of *Being and Time*), presence-at-hand is correlated with a vision akin to staring – the 'hammer "merely stared at", for example' – and that, apart from *this* experience with vision, all experiences are assumed to be essentially instrumental:

> There are no simply sheerly present things; for everything is openly or concealedly ready-to-hand, and what is ready-to-hand – the hammer when one takes hold of it and uses it – is not sheerly present as a self-contained positivity. Rather, it is extended beyond itself into the referential totality by which it is determined. . . . There is no pure presence ['no simple sensory presence']; for in whatever presents itself there is already in play the operation of signification.[186]

The metaphysical sense of 'presence' contested by Derrida comes through very clearly in a recent interview. What Derrida has to say indicates that we cannot continue to suppose that the sense of 'presence' involved in *Gelassenheit* is threatened by the deconstructive critique. Here is what Derrida says:

> The priority of spoken language over written or silent language stems from the fact that when words are spoken, the speaker and the listener are supposed to be simultaneously present to one another; they are supposed to be the same, pure unmediated presence. This ideal of perfect self-presence, of the immediate possession of meaning, is what is expressed by the phonocentric necessity. Writing, on the other hand, is considered subversive

> insofar as it creates a spatial and temporal distance between author and audience; writing presupposes the absence of the author and so we can never be sure exactly what is meant by a written text. . . .[187]

The presence involved in *Gelassenheit*, however, is not a 'perfect self-presence'. Nor is it an 'immediate possession of meaning'. Nor is it incompatible with spatial and temporal distance: it is not a pure 'immediacy' in the sense of a 'fusion' or 'coincidence'. Nor does it fail to acknowledge alterity and difference. Nor is it ever 'a mutually intuitive correspondence between two human presences'.[188] Nor does it suppress or occlude the invisibility of the invisible, or the absence of concealed depth and dimensionality. Nor does it constitute a situation in which there can be absolute certainty, absolute finality, or closure of meaning. Nor does it require a metaphysical abstraction from its existential situation. On the contrary, *Gelassenheit* is essentially 'circumspective': it is situated in a field of practical relations, in a world; and it is inseparable from its circumstances.

After the *Kehre*, the 'turning point' in his thinking, Heidegger realized that we need to recognize a *third* sense of 'presence', different from both *Vorhandensein* and *Zuhandensein*. This sense he called *Anwesenheit*. But Sallis follows Derrida in not recognizing – or anyway, remaining silent about – this third sense. About 'presence' in *this* sense, they apparently have nothing to say. Is this sense of 'presence' metaphysical, and therefore subject, like the senses examined in *Being and Time*, to the critique which is aimed at the 'metaphysics of presence'? I do not think so. Rather, the third is a sense of 'presence' which comes to the fore only *after* the impact of this critique. And it is, I submit, this other sense of 'presence' which explicates the attitude of *Gelassenheit*. Letting-be is unquestionably *not* an attitude congenial to the reification and totalization characteristic of the 'metaphysics of presence'. Therefore, the 'presence' – the being-with – involved in *Gelassenheit* is not an experience with the Being of beings totally determined by the history of metaphysics.

We should note that, when an *example* of presence-at-hand is chosen for our reflection, it is an act of staring. Thus, we are being asked to recognize in our experience with vision only *two* modes of visual presence: (i) the 'presence' which correlates with 'pure seeing', i.e., with disinterested theoretical staring and (ii) the 'presence' which correlates with instrumental rationality. But what about aesthetic perception? Surely it must be conceded that the kind of looking that takes place in the presence of artworks is neither of these.

Anwesenheit, the sense of 'presence' demonstrated in *Gelassenheit*, does not fall under the spell of the 'metaphysics of presence'. *Gelassenheit*, a third mode of presence, is in fact Heidegger's *answer* to the history of this metaphysics. It manifests an historically new way of being-with, and therefore also, a new epistemology, a new ontology.

Of decisive importance for this argument is the *difference* between the 'presence' involved in *Gelassenheit* and the 'presence' involved in *Vorhandensein*. The latter is described (in our second passage from Sallis's text) as involving 'a self-contained positivity'. Thus, when Sallis argues *against* the possibility of 'presence' *in this sense*, he observes – accurately and appropriately – that the hammer, like all other things, is 'extended beyond itself into the referential totality'. In other words, his argument against the 'presence' of *Vorhandensein* is that everything we encounter belongs to a referential field: nothing exists as a 'self-contained positivity', or as an isolated substance entirely independent of its situation within the world. But the sense of 'presence' which is manifest in the attitude of *Gelassenheit* is *not at all* a sense defined in terms of traditional substantiality and field-independence. On the contrary, *Discourse on Thinking*, Heidegger's work on *Gelassenheit*, makes it very clear, through an abundance of descriptive formulations, that the being-with of *Gelassenheit* constitutes a 'presence' which always takes place in – and with an awareness of – a referential field, a field or region of Being.

We may therefore conclude that the 'presence' of *Gelassenheit* is *not* called into question by the critique of the 'metaphysics of presence'. Letting-be simply constitutes *a radically different relationship* with beings. In vision, for example, letting-be is certainly not a 'simple sensory presence', a meaningless, worldless sensuous abstraction, an encounter reducible to the philosopher's sense datum. Moreover, it is neither a 'cold', theoretically disinterested staring-at, nor a looking-at totally possessed by instrumental calculations. Letting-be is an interested looking which cares; it is a being-with which cares; it is a response-ability to the presencing of Being which lets it come forth, lets it be present, *without* needing to master and dominate its presence. Letting-be is a being-with which understands the Being of the ground, and therefore cares *very much* for the field, the ground, of our visionary life.

Far from undermining the claim that *Gelassenheit* exemplifies a valuable experience with 'presence' (*Anwesenheit*), the post-modern critique of the metaphysics of presence *first makes it possible for us* to conceive a gaze belonging to a new epoch of history: a gaze whose *caring* is what fulfills its hermeneutic capacity, its capacity to open into unconcealment. I agree that our historical plight is connected to the dominance of a metaphysics of presence, and that we must radically deconstruct its operations in the structuring of our visionary situations. But, beyond this negative stage, we must attempt, more positively, to think a gaze which does not position itself in opposition to Being – and does not 'posit' Being as a constant, permanent, fixed presence, an immutable substance, or a reified totality.

'Gelassenheit' is a very radical postmodern concept, because it articulates a relationship with beings, and consequently to the Being of beings, which is not totally determined by the modern metaphysics of presence.

Perhaps what we need now are some concrete examples. I shall suggest four. (i) Normally, when we are walking outdoors in the most bitter cold of a northern winter, we tend to oppose and resist the environmental mood. And if, at that time, we become aware of (the character of) our bodily way of being-in-the-world, we will realize, with a bodily felt certainty, that, instead of relaxing and yielding to the cold, we have become tense and tight, bodily fighting off the hostile, undesired cold. But if, instead, we should let ourselves accept and welcome it, we would find that our breathing has improved and the circulation of blood has been greatly facilitated. As a result, of course, we would feel, and would be, less cold. *Gelassenheit* is *healthy* for the physical body: it is in fact a *norm* for our physical well-being. We might even say that *Gelassenheit* involves shifts in our patterning of energy and activity which fulfill an *organismic* condition and function. The fact that, when we relate to the winter environment in an attitude, a *Stimmung*, of letting-be, letting go of our representations, we suffer less the objective cold, certainly constitutes reasonable evidence for this perhaps surprising conjecture.

(ii) What do we do when we have forgotten something?

We always try very strenuously to remember; and typically, we persevere for awhile, even though we usually get the sense, a quite distinct sense, that the more unremittingly we pursue and try to grasp it, the longer we will fail. We all somehow know, knowing through our bodies, that the only way to succeed is to yield to the forgetting, to surrender the ego's need to recall. In order to remember, we must first forget: we must forgo the getting; we must neutralize the desire to remember; we must let go and let be. *Gelassenheit* is a recollection whose non-duality finally brings forth a presence here.

(iii) In the late spring, when all the flowering trees and shrubs are in full blossom, the air is heavy with their fragrance – many different fragrances, gathered into one bouquet. And yet, if we sniff the air very intensely, anxious to capture the totality of the gift, we find ourselves denied gratification. Intense olefactory activity, an arrowlike intentionality narrowed to a greedy, possessive form of desire, will miss the gift; by contrast, a more relaxed approach, simply lingering circumspectively and receptively in the atmosphere of that fragrance, our senses alive and open, being *with* it in a way that lets go of the need to possess it, to bottle it up in our nostrils, being with it in a way that lets it be as an enveloping presence, will be amply rewarded with a richness of olefactory sensations. This point can be tested. If, as we saunter through a garden, we suddenly become aware of an especially seductive, especially delightful fragrance, we normally tend to *locate the source* (*archē*) of our pleasure and move closer to it – getting as close, in fact, as we can. (This tendency to master the source, the *fons et origo,* is of course at work in our metaphysics.) But if we then try to inhale this fragrance with a hungry intensity, no fragrance comes forth. The fragrance was not to be 'had' in the source, but only in the atmosphere: it is a presence which can only be enjoyed without driving exertion, without willfulness, without aggressive desire. When we go *directly* to the 'source' to inhale a more concentrated scent, the presence withdraws. Only in *Gelassenheit* is there a holistic structuring of energy and activity which can receive this presence.

(iv) Most important of all, *Gelassenheit* constitutes a radical transformation of sociality. In interpersonal relationships, the difference between the 'presence' of *Gelassenheit* and the 'presence' of *Zuhandensein* and *Vorhandensein* is of great consequence. 'The genuine greeting', writes Heidegger, 'offers to the one greeted the harmony of his own being'. (*Der echte Gruss schenkt dem Gegrüssten den Anklang seines Wesens.*)[189] The 'presence' of *Gelassenheit* is to be contrasted with ways of being with others characterized by distraction, divided attention, conscious or unconscious hostility, prejudice and stereotyping, exclusively instrumental preoccupations, and interests or representations which direct our attention away from the individuality of the other. No meeting, no contact with others, no communication with others, is ever absolutely or totally free of all such mediations. Nevertheless, we all know the difference very well, often as very young children, between the representational 'presence' of *Zuhandensein* and *Vorhandensein* and the entirely different 'presence' which approximates, or shifts in the direction of, *Gelassenheit*. Parents sometimes try to feign an attentiveness to what their children are saying or doing. Presumably, they hope, or believe, that their children won't sense the difference. But children, being extremely sensitive to parental moods, can always immediately *feel* their parents' withdrawal or distraction – their unmistakable absence. The 'presence' of *Gelassenheit* does not mean a fusion or merging of identities. It does not mean total coincidence, or total possession. It does not mean reciprocal transparency, or total disclosure. It means being genuinely *with* the other, in a contact or communication which feels, not of course total, but distinctly whole: 'complete' in the sense that what needs to be, or wants to be, brought forth, what is there in the situation, needing or wanting to be brought out, is reciprocally brought out and unfolded into the socially constituted openness.

The gaze of *Gelassenheit* is a neutralized gaze: a gaze in which the metaphysical dualisms rooted in our egocentric patternings of desire, our egocentric intentionalities, have been transformed into ways of caring, of being with others in 'circumspective concern'. The neutralization does *not* destroy feeling; it does not encourage neglect or indifference. On the contrary, it is this process of neutralization which first makes real caring possible. *Gelassenheit* is not willful; but it is a style of agency: it is not passivity, but rather a mode of comportment different from that which takes place according to the tensions in the opposition between activity and passivity. Receptiveness, openness, do require effort; and therefore, they are achievements – achievements of character.

Translated into the world of interpersonal relationships, the two metaphysical senses of 'presence' – *Zuhandensein* and *Vorhandensein* – involve an oppositional structure, a structure of subject and object. 'Gelassenheit', however, brings to language the ideal of a different structuring of positions: one where there would be a reciprocated respect for the differentness of being, the untotalizable otherness, of the other. 'Gelassenheit' articulates our sense that there is an experience of 'presence' in our being with others which is radically different from the instrumental and theoretically mediated experiences with others in which our 'metaphysics of presence' predominates.

In closing, I would like to argue that the development of *Gelassenheit* is potentially a practice with great *political* significance. As Merleau-Ponty reports in 'The Child's Relations with Others', a rigid style of perceptiveness is characteristic of a rigid personality.[190] It also turns out, not surprisingly, that such rigidity, extremely intolerant of ambiguity, of uncertainty, of difference, of pluralism, is precisely what constitutes the essence of the 'authoritarian' personality, the type of personality which forms the basis of, and is in turn formed by, totalitarian systems of power. If we see that, in the 'metaphysics of presence', what is called 'presence' is reified and totalized, we need also to see that the same may be said of 'presence' insofar as it is determined by the conditions of a totalitarian regime. The political significance of *Gelassenheit* as a *practice* is that letting-be would overcome positions of rigidity, dogmatism, and intolerance; that it undoes the will to coercion, manipulation, mastery and domination; and that it transforms the pathological compulsion to secure, to make certain, to seize, capture, and possess: in *Gelassenheit*, there is a neutralization of the desire to totalize.

However, I must emphasize that, since vision takes place in the world, and since this world is a place of oppression and socially produced suffering, *Gelassenheit* is only an *ideal* of vision, a 'practical postulate' for our visionary being; though allowed some measure of actualization – provided that there are correspondingly radical changes in the body politic, our social environment – *Gelassenheit* is not a way of life which can ever be completely achieved. Beyond a certain point, the individual's development of this capacity requires collective support: the support of society as a whole; the support of the political economy. Individual practice is necessary, and much can be accomplished, both for the individual and for society, in this way. But as long as there are social conditions – injustice, unfreedom, fear and hate – which promote reification and totalization, effects of the hegemony of the metaphysics of presence, it must be understood that the ontological attitude of *Gelassenheit* will remain deeply corrupted, pathologized, and incomplete.

There can be no doubt that *Being and Time* decisively set in motion a critical rupture with the history of metaphysics. Nevertheless, as Heidegger himself later realized, that work also continued the metaphysical tradition. The *Kehre*, the so-called turning point in Heidegger's thinking, accordingly marks his attempt to effect a further, more radical rupture. Of course, as Derrida helps us to see, Heidegger continued to think within this tradition, in the sense that his thinking remained oriented, as before, by the Question of Being. However, it is crucial to understand that, although this question was still taken to concern 'presence', the significance of the *Kehre* is that Heidegger finally began to think out the possibility of an historically new (or different) sense of 'presence'. The concept of 'Gelassenheit' is his attempt to articulate a way of being with other people and things which lets the presencing of Being happen (*ereignet*, *apophainesthai*) in a different way. 'Gelassenheit' conceptualizes a human *presence-to-beings* in which the traditional metaphysical relationship to the Being of beings would be 'overcome'.

What is the *Kehre*? The *Kehre* manifests a further phase in Heidegger's project of 'overcoming' metaphysics; and *Gelassenheit*, a radically different way of *relating* to the historical presencing of Being, is at the very heart of this further step. The *Kehre* embodies Heidegger's attempt to let the Question of Being itself motivate our thinking. Instead of responding to this question by turning immediately to the presumed ontological structures of Dasein and letting our thinking be further conditioned by the historical determinacy of those presumed structures, what he proposes is that we let our thinking be directed solely by the Question itself. That is to say, we try to let ourselves be questioned in an attitude of *Gelassenheit*: our capacity for openness, receptiveness, and responsiveness to the presencing of Being.

To the extent that we can achieve *Gelassenheit*, we can break out of our metaphysical history; correlatively, to the extent that we have not achieved it, we remain situated within this ancient history. I firmly believe that, both as individuals and as collectivities, we are capable of taking significant steps in this direction. But I am also convinced that *Gelassenheit* is an ideal we cannot completely actualize, and that, for this reason, the history of metaphysics is not something we could ever completely overcome. I would only add my hope, here, that this impossibility will not be taken as cause for despair. On the contrary, I deeply hope that it will motivate us to further thought and exertion.

In *Daybreak*, Nietzsche writes of the 'quiet eye of eternity'.[191] If such a vision be possible in this world, it is surely to be manifested in the gaze of *Gelassenheit*, the gaze with the wisdom of the Ainu Spider Goddess, the gaze for which the presencing of Being, the Being of beings, is not thoroughly determined by the metaphysics of presence in its historical life. Thus we can perhaps understand why, in the fourteenth century, one of Tibet's greatest teachers and scholars, the monk Klongchen rab-'byams-pa, recommended for his students, and for all who strive to embody enlightenment, this ongoing practice:

> The subject dealing with its object is like a dream:
> Although there is no duality of subject and object,
> ingrained tendencies [intentionalities] cause duality
> to appear. What is postulated by the intellect is
> nothing [metaphysically] self-sufficient. – Know
> non-duality straight away.[192]

4 COMPASSION: THE GREAT BEHOLDING

Opening Conversation

(1) I am the victim of a double infirmity: what I see is an affliction to me; and what I do not see, a reproach.
Claude Levi-Strauss, *Tristes Tropiques*[193]

(2) Documentaries don't necessarily betray the fact of the Holocaust, but too often they betray its truth: the grainy footage of Nazi atrocities stimulates only nausea and revulsion – the choice is either to look

away, or to look and not see.
David Kehr, 'Shoah: The End of the Line'[194]

(3) I not only saw but felt in my body all that I saw.
Edward Bellamy, *Looking Backwards*[195]

(4) Becoming able to deal skillfully with schizophrenic communication requires one, more than anything else, to become able to endure seeing, and at least momentarily sharing at a feeling level, the world in which the schizophrenic individual exists.
Harold Searles, *Collected Papers on Schizophrenia and Related Subjects*[196]

(5) . . . the gaze of compassion . . .
Foucault, *The Birth of the Clinic*[197]

(6) . . . seeing – an activity that is ultimately derived from touching.
Freud, *Three Essays on the Theory of Sexuality*[198]

(7) In what the senses of sight, hearing, and touch convey, in the sensations of color, sound, roughness, and hardness, things move us bodily, in the literal meaning of the word. The thing is the *aisthēton*. . . .
Heidegger, 'The Origin of the Work of Art'[199]

(8) In so far as I have sensory functions, a visual, auditory and tactile field, I am already in communication with others. . . .
Merleau-Ponty, *Phenomenology of Perception*[200]

(9) I live in the facial expressions of the other, as I feel him living in mine.
Merleau-Ponty, 'The Child's Relations with Others'[201]

(10) [Human beings] have in common a natural light or opening to being. . . .
Merleau-Ponty, *Signs*[202]

(11) All flesh . . . radiates beyond itself.
Merleau-Ponty, 'Eye and Mind'[203]

> (12) Noble, then, is the bond which links together sight and visibility, and great beyond other bonds by no small difference of nature; for light is their bond. . . .
>
> Plato, *The Republic*[204]

I have argued in this chapter that crying is the root of human vision. This suggests that compassion, the capacity to feel what others are feeling and take their suffering to heart, is an essential factor in the fulfillment of our vision: a fulfillment already manifest in our capacity to be so moved by what we see that we break down and cry. Compassion is our visionary fulfillment because the sociality of feeling-with is always already inscribed into the flesh of our visionary organs.

The medium of crying is water, principle of continuity, cohesion, and change. When the heart opens, the being of vision is effectively rooted: rooted, if we may rely, here, on some words from Emerson's *Threnody*, in 'waters with tears of ancient sorrow'. According to the depth of feeling in which our vision is rooted, the depth of awareness from which our vision is drawn, such is the extent of our capacity to see with eyes of compassion – and make visible, in the world of our living, what those eyes see a need to accomplish.

When we are moved and motivated by compassion, our vision is no longer completely determined by the boundaries of the ego-logical body, the encapsulated body that corresponds to the metaphysics of an egocentric, logocentric consciousness. It issues, rather, from what Arnold Mindell calls the 'dreambody':

> The existence of the dreambody [i.e., the body which bears the 'collective unconscious'] means that when we are with others, they are parts of us. We are individuals, and we can act as if we are alone, but there is no direct and clear division between inner and outer. Thus, regardless of how alone we feel, we are parts of the entire world, and it influences the world. [Recognition that our existence is ultimately drawn from the] existence of the universal dreambody should give us a more human, more intimate relationship to both those who are near us, and to those whom we will never see. . . .

> It's our ability to get beyond looking at ourselves as the center of the universe, and our ability to change viewpoints so we can see ourselves as parts of a larger 'personality' that will determine whether or not we continue world war, destructive ambitiousness and annihilation.[205]

The dreambody is a vision; it is the vision which Merleau-Ponty wants to encourage with his articulation of the flesh, the dimension of our intercorporeality, our intertwining.

(a) Vision in Touch

As the intertwining of radiant energies in the field of sheer lucency begins to take the shape of a pre-personal field of light belonging to the body-subject, it manifests in the interweaving of sensory textures, and as the vectorial crisscrossing of fields of sensibility. Thus, it is true that, as Merleau-Ponty says, 'each organ of sense explores the object in its own way. . . .'[206] But the focal object (*Gegen-stand*), standing out from the field of sensory exploration in the manner of a pattern of interference, always presences as a *polarity* around which, simultaneously, all five of our senses pivot: 'The senses intercommunicate by *opening* onto the structure of the thing'.[207] Because of this inter-communication,

> Not only do I use my fingers and my whole body as a single organ, but also, thanks to this unity of the body, the tactile perceptions gained through an organ are *immediately translated* into the language of the rest. . . .[208]

Now, what makes this of special interest to us is the fact that 'the palpation of the eye is [to be understood as] a remarkable variant [of tactile palpation]'.[209] For we cannot deny, if we have attended closely to our experience of vision, that the look 'envelopes, palpates, espouses the visible things'.[210] As Merleau-Ponty argues in *Phenomenology of Perception*:

> we cannot withhold, from the sense of touch, spatiality in the sense of a grasp of co-existences.

> The very fact that the way is paved to true vision through a phase of transition and through a sort of touch effected by the eyes would be incomprehensible unless there were a quasi-spatial tactile field, into which the first visual perceptions may be inserted. Sight would never communicate directly with touch, as it in fact does in the normal adult, if the sense of touch, even when artificially isolated, were not so organized as to make co-existence possible.[211]

Touch is 'the way' to 'true vision' because the field's primordial *hold* on our gaze yields the lesson of *tact*, and this is an essential development in our capacity for 'true vision'. The response-ability of vision is already made possible through the inherent tactfulness, often seriously disturbed, of our visionary being. In 'The Intertwining – The Chiasm', Merleau-Ponty goes, I think, more deeply into the tactile character of our vision. For it is only a beginning to recognize that 'every vision takes place somewhere in the tactile space'.[212]

Stanley Cavell asks us, in *The Claim of Reason* to consider: 'If the body individuates flesh and spirit, singles me out, what does the soul do? It binds me to others.'[213] But what if the body is, as Wittgenstein says, the 'best picture' of the soul? I would say that the 'soul' of which Cavell speaks is the *depth* of the body. Vision is not essentially ego-logical; nor is it inevitably, for all times, ego-logical. We are always in fact already in touch, by way of our body of feeling, with much *more* than we can explicitly know and much more than we can see by the light of reason alone. We are in touch with others, other beings, by grace of a radiance, an intertwining of visionary presences, which precedes volition, choice, and the influences of socialization. Yet we are also never *fully* in touch. In fact, as socially constituted adults, we at once get *out* of touch in the very ecstasy of sensation, since that moment initiates the arising of the egocentric vision, and we separate ourselves from the rich aesthetic texture of the sensory fields. But if we take the time to get in touch, to renew this rudimentary experience of contact, we may find ourselves enjoying a deeper, more extensive, more sympathetic vision:

> [One] feels that he is the sensible itself coming to

> itself and that, in return, the sensible is . . . as it were his double or an *extension* of his own flesh.[214]

As we get in touch with our felt sense of the visual field, we will begin to discover a treasure of implicit knowledge. For it is this tactile extension of our visual being, this primordial contact with what is given by the presencing of light, which first sets down our destiny as seers and holds out to us the 'promise' of a new historical enlightenment:

> the visible spectacle belongs to the touch neither more nor less than do the 'tactile qualities'. We must habituate ourselves to think that every visible is cut out in the tangible, every tactile being in some manner *promised* to visibility, and that there is encroachment, infringement, not only between the touched and the touching, but also between the tangible and the visible. . . .[215]

The *Phenomenology of Perception* contributes the light of its understanding to the visibility of this tactile vision:

> One sees the hardness and brittleness of glass, and when, with a tinkling sound, it breaks, this sound is conveyed by the visible glass. One sees the springiness of steel, the ductility of red-hot steel, the hardness of a plane blade, the softness of shavings.[216]

But this kind of vision would not be possible unless we were always and already in orientating touch with what we see, and open, in this primordial sense, to being correspondingly touched, visibly touched and visibly moved, moved, at times, to tears, even at an immense physical distance. The tactile vision that Merleau-Ponty describes, therefore, is only the most *rudimentary* level of our experience, foreshadowing the fulfillment of a 'true vision', an enlightened vision, for which our *sense* of being in touch is the way. Thus we ask: Could our vision be even more closely *in touch*, and even more *open* to being visibly touched and oriented? If we read *The Monadology* of Leibniz with care, I think we will find an affirmative answer that anticipates the work of thought we are undertaking here:

> For since the world is a *plenum*, rendering all matter connected, and since in a *plenum* every motion has some effect on distant bodies in proportion to their distance, so that each body is affected not only by those in contact with it, and feels in some way all that happens to them, but also by their means is affected by those which are in contact with the former, with which it itself is in immediate contact, it follows that this intercommunication extends to any distance whatever. And consequently, each body feels all that happens in the universe.[217]

When vision is in touch with itself, it is also in touch with others, because our visionary being, and the ontology which it brings forth, are grounded in an elemental flesh, a universal matrix of intertwining gazes, fields, zones, horizons, clearings. Vision in touch is vision moved by feeling, and not just the philosopher's reason. By feeling, and not emotion or desire: feeling is global, syncretic, synergic; it is a sense of wholeness and continuity which precedes the more differentiated formation of desire, intentionalities with a more egocentric focus. By feeling and not passion; for feeling, here, is a basis for moral agency, and is not a condition of passive suffering.

(b) In Beholding, We Are Beholden

> The eye by which I see God is the same eye by which God sees me.
>
> Meister Eckhart, *Sermons*[218]

> . . . the seer is the seen and the seen is the seer.
>
> Suzuki Daisetz, *Studies in Zen*[219]

In *The Claim of Reason*, Cavell observes that,

> what the philosophers call 'the senses' are themselves conceived in terms of this idea of a geometrically fixed position, disconnected from the fact of their possession and use by a creature who must act.[220]

Our reflections on the sensational emergence, in the

Renaissance, of a new historical relationship to linear perspective – an event heralding the beginning of modernity – should of course be recalled here. Fixed position and total geometric intelligibility are a requirement of the utmost importance for the gaze of modern reason. Eventually, though, and as an irony of history, this same perspectivism dislocated us, decentering the ego-logical and logocentric subject. It is now possible for us to see that the gaze which appears in the texts of our metaphysics is a gaze which does not move: a gaze, therefore, which is never moved to act, never motivated by the calling, the appeal, the claim, of what is seen and beheld. It is a gaze whose beholding is without any sense of its corresponding beholdenness: without any sense of its response-ability.

We, of the modern epoch, experience our world as a picture (*Bild*). Now, a picture is something we 'view'. The viewpoint of a view, however, is a distinctive structure of vision. Whatever the tragedy, however intense the pain, we can turn it into a 'picture' – something we can 'show' and 'watch', *without being touched*, in that paradoxical medium of remoteness we call 'tele-vision'. It was inevitable that the communication, on television, of matters of common concern should be reduced, first, to the reporting of 'news', and then, finally, to a kind of 'show' that must compete for attention with our entertainments.

By contrast, 'in the age of the Greeks, the world *cannot* become picture',[221] since it was understood, even if only implicitly and, in a sense, obscurely, that, 'That which is *does not come into being at all* through the fact that man first *looks* upon it. . . .[222]

In earlier epochs – for example, among the Greeks, –

> *man is the one who is looked upon* by that which is; he is the one who is . . . gathered toward presencing, by that which opens itself. *To be beheld by what is,* is to be included and maintained with its openness.[223]

To behold is to be *held* by what one sees. To behold is, in this sense, to be also *beheld*. Conversely, since the beheld is that which *holds* our gaze – holds it, sometimes, and binds it under a spell, it is also true to say that the beheld is also the one beholding. In beholding, though, we are held not only by what we have beheld; we are held at the same time by the entire world of visibility; and ultimately, by the field of its lighting. It is only *by grace of* this world-clearing illumination that the beings which we behold shine forth. Sustained by this binding field of illumination, our vision, *zugehörig*, is beholden: beholden to the lighting and clearing of Being. Being held well, we are kept well, and are always already beholden, long before we are conscious of our situation, our *Befindlichkeit*.

But in what sense are we beholden? It would seem that we have 'forgotten' this primordial beholdenness, 'forgotten' that, in this grounding of our gaze, we are unavoidably beholden. (If we were to think, here, in the vocabulary of *Being and Time*, we should speak, I believe, of the *Schuldigkeit* of vision. But of course I want to purge our words of their onto-theological interpretation.) What would a vision be like, if it 'remembered' its ontological beholdenness? How could our vision make its beholdenness visible? What would be the *character* of a gaze which is open to being touched and moved by its 'recollection' of this beholdenness? Since Merleau-Ponty has written of the 'consecratory gesture', we might find it illuminating to envision a gaze visibly *moved to act* by this process of recollection.[224] Since our visionary being is held open by the field into whose intertwining we are bound, we cannot remain indifferent to our beholdenness without being adversely affected. In this sense, then, we are responsible for the development of our response-ability.

In his essay on 'Education', Emerson writes:

> Consent yourself to be an organ of your highest thought, and lo! suddenly you . . . are the fountain of an energy that goes pulsing on with waves of benefit to the borders of Society, and to the circumference of things.[225]

Emerson's words should remind us that we, as visionary beings, are not beholden *only* to the lighting of Being as such. Living in a world with others, we are by necessity inscribed into the social constitution of vision. Even our individual relationship to the lighting which surrounds us, our relationship to the presencing of Being, is mediated by social, cultural, and historical conditions. Thus, for example, intersubjectivity is constitutive of our subjective identity; it is confirming; it establishes our basic sense of reality, our basic trust in this sense. When we reflect on our ontological beholdenness, we must not forget that we are also beholden to others; that, without them, we should not, in an important sense, be 'seeing'. It is consequently in the gaze of compassion, the gaze touched and moved by its sense of being with others, that vision shows and enacts its beholdenness. *A fortiori,* it is in the caring gaze of compassion, lighting the way for actions and activities which embrace the being of others, that the tactile nature of our vision – its 'primordial contact' with the visible – actualizes the tact which is already laying claim to our responsiveness and making us responsible for its deepest potential.

I said, just before quoting from the text by Emerson, that we are responsible for the development of our response-ability. And what I mean by this is that, because we are sentient and responsive beings, beings whose visionary existence is always already inscribed into the intertwinings of our being-with-others, a sense of response-ability is inherently conceded by our vision from the very first moment on. For vision to take place at all, our eyes must be open and receptive, responsive to the solicitations of the visible. This immediate responsiveness enacts, and accordingly attests, our primordial sense of beholdenness. From its very inception, as moved to respond, our vision concedes its response-ability and acknowledges a task.

The social beholdenness of vision, its motivational character, and its responsibility for tact and respect, are already set down for us in the very nature of its formation. Self-esteem, or self-respect, is a necessary condition for the individual's development of a sense of self. People without respect for themselves feel empty and do not enjoy a sense of who they are: lacking self-esteem, their identity is alienated, entirely dependent on others. Self-respect is also necessary as a condition for the respect due to others: if we do not respect ourselves, we shall not respect others.

But how do we develop self-respect and learn to show respect for others? The social experiences of early childhood are basically decisive. Of great importance, here, are the child's early experiences with vision. Our words themselves – 'respect', 'self-respect', 'regard' and 'self-regard' – should remind us of the fact that visionary experience is normally constitutive of moral character, and even our primal sense of self-identity, since self-respect and respect for others are essential to the formation of a strong sense of individual identity. It is in the intertwining of gazes and in the character of their mirroring that the child gifted with vision – a visionary response-ability – must learn the meaning of respect. Seeing and feeling a gaze of respect tactfully directed her way, the child simultaneously learns the meaning of respect for others and develops, within herself, a sense of identity: an identity centred in self-respect and grounded in a sense of being a being with unconditional value.

As Carl Rogers has argued, with independent support from Melanie Klein, Harry Stack Sullivan, Heinz Kohut, Jacques Lacan, and Merleau-Ponty, the child needs to see, mirrored and made visible in the gazes of those who surround him, an 'unconditional positive regard.'[226] Thus, it can be seen that we are beholden to others, beholden to the *being* of others, for a vision of our own identity. But it may also be seen that the gaze learning respect is first of all a gaze moved by an experience of compassion: a gaze which has been granted some experience of mirrored and reciprocated responsiveness. (Sullivan points out that the *face* is shaped and moulded by social, cultural and historical processes, and that the child's earliest visual experiences with the gazes of others are extremely important for the composure of the face. But, since the face is the *persona*, the most expressive manifestation of the *person*, we return once again to the constitutive role of the gazes of others in the formation of personal identity.)

In a study on the psychological dimension of visual problems, Marilyn Rosanes-Berret notes that,

> Most studies involved with visual afflictions stress anatomical and structural deviations without regard to psychological factors. However, psychosomatic medicine has shown how artificial is the dividing line between psyche and soma, and the eyes are part of the soma. The content of speech, motoric behavior, breathing processes, voice and verbal behavior, sensory awareness, are all affected by feeling states such as embarrassment, expectancy, dread, and excitement. Visual processes are no exception to the rule.[227]

Thus, for example, she points out that, in his early work with 'conversion hysteria', as well as in his subsequent work on anxiety and the formation of symptoms, 'Freud recognized that some forms of blindness might be caused by emotional states.'[228] She accordingly challenges thinking which continues to ignore, neglect, or deny the possibility of significant existential relationships between personality difficulties and disturbances in visual functioning:

> Deviant visual functioning and the structural malformation which develops from it are not considered in the light of the possibility that the individual is expressing a psychological disturbance through the visual system.[229]

The principal point she then attempts to argue is that the healthy life of our visionary being requires integrity of character and the integration of personality, i.e., the integration of split-off parts of the psyche. If she is right, however, then healthy vision also depends on self-esteem – and therefore, on the existence of social conditions which encourage visual interactions that develop it. Responsibility is rooted in responsiveness. Therefore, taking responsibility for our responsiveness – for our response-ability – is always essential for the adult:

> When the patient can take responsibility for his experiences, needs, desires, and wishes, he can perceive how he uses his various organ systems, including the visual apparatus.[230]

According to Rosanes-Berret,

> To a great extent, those with the commonest eye difficulties are unaware of how they use vision. When the patient recognizes how and why he stares, pushes, and distorts his eyes in the process of seeing, he may allow himself to experience looking differently; that is, he may look casually, effortlessly, and yet be alert and see. In myopes, [for example,] the individual is prone to limit motoric expression. . . . It might be said that he is afraid movements might be seen and be dangerous, and so he tends to hold back activity.[231]

To what degree is the vision we consider to be 'normal'

motivated and structured by socially produced – and by no means inevitable – anxiety? To what degree could this conditioning be modified by changes in the character of social interactions?

> In psychotic states, anxiety is blown up to panic proportions, and this too can have an effect on vision. Some psychotic patients, particularly those with paranoid delusions, frequently mention that their distance vision blurs more markedly as their fright increases, and clears again as they grow calmer.[232]

Does this not suggest that, in a society which encouraged visionary practices, practices to develop relaxation, calmness, and the neutralization of desire, even our so-called 'normal' vision might be given an opportunity to achieve a new degree of responsiveness, a new kind of clarity?

In other cases she reports, patients suffering blurred vision, blind spots, and fluctuating myopia found their vision restored to clarity once they could accept their erotic feelings and take responsibility for them. She also reports the story of a nine-year-old child, diagnosed with myopia, whose visionary problems in school could be traced back to feelings of inadequacy and insecurity he developed in his family. These feelings made it impossible for him to look with normal skill at the lessons written on the blackboard. 'The more he stared', she writes, 'the greater the blur. When asked to allow himself to look casually, he became frightened and said he could not learn; he could never please his parents; he was [he insisted] a really wicked child and did not want to see himself or let himself be seen.'[233] His vision did not improve until he gained some self-esteem – first in the eyes of significant others, then, of course, in his own eyes.

Other cases are no less thought-provoking. 'Frequently, when a patient who is not seeing well is asked to experience himself, he relates that he feels empty.'[234] Now we know, from ample clinical documentation, that the experience of emptiness, a typical aspect of depression and a symptom, as I have argued elsewhere, of the nihilism raging in our epoch, is related to the felt absence of self-esteem, a strong sense of self-identity. Is it possible that the pathologies of vision which we have been explicating in this book – pathologies of a vision which is commonly considered to be 'normal', and which only a critical hermeneutics, a hermeneutics of 'suspicion', can call into question – are epidemiologically related to the conditions of our historical situation?

It is of course widely acknowledged that ours is an alienated, alienating society. And it is no less widely acknowledged that this alienation makes an impact on the individual. If individual psychology can affect visual perception, is it therefore unreasonable to speculate about how social alienation might circulate throughout the system of intertwining gazes and affect the character of our vision?

> When a patient speaks for the organ in distress – in this case the eye – he becomes aware of what the eye is expressing through distorted functioning and/or structure. The alienated affective expression can then be experienced. 'I (the eye) dare not see me or know how I feel. When I limit my sight, I deny that which is threatening and so avoid uncomfortable experiences or recognition of that which should not exist for me'. To identify with the alienated feelings is to see again. This is necessary before lasting improvement of vision is possible.[235]

I submit that alienation is always a socially constituted phenomenon; that it is never only an internal psychological experience, something which happens apart from processes of social interaction; and that, when it occurs, it must therefore be true that a visionary network of other gazes – perhaps the family – is partially responsible for the alienating character of its responsiveness. The infant's sense of self, of being-a-body, of being a being of value, is constituted in the intertwining of gazes: living within their intersection, she acquires a firm sense of who she is. Since the very act of beholding other people is always already a mode of responsiveness which acknowledges their claim to a gaze of respect, it needs to be observed that, when we behold other people, we necessarily 'consent' to being held by them in a beholdenness for which we are partly responsible. In other words, we have consented to being held responsible for their response-ability – if only for the time our gazes are intertwined. When those with whom we are in visual contact suffer alienation, our beholding makes us to some degree responsible for the course of their suffering.

Rosanes-Berrett concludes her study by declaring that,

> Genuine change occurs only when one is in touch with self. Then ears that dared not hear, eyes that dared not see, and a heart that dared not feel can come to life again and function normally.[236]

Getting in touch with our visionary self is not a process, however, which can happen in social isolation. Since it involves self-respect, it must at some stage be encouraged by experiences with others which communicate a vision of respect, a respectful gaze. There are always enabling conditions of a social nature which surround and continue such a process. Vision in touch, in touch with itself and in touch with others, is therefore always a vision moved by what it has seen, a vision moved by its sense of being beholden. Because it decentres the visionary subject, this sense of beholdenness is both morally and ontologically important. It constitutes an affective basis for the development of *Gelassenheit*, vision attuned by its guardian awareness of Being.

(c) Being of their Family

Opening Conversation

> (1) Local disease . . . occurs only because some entity whose freedom or life exists only in so far as it may remain in the whole, strives to exist for itself.
>
> F. W. J. Schelling, *Of Human Freedom*[237]

> (2) I have been struck . . . with the highly significant circumstances in which fatal or near-fatal cardiovascular accidents, quite beyond the much more frequent upsurges of anxiety in sometimes psychotic proportions, have occurred among parents of recovering schizophrenic patients [who are 'given to feel that the resolution of the symbiotic mode of relatedness would mean the murder of the parent', and that their 'own potentially individual self is . . . an inherently murderous self.'] Several such incidents have occurred in the families of patients with whom I have been working; but this

> statistically insignificant sampling grows to sobering proportions when one sees it supported by similar incidents which have occurred in the course of my 14 years of work at Chestnut Lodge, in the collective experiences of, in cumulative total, perhaps 50 therapists who have worked with hundreds of patients here over the years.
>
> Harold Searles, 'Transference Psychosis in the Psychotherapy of Chronic Schizophrenia'[238]

(3) The mechanism of 'pathic projection' determines that those in power perceive as human only their own reflected image, instead of reflecting back the human as precisely what is different. Murder is thus the repeated attempt, by yet greater madness, to distort the madness of such false perception into reason: what was not seen as human and yet is human, is made a thing, so that its stirrings can no longer refute the manic gaze.

Theodor Adorno, *Minima Moralia: Reflections from Damaged Life*[239]

(4) . . . I shut my eyes to others. . . . [In consequence] the other remains as unacknowledged, that is, as denied. I have shut my eyes to *this* other. And this is now part of the other's knowledge. To acknowledge him now would be to know this. To deny him now would be to deny this, deny this denial of him: to shut his eyes to me. Either way I implicate myself in his existence. There is the problem of the other. – The crucified human body is our best picture of the unacknowledged human soul.

Stanley Cavell, *The Claim of Reason*[240]

Our title returns us to Merleau-Ponty, who writes:

> being of their family, itself visible and tangible, it [our vision] uses its own being as a means to participate in theirs.[241]

The gaze which responds with care, with the response-ability of compassion, is vision's most visible avowal of the truth that it belongs, inextricably, to the family of beings by which it is touched and moved. And the acknowledgement of this kinship of flesh, spanning the field of the visible, demonstrates as no proofs of reason possibly could that the character of the gaze, its moral or nomological sense of identity, is deeply beholden to others, and that this character is correspondingly fulfilled, phenomenologically speaking, by virtue of our enactment of the beholdenness which comes from such awareness and acknowledgement. It is not only that the co-existence of others is a knowledge beyond every shadow of doubt, but that their co-existence is necessary for our own.

Sociality, i.e., the network of interactions philosophers call 'intersubjectivity', is *prior* to individuation and monadological consciousness. We emerge as egocentric subjects, as ego-logically shaped and informed bodies, in interdependent co-emergence with others. I am, first of all, intercorporeality. In more mythopoetic words, we might say, using Lévy-Bruhl, that *Dasein* begins, as an infant, in a condition of 'participation mystique', taking part in a symbiotic interaction and interpenetration of lives. Thus, when Yeshé Tsogyal, a Tibetan woman of wisdom living in the ninth century, wrote that 'the feeling of love perfects the glance', we need to understand that the reciprocity of love means the perfection of *two* glances: that love perfects the glance of the one who gives it; and that it is also the glance of love which perfects the glance of the one who receives it.[242] Love puts us in touch with a light 'which first brings to its radiance what is present'.[243] In the light of this feeling, perhaps we can see why the metaphysical eye of reason is blind and how this blindness governs the human world. In the unconscious responsiveness of intercorporeality, the response-ability of our gaze, and the ethic of responsibility which fulfills it, is always already bodily schematized, a rudimentary acknowledgement of others visibly like ourselves, and of our being of their family.

We should perhaps recall, here, what Descartes has to say, in his second *Meditation*, about the eye's participation in a social world of visible kinship. Descartes needs to know beyond all possible doubt that other human beings really exist. But how could he possibly 'know' this when he has deliberately suspended all *evidence* of co-existence – that spontaneous gift of kinship, of which, thanks to the deep tactility of vision, he is always and already sensible?[244] Before co-existence has been cognized, before it can be 'known', it has already been acknowledged. Cartesian reason must begin to recognize the body's gift of a primordial wisdom. In skepticism, the metaphysical eye of reason suffers in a self-imposed solitude; but it can be argued that it accurately reflects historical reality: a world of social conflict, a world in which our modern interpretation of individualism carries us closer and closer to historical tragedy.

In the plan of Bentham's Panopticon, the guards can look without being seen. When looking at others in this way, though, we do not see the being of the other. To see the being of the other, we must be open, vulnerable, touched and moved by the living presence of the other: we must consent to being seen.[245]

In 'The Child's Relations with Others', Merleau-Ponty reviews various philosophical attempts to understand our perception of other human beings. He shows why Husserl, for example, unwilling to entrust his thinking to the wisdom of the body of feeling, utterly failed to resolve the 'problem of intersubjectivity'.[246] Eventually, as we know, Husserl began to recognize and approach the intercorporeality of an elemental flesh, introducing into his transcendental analysis an account of 'intentional transgression'. But even this, together with an account of 'coupling' (*Paarung*), could not overcome the solipsism in his ego-centred transcendentalism. The problem with Husserl's phenomenology of the intersubjective constitution of our social world is that it overlooks the dynamics of social interaction. It does not see intersubjectivity in the making – through social interactions. And it persists in seeing intersubjectivity from the standpoint of the ego-logical monad, whose transcendental 'life' is essentially prior to, and independent of, all forms of participation in social dynamics.

Henri Wallon therefore took an important step beyond Husserl when he embodied the 'intentional transgression' towards which Husserl was pointing: basically he reformulated it in terms of a process he called 'postural impregnation'. Merleau-Ponty does not disagree. However, he argues that this phenomenon needs to be seen less 'passively', as a mode of 'pre-communication': a preverbally reciprocated acknowledgement of some primordial kinship. In the 'contagion of smiles' which gathers the child and its mothering one, there is a field of gazes, mirroring, touching, moving, and interpenetrating one another; and when we see their faces lit up with joy, we cannot easily deny that their symbiotic relationship expresses the truth of a shared flesh, a dynamic of co-constitution. Thus we see that, even before we are named, or know others by name, we belong together as interacting subjects:

> There is . . . no problem of the *alter ego* because it is not I who sees, not he who sees; because an anonymous visibility inhabits both of us, a vision in general, in virtue of that primordial property that belongs to flesh, of being here and now and of radiating everywhere and forever.[247]

Others are, then, 'of my own flesh', and we are deeply gathered, all of us, in the *Legein* of a lighting which is the elemental flesh of the visual matrix. Belonging, as we do, to this luminous flesh, we are, in the words of Merleau-Ponty, 'together gathered into a single world, in which we all participate as anonymous subjects of perception.'[248] Consequently, 'if I wanted to render precisely the perceptual experience, I ought to say that *one* perceives in me, not that I perceive'.[249] And he adds, disconcertingly perhaps, that 'Every perception contains within it the germ of a dream of depersonalization'.[250] This allusion to a dream of depersonalization is of course disconcerting, because we live in a culture which values personhood, the individuation of Dasein. But individuation of personality is not in question here. What Merleau-Ponty means is that, inherent in the prepersonal dimension of our visionary life, there is a tremendous capacity for responsiveness which normally remains incompletely developed and essentially unfulfilled, even with, and in a sense precisely because of, the rise to power of an egocentric, ego-logical personality. The 'dream' which inhabits our prepersonal existence is a dream, a vision of being, which *transcends* the boundaries of responsiveness set down by the socialized ego. The dream, however, is not just infantile; it is archaic, archetypal; and it suggests the possibility that the prepersonal anonymity of the infantile gaze, a 'selflessness' which is only *aufgehoben*, and not entirely destroyed in our subsequent growth, could be the tangible, felt basis for the achievement of a 'higher' anonymity, a 'depersonalization' which would, in fact, realize the deeply incarnate dream of a *transpersonal extension* of our visionary capacity: a truly *selfless* vision, selfless not in the sense that it is not yet individuated, like the vision of an infant, but selfless, rather, by virtue of an authentically individuating commitment to an ethics of responsibility and caring. The 'dream of depersonalization' is not a dream of regression, but a vision, rather, of community, of inter-subjective co-existence, grounded in the spontaneous responsiveness of our prepersonal life. It is a dream which summons us to action; it is a dream which challenges us to see *beyond* the selfishness of merely personal existence.

In his 'Nature and Logos' lecture at the Collège de France, Merleau-Ponty speaks of 'an ideal community of embodied subjects, of an intercorporeality. . . .'[251] The 'original ecstasy' of the flesh, visibly manifesting as the radiance of the gaze of compassion, is the basis for this dream of a new historical community.

Recalling Merleau-Ponty's observation that the flesh is radiant, that it radiates beyond itself, we are adding, here, that compassion is a visible extension of this radiance, an ecstasy of the visionary flesh, gathering all sentient beings into a planetary 'family'. As thoughtful organs of Being, as organs of a flesh through which the lighting of Being is taken up, treasured, protected, and shared, eyes of compassion extend the original gift of radiance, letting it be tangible and visible in the world of our living.

The gaze of compassion is moved by a vision of the Being of beings which *brings out* the radiance, the 'natural light', inherently present in the other; and in the community of beings it gathers into its embrace, this, then, is true: that, as Yeshé Tsogyal says, 'the feeling of love perfects the glance'.[252]

5 THE CHILD OF JOY

Opening Conversation

(1) . . . and a little child shall lead them.

Isaiah, xi, 6

(2) . . . Child of the pure unclouded brow/And dreaming eyes of wonder. . . .

Lewis Carroll, *Through the Looking Glass*[253]

(3) . . . the reasonable man: there are things he does not see which even a child sees; there are things he does

not hear which even a child hears, and these things are precisely the most important things. . . .

Nietzsche, 'On the Uses and Disadvantages of History for Life', *Untimely Meditations*[254]

(4) L'enfant abdique son extase. . . .

Stéphane Mallarmé, *Prose*[255]

(5) It is only through such truthfulness [about oneself and about others] that the distress, the inner misery, of modern man will come to light, and that, in place of that anxious concealment through convention and masquerade, art and religion, . . . a culture [will emerge] which corresponds to real needs and does not . . . deceive itself as to these needs and thereby become a walking lie.

Nietzsche, 'On the Uses and Disadvantages of History for Life'[256]

(6) . . . in a system of discipline, the child is more individualized than the adult, the patient more than the healthy man, the madman and the delinquent more than the normal and the non-delinquent. In each case, it is toward the first of these pairs that all the individualizing mechanisms are turned in our civilization; and when one wishes to individualize the healthy, normal, and law-abiding adult, it is always by asking him how much of the child he has in him, what secret madness lies within him, what fundamental crime he has dreamt of committing.

Foucault, *Discipline and Punish: The Birth of the Prison*[257]

Mallarmé's words are difficult to translate. My translation is already, of course, an interpretation: The child of joy renounces – leaves behind – its experience of ecstasy. In the ecstatic moment of vision, the gaze is open to enchantment. But socialization begins at once, and this moment is abandoned – forever. It will never be retrieved exactly as it was. The child grows up and renounces, or is expected to renounce, all 'childish things'. But, in a curious way, the philosopher is one who remains, or dreams of remaining, a child of vision. I am thinking, here, of Herakleitos, Sokrates, Nietzsche, Wittgenstein: of their lives and their words.

Dasein, for Heidegger, is an 'ekstatic openness-to-Being'. Is it? Heidegger's 'is', here, is tenseless: it repeats the timelessness of the metaphysical tradition. To be sure, Heidegger's *Dasein* is historical; but it is, curiously, a being without biography, without any narrative of personal history. Reading between the lines, we must presume from his account that the *Dasein* he describes is an adult. But, unlike Athena, we human beings do not enter the world as adults. Heidegger's *Dasein* is, as Derrida has noted, without gender; but it is also without personal development: without infancy, childhood, adolescence; without any lifetime process of individuation.

I want to call attention to this omission and correct his inaccurate representation. For, Heidegger's silence in regard to personal development is not without philosophical implications. I do not want to repeat what I have said earlier, beginning with the Introduction, concerning the significance of the child's experience with vision. Here, then, I would like only to emphasize that I am not arguing for infantile regression. Rather, what I am proposing is a hermeneutical movement in response to our present needs, both individual, or biographical, and collective, or historical: a movement which 'goes back' in order to 'go forward'.

There is an ecstatic experience with Being – a joyful experience with Being – of which *Dasein* is capable: phenomenologically considered, it would be an experience of individual actualization and fulfillment in relation to that primordial existential structure of human being which constitutes what Heidgger calls 'ekstatic temporality'. Heidegger's ontological concept *prefigures* a possible direction for *Dasein* to take up in everyday life as a process of authentic individuation. The ekstatic structure is a potential which can be, and needs to be, developed; it is a structural potential through which we live in, and experience, the world: first as a child, later as an adult. The child's way of living and experiencing this ekstatic structuring of his being-in-the-world is not the same as the way characteristic of the adult: it is a different aesthetic, a different sensibility, a different moodedness; it involves a different epistemology and a different ontology. The child inhabits a *different* pre-ontological understanding of Being.

The project I want to propose, here, namely, the 'retrieval' of the child's experience with the ekstatic structuring, is not an infantile regression; nor is it a continuation of the metaphysical obsession with an absolute origin.

As adults, we carry within us, each one, a dream, a vision, of childhood. (In 'Anima and Animus', Jung speaks of innately embodied 'virtual images'.)[258] Not only memories of our actual childhood, but also memories of childhood dreams and phantasies, still evocative representations of the childhood we always wanted. We carry within us, by grace of our embodiment, a retrievable sense of the 'emotional essence' of childhood: an 'emotional essence' derived as much from our childhood dreams as from our actual experiences at that age. We can make our present *sense* of what 'childhood' means to us, in our lives today, into an 'intelligible essence'.

If we reflect together on this 'emotional essence', this bodily carried, bodily felt *sense* of what 'childhood' is and means, I think that we will find considerable agreement. According, then, to this shared sense of the meaning of the child's early life-experience, there is an initial openness, an open attunement by and to Being. Unless, or until, the circumstances of the child's socialization compel the child to close. Obviously, life-threatening situations, causing intense fear, anxiety, and mistrust, will compel such closure. But if the child is genuinely cared for, cared for well, this closure will never be extreme, autistic.

The early experience of the child well cared-for is very precious, for it gives birth to a sense of the 'essential meaning' of childhood that stays with us into our adulthood, sheltering within it a vision of the presencing of Being, and therefore a pre-ontological understanding of Being, which we of the modern world very much need to contact, recollect, and integrate into our living.

It is not a question of infantile regression. Nor is it a question of restoring our childhood as it was in the past. The process of thought I am suggesting is, rather, a way of moving forward: integrating into *present* living a sense of being, and of life's possibilities, from which we have become detached. What we 'retrieve' is not so much our *actual* childhood past, but rather more our *present* sense, our presently embodied, living *sense*, of what it signifies in the depth of our dreams, our reveries, our phantasies: the childhood we would have liked, if only. . . . As our socialization proceeds, children forget this sense; as adults, we lose touch with it; we let it sediment in our muscles; we bury it our flesh. If now we should begin to retrieve it, we might bring something of its distinctive ontology into critical and creative opposition to our prevailing metaphysics. Our presently retrieved sense of the child's enchantment – a sense both individual and collective, and which we make present through the work of retrieval – could perhaps open us to a different visionary attitude – and to a correspondingly different ontology of presencing.

There is something universal in the ontology that corresponds to the development, in the child, of its visionary capacity. All adult human beings have passed through childhood. This much, at least, is biologically inwrought, preprogrammed. But it is not only our biology which is to some degree universal; because of this biological condition, our *experience* of life is also to some degree prestructured, universal, archetypal, transcending cultural, historical, and even socio-economic kinds of difference.

I want to characterize the 'universal childhood' whose sense I am speaking of retrieving in terms of the 'child of joy'. Thus I want to say that every adult human being carries within, sometimes deeply buried in a flesh of pain and torture, a child of joy: a bodily carried sense, determinately indeterminate, but always further determinable, of what that means, what it would be like; an implicit or latent sense of childhood that somehow survives, though perhaps only in fragments: fragments of dreams, phantasies, visions. And I want to emphasize the importance of retrieving this sense, this child of joy, not only for our own individuation, but also, since we are intercorporeally socialized beings from the very beginning, for our collective vision, our collective historical life. When this vision of childhood passes before the eyes of an adult, there is a creative conflict between two visions, and between their corresponding ontologies. At that moment, our prevailing metaphysics is challenged by a different pre-ontological understanding (ontological pre-understanding) of Being. And this challenge, which has yet to happen as a collective experience, could have historical effects.

I am suggesting that this retrieval of the child of joy constitutes a practice of the self, a practice of caring for the self. But, since the self is not a metaphysically isolated subject, such a practice is also, necessarily, a practice of caring for all that the self beholds and encounters through vision. Consequently, we are concerned, here, with a practice of historical relevance.

How can I speak, however, of retrieving our *universal beginning* as children of joy? After all, we cannot ever *return* to this 'origin', this *archē* and archetype. Nor could we ever be absolutely certain about the essential nature of this 'original' child, certain with transcendental knowledge. Isn't the 'original child of joy' just a myth? A conceit of nostalgia? But I am not proposing a process of retrieval about which it would be appropriate to say that it is tied to a metaphysics of origins. My proposal is less grandiose. We simply make contact with whatever sense of this child of joy we happen to carry with us by grace of our sedimentary embodiment. Whatever we contact, when we 'look into ourselves' to find a child of joy; whatever we feel and sense and see and learn about ourselves when we attempt to retrieve our embodied, sedimented *sense* of this child; whatever is evoked, whatever we discover, whatever we see, as we search within ourselves for hints and traces of a meaning carried by the body and resonating in a powerful way in response to the words of this concept: whatever emerges in that way, *that* is 'the child of joy'.

The concept of the 'child of joy' is therefore a phenomenological concept, and it functions hermeneutically, like a metaphor, referring us to our body of experience and directing our attention to it in a way that enables us to make contact with a sense of our past, which, however vague and indeterminate, is nevertheless a *meaningful* feeling; and it can accordingly be raised into consciousness in a process of ongoing self-determination. The concept does not refer to an original, universally fixed, completely predetermined content of experience, but to a vital *sense* of our past which can be brought into presence in a process that has to achieve it, that has to make it happen. The retrieval uses the concept as a 'handle' for making something happen in our visionary life: a heuristic metaphor for opening us to a vision whose future lies before us. The concept gathers us into an ontological project.

In *Being and Time*, Heidegger speaks of this project in terms which, in my view, would seem to admit of an interpretation that points to a retrieval of the 'child of joy' as a part of the task he later called the 'recollection of Being':

> We understand this task as one in which, by taking the question of Being as our clew, we are to destroy the traditional content of ancient ontology until we arrive at those primordial experiences in which we

> achieved our first ways of determining the nature of Being. . . .[259]

I am suggesting that the 'child of joy' is an ancient cultural symbol for 'those primordial experiences in which we achieved our first ways of determining the nature of Being'. I am also saying that we can see the task in *reverse* order: it is by retrieving our 'primordial' experiences of Being that we will see a way to deconstruct the ancient ontology. What Heidegger's ordering emphasizes is, rather, the fact that the traditional ontology is an ideological system which blocks off our access to such experiences, and that we can only gain access to the primordial insofar as we are no longer entangled in the prevailing ontology. We need, I think, to put both strategies in motion.

Let us now, in conclusion, review the steps of the interpretation I am proposing for the retrieval of our inner child: the child, that is, of joy. The interpretation depends, I believe, on our seeing in this joy the ontological significance implicitly adumbrated in the 'analytic' of Heidegger's *Being and Time*. Here are the most crucial hermeneutical steps. (i) According to this analytic, *Dasein* is an *ekstasis*. (ii) This *ekstasis* is an *ontological* determination of human existence. (iii) But this determination is not a totally pre-determined fate; the ekstatic structure only determines a field of existential possibilities; and it accordingly constitutes for us an existential task: What shall we make of our *Befindlichkeit*, this *ekstasis* into which we have been thrown? (iv) This question is a questioning of our being; it points toward our potential, and lets us sense its call, its need, for fulfillment. (v) The *ekstasis* is the ontological gift of our potential; and in this sense, it is not fulfilled: it is really, i.e., more accurately described as, our *pre-ontological* understanding of Being, awaiting fulfillment through authentic existence. (vi) What constitutes the realization and fulfillment of this potential, however, is a *retrieval* of the joy of the child, the child's vision of enchantment, the child's pre-ontological understanding of Being. (vii) The child of joy is a possibility we can reflectively *sense* through our embodiment. (viii) The recognition of this possibility brings us closer to retrieving it. (ix) The retrieval of (our sense of) the child of joy is a retrieval of the *ekstatic* structuring of human existence which further fulfills the needfulness of this ontological condition. (x) What most deeply fulfills the *ekstatic* possibility of *Dasein*'s existence is an experience of joy, because it is the fulfillment of *this* possibility which most deeply fulfills the human being as a whole. Many authorities have argued this well: Dewey, Erikson, Piaget, Harold Searles, Harry Stack Sullivan; and perhaps Aristotle. (xi) Since the child of joy is the child who enjoys a pre-ontological understanding of Being which we need to integrate into our adult world, in retrieving the child of joy, we take a further step in our historical self-determination – in the realization of our ontological potential and in the fulfillment of its predisposition.

'L'enfant abdique son exstase'. Mallarmé is right. But Nietzsche, too. We, as adults, as heirs to the child's archetypal joy, can turn that experience of Being, that vision of its enchantment, into a more mature, more civilized and civilizing vision, motivated to worldly action by its being in touch with a body of joyful wisdom. In this sense, too, we are indeed renewed, historically, through the 'immortal' child.

Gold ring of Isopata, ca 1500 B.C., recovered near Knossos, showing four whirling females, perhaps Maenads, with insect heads and hands. Goddess on right. Note disembodied eye. Ring depicts a ceremony related to the visionary ecstasies of the Eleusinian Mysteries. From Marija Gimbutas, *Gods and Goddesses of Old Europe* (University of California Press, 1974).

6 THE FEMININE ARCHETYPE: A VISION OF CULTURAL CHANGE*

(1) It is to be suspected that the essential repressed element is always femininity.

Freud, Letter to Wilhelm Fliess[260]

(2) Woman always stands just where man's shadow falls. . . .

Jung, 'Woman in Europe'[261]

(3) Women . . . limping on the edges of the History of Man. . . .

Rita Mae Brown, 'The New Lost Feminist'[262]

(4) Women's moral judgement is more contextual, more immersed in the details of relationships and narratives. It shows a greater propensity to take the [concrete] standpoint of the 'particular other,' and women appear more adept at revealing the feelings of empathy and sympathy required by this. Once these cognitive characteristics are seen not as deficiencies, but as essential components of adult moral reasoning . . . , then women's apparent moral confusion of judgement becomes a sign of their strength.

Seyla Benhabib, 'The Generalized and the Concrete Other: Toward A Feminist Critique of Substitutionalist Universalism'[263]

(5) . . . the [contextual] Background is the realm of the wild reality of women's Selves. Objectification and alienation take place when we are locked into the [universalism of an abstracted,] male-centered, monodimensional foreground.

Mary Daly, *Gyn/Ecology*[264]

(6) . . . this [masculine] ego: the more it learns, the more words and honours it finds for body and earth.

Nietzsche, *Thus Spake Zarathustra*[265]

(7) . . . to understand and to combat woman's oppression, it is no longer sufficient to demand woman's political and economic emancipation alone; it is also necessary to question those psychosexual relations in the domestic and private spheres within which women's lives unfold, and through which gender identity is reproduced. To explicate woman's oppression it is necessary to uncover the power of those symbols, myths, and fantasies that entrap both sexes in the unquestioned world of gender roles. Perhaps one of the most fundamental of these myths and symbols has been the ideal of autonomy conceived in the image of *a disembedded and disembodied male ego*. This vision of autonomy was and continues to be based upon an implicit politics which defines the domestic, intimate sphere as ahistorical, unchanging, and immutable, thereby removing it from reflection and discussion.

Seyla Benhabib, 'The Generalized and the Concrete Other'[266]

(8) As a woman thinking, I experience no . . . division in my own being between nature and culture, between my female body and my conscious thought. In bringing the light of critical thinking to bear on her subject, in the very act of becoming more conscious of her situation in the world, a woman may feel herself coming deeper than ever into touch with her unconscious and with her body [i.e., with all that 'the History of Man' has marginalized, pushed to the edges, pushed into the forgotten background]. Woman-reading-Neumann, woman-reading-Freud, woman-reading-Engels or Levi-Strauss, has to draw on her own deep experience for strength and clarity in discrimination, analysis, criticism. She has to ask herself, not merely, 'What does my own prior intellectual training tell me?' but 'What do my own brain, my own body, tell me – my memories, my sexuality, my dreams, my powers and energies?'

Adrienne Rich, *Of Woman Born*[267]

(9) Woman's psychology is founded on the principle of Eros, the great binder and loosener, whereas from ancient times the ruling principle ascribed to the man is Logos. The concept of Eros could be expressed in modern times as psychic relatedness, and that of Logos as objective interest. . . . [I]t is the function of Eros to unite what Logos has sundered. The woman of today is faced with a tremendous cultural task – perhaps it will be the dawn of a new era.

Jung, 'Woman in Europe'[268]

(10) The 'Apollonian' rational control of nature, as opposed to the instinctual excesses of the cult of Dionysus, the power of consciousness as opposed to the unconscious, the celebration of father-right over mother-right, come together in this mythology. Why the sun should have come to embody a split consciousness, while the worship of the moon allowed for coexistent opposites, a holistic process, is an interesting question. The fact that the moon itself is continually changing, and is visible in so many forms, while the sun presents itself in one, single, unvarying [authoritarian] form, may account for the kinds of human perceptions which would be powerfully drawn to one or the other. At all events, with the advent of solar religion, the great Mother, in her manifold persons and expression, begins to suffer reduction; parts of her are split off, some undergo a gender change, and henceforth woman herself will be living on patriarchal terms, under the laws of male divinities and in the light of male judgments.

Adrienne Rich, *Of Woman Born*[269]

(11) The contextuality, narrativity, and specificity of women's moral judgment is not a sign of weakness or deficiency, but a manifestation of a vision of moral maturity that views the self as a being immersed in a network of relationships with others. According to this vision, the respect for each others' needs and the mutuality of efforts to satisfy them

sustain moral growth and development.

Seyla Benhabib, 'The Generalized and the Concrete Other'[270]

(12) What the upheavals of the last twenty or thirty years mean for man's world is apparent to everyone; we can read about it every day in the newspapers. But what it means for woman is not so evident. Neither politically, nor economically, nor spiritually is she a factor of visible importance. If she were, she would loom more largely in man's field of vision and would have to be considered a rival. Sometimes she is seen in this role, but only as a man, so to speak, who is accidentally a woman. But since as a rule her place is on man's intimate side, the side of him that merely feels and has no eyes and does not want to see, woman appears as an impenetrable mask behind which everything possible and impossible can be conjectured – and actually seen! – without his getting anywhere near the mark.

Jung, 'Woman in Europe'[271]

(13) [A] certain degree of anatomical hermaphroditism occurs normally. In every normal male or female individual, *traces* are found of the apparatus of the opposite sex. These either persist without function as rudimentary organs or become modified and take on other functions. . . . These long-familiar facts of anatomy lead us to suppose that an originally bisexual physical disposition has, in the course of evolution, become modified into a unisexual one, leaving behind only a few *traces* of the sex that has become atrophied.

Sigmund Freud, *Three Essays on The Theory of Sexuality*[272]

(14) I am Protennoia, the Thought that exists in the Light. . . . She who exists before the All. . . . I am the Invisible One within the All. . . . I am perception and knowledge, uttering a voice by means of Thought. I cry out in everyone, and they know that a seed dwells within. . . . I am the Voice . . . it is I who speak within every creature. . . . Now I have come a second time in the likeness of a

female, and have spoken. . . . I have revealed myself in the Thought of the likeness of my masculinity. . . . I am androgynous. . . . I am both Father and Mother. . . .

Gnostic vision of 'Trimorphic Protennoia', *Nag Hammadi*[273]

(15) I remember the vision quest of a man who constantly dreamed about running. After he began to train in the art of long-distance running, his dreams still insisted that there was much to learn from this and that if he ran far enough he would find a magical woman. We decided to run together once and to remain watchful and see what we could discover. After a while he experienced stiffness in his back because, he explained, it was not rotating at the base of the spine in conjunction with the stride of his legs or the swinging motion of his arms. He felt stiff and compelled to rigidly face forward while running. But when he allowed his back to rotate he realized why he was stiff. When he swung with his running rhythm he felt he moved like a woman with breasts, and something [coming from his head, his intellect] said, 'This is wrong, you must remain a man!' However, his body said that the new swing was a more restful and relaxed way of running. This man's body was saying that he needed more anima, more swing, more softness and less strength. Developing the art of long-distance running required developing femininity and loosening up his inhibitions. His vision quest revealed a magic woman. Women suffer analogous inhibitions insofar as culturally accepted masculine characteristics are forbidden in female behavior.

Arnold Mindell, *Dreambody: The Body's Role in Revealing the Self*[274]

(16) . . . the man will be forced to develop his feminine side, to open his eyes to the psyche and to Eros. It is a task he cannot avoid. . . .

Jung, 'Woman in Europe'[275]

(17) The will, the male power, organizes, imposes its own thought and wish on others, and makes that

military eye which controls boys as it controls men.

Ralph Waldo Emerson, 'Education'[276]

(18) . . . how accustomed we have become to seeing life through men's eyes. . . . [Women are more practised at] seeing a world comprised of relationships rather than of people standing alone [and in conflict with one another], a world that coheres through [the imposition of] systems of rules. . . .

Carol Gilligan, *In a Different Voice: Psychological Theory and Women's Development*[277]

In our culture, a binary logic of oppositions has long differentiated men and women, and their respective positions within the prevailing sex-gender system have been pervasively determined according to dualisms established long ago. These dualisms are codified in our metaphysics, and veiled in false justifications. The one is identified with activity, the other with passivity; the one with mind, the other with body; the one with sky, the other with earth; the one with ego, the other with libido; the one with order, the other with disorder; the one with maturity, the other with its absence; the one with reason, the other with passion; the one with clarity, the other with obscurity; the one with the light, the other with the dark; the one with culture, the other with nature; the one with spirit, the other with matter; the one with forms of consciousness, the other with the mysteries of the unconscious; the one with the making of history, the other with fate. Since men have occupied the dominant positions in these bipolar structures, the institutionalization of the dualisms has functioned to subordinate and exploit women. It can no longer be denied that women have suffered terribly under the oppressive rule of men. They have survived; but their being has suffered under this system of patriarchy, and it is time to bring this ancient rule to an end.

But the history of Western metaphysics, the very tradition which Heidegger himself summons us to 'destroy', is a reflection of this patriarchy; it is a manifestation of the fact that it is men who have dominated our society and culture. Thus we are compelled to acknowledge, today, that since it is men who have presided over the telling of metaphysics, the history of Being, the very presencing of Being itself, must be a production of the patriarchal system. In other words, it is not only in order to emancipate women that we must overcome the tradition of metaphysics, although this would certainly be reason enough, but also in order to release Being as such, the dimensionality of beings as a whole, from a history determined by the masculine will to power and its modern form of technology. If the advent of nihilism is related to the rise of the modern will to power, then it is also related to the history of male domination. And that means that there can be no historically significant break with the history of male domination without a break with the history of metaphysics; and, conversely, that there can be no significant break with the history of metaphysics without a corresponding break with the history of male domination.

The historical possibility of a new vision of Being, and therefore, a new vision of the *human* being, a vision which would bring with it the liberation of all human beings from the prevailing forms of domination and oppression, requires that we reflect in a critical way on the character of our visionary being in a world whose archetypal principles have been organized by men into a sex-gender system which ensures the continuation of their own power . . . and their own interpretation of 'power'.

Since women have traditionally been identified with the body, with earth and the elemental, with the unconscious, with instinct, passion, feeling, ground and background, darkness and obscurity, how can it possibly be emancipatory for them if, in this book, we emphasize the importance of these things? How can our project be liberating for them, when it associates their being with these traditional sources of identity? These questions need to be answered; otherwise, an essential aspect of our project will not be understood.

To begin with, we need to see the difference between *archetypes* and *stereotypes*. The archetypes are primordial and transhistorical patterns of experiencing. Stereotypes are degraded archetypes, socially instituted and majority-validated. In the discourse of this book, 'body', 'earth', 'background', 'darkness', 'obscurity', and 'matter' (I mention only a few examples) designate transhistorical archetypes, not historically oppressive stereotypes. Under patriarchal supervision, the dimensionality of archetypes manifesting in our civilization has been systematically reduced, and some, such as those I just named, have been restricted to the lives of the women. Thus restricted, the archetypes do indeed function oppressively, and not only in the lives of women, but also, though in a different way, in the lives of men. For women, the oppression consists in the fact that they are limited to the enactment of these 'subordinate' archetypes; in other words, they are excluded from participation in the enactment of other archetypes, archetypes no less essential for their individuation and fulfillment as human beings. For men, the oppression consists in the fact that they have instituted a society which will not allow them to take part in the enactment of archetypes exclusively assigned to the psychology of women. Men may be the masters in relation to women; but, as Hegel and Marx understood, the master's position of mastery is always, nevertheless, a position of degradation and enslavement. The archetypes which our patriarchal society has stereotyped, and which it continually reproduces as 'essentially' and 'inherently' feminine, must be developed, now, by our society as a whole, and by men in particular. At the same time, society must allow women to free themselves from the bondage to degraded and degrading stereotypes, so that they can begin to develop themselves in relation to other creative archetypes.

As a society, as a culture, we need to appreciate that the historical forms in which the archetypes have manifested are not necessary and eternal, but contingent and changeable; we need to understand that the forms within which women – and men, too – have been confined are not pre-ordained fates, destinies to be suffered and endured generation after generation. The historical forms ought rather to be seen as challenges to our imagination, our capacity to envision new possibilities for social existence. They also challenge our willingness to share in the difficulties of the task, and demand from us a real commitment to improve communication between the two systems of sex and gender. This willingness and commitment must of course be grounded in a mutual recognition of the need for men and women to work together, moving step by step toward greater reciprocity, greater respect, and greater equality.

What our present historical situation calls for is a pervasive change in social attitudes, and in the institutions which reproduce them: for example, acceptance of archetypes which manifest at the present time mostly in association with the feminine psyche. This is why we have focused so much attention on the archetypes. Not in order to perpetuate an oppressive essentialism, to fixate the historically contingent identity of the feminine in terms of body, earth, matter, emotionality, darkness, holistic awareness, and the unconscious, but in order to affirm their enduring value, their supreme importance, for each and every one of us, men as well as women.

Archetypes which for too long have been constitutive only of femininity must be seen and valued as essential for the fulfillment of masculinity. Men must begin to recognize and value the spirit of femininity within themselves. They will not truly love and respect themselves, nor love and respect women, until they do.

What I want to say, then, is that these traditional associations – associations, for example, between women and body, women and earth – are in fact archetypal, and should be recognized and appreciated as such: they should not be seen stereotypically, in association only with women; nor should they be devalued in consequence of this presumed association. Since men are (still) in control, it is imperative that they change their traditional attitudes, their beliefs, and their modes of comportment. As men change in the required ways, the archetypes traditionally associated with women will be increasingly accepted within society as a whole. They will be integrated into its network of practices and institutions, transforming them accordingly. And of course, as these changes occur, the oppression of women will continue to diminish. Eventually, perhaps, the traditional connections between women and body, earth, matter, emotionality, darkness and the unconscious will no longer be regarded with a contempt that places women in subordinate positions; and they will find themselves increasingly free to change: freer to change themselves, freer to change the nature of these archetypal connections, and freer to develop historically unprecedented connections with other equally important archetypal complexes.

I would like to retrieve the creative, emancipatory power of the archetypes concealed within the patriarchy's historical imposition of stereotypes injurious to the feminine pole of human experience. Affirming archetypal connections – the connection, for example, between women and the body – is not necessarily a reactionary move: it all depends on how this affirmation functions in the discourse where it takes place. In the discourse entrusted with our project, the affirmation has two equal objectives: to draw the lives of men into the dimensions of their embodiment and to release women from their ancient confinement within those same dimensions. In the postmodern situation, there is an important sense in which we men and women do not know who we are. This 'not knowing' gives to our lives a new ontological question: difficult, of course, and not without pain; but also, I believe, enriching beyond known measure.

Knowledge: A Critique of the Patriarchal Paradigm

In *The Genealogy of Morals*, Nietzsche declares that:

> It is of the greatest importance to know how to put the most diverse perspectives and psychological interpretations at the service of intellection. Let us, from now on, be on our guard against the hallowed philosophers' myth of a 'pure, will-less, painless, timeless knower'; let us beware of the tentacles of such contradictory notions as 'pure reason', 'absolute knowledge', 'absolute intelligence'. All these concepts presuppose an eye such as no living being can imagine, an eye required to have no direction, to abrogate its active and interpretive powers – precisely those powers that alone make of seeing a seeing of *something*. All seeing is essentially perspective, and so is all knowing. The more emotions we allow to speak in a given matter, the more different eyes we can put on in order to view a given spectacle, the more complete will be our conception of it, the greater our 'objectivity'.[278]

In an essay called 'Feeling and Knowing: Emotion in Feminist Theory', Alison Jaggar sharpens and develops this critique, turning it very specifically against the logocentric epistemology of the patriarchal tradition. Moved by her vision of an 'epistemology of care or love', she sees what Nietzsche, unfortunately, could never have seen.[279] Moreover, part of what Nietzsche did not see, namely, that the paradigm of knowledge against which he was rebelling is inextricably tied up with the historical domination of men,

Jaggar sees at the very heart of the postmodern critique he inaugurated. Jaggar writes:

> . . . just as earlier I challenged the prevailing view that reliable knowledge should be value-free or disinterested, I now want to argue that reliable knowledge need not, cannot, and should not be dispassionate. Just as the disinterested observer is a myth, so is the dispassionate observer – a myth, moreover, with a powerful ideological function. Feminist practice disproves this myth and feminist epistemology must demystify it.[280]

According to the paradigm imposed by our patriarchal tradition, 'knowledge' must be disinterested and dispassionate, a product of value-free enquiry. The patriarchal pursuit of 'knowledge' requires the pure objectivity of a disengaged, unmoved observer. The patriarchal ideal of 'knowledge' excludes or overcomes its relationship to our sensibility: even if it originates in sensation, in the 'passivity' or 'receptivity' of sensuous awareness, it must detach itself, must abstract itself, from the body of felt experience; it must overcome all 'passivity' through its 'active' reworking of the material it is given. The patriarchy is logocentric; as Jung says, it values *Logos* and rejects *Eros*. Patriarchal 'knowledge' is detached, abstract, universal, and totally committed to the ideal of objectivity. Patriarchal 'knowledge' is also essentially hierarchical and is built on a sturdy foundation of unquestionable, absolutely authoritative propositions.

In her essay, Jaggar emphasizes the fact that our traditional paradigm of knowledge requires the exclusion of all feelings, or at least their subordination and control by a rationality which is purely theoretical, and which lays claim to the power of an impartial universality. But she sees cause for some hope in the fact that 'the hegemony which our society exercises over our emotional constitution is not total.'[281] Arguing against the patriarchal dualism of reason and emotion, a polarization which turns emotions into mere sentimentality and irrationalism, she maintains that,

> Critical reflection on our emotions is not a self-indulgent substitute for political theory and political action. It is itself a kind of political analysis and

> political practice, indispensable for an adequate feminist theory and social transformation.[282]

And she reflects on the fact that, as she says,

> We can now see that women's subversive insights owe much to women's outlaw emotions, which themselves are inevitable responses to women's social subordination.[283]

To support her contention, she examines the critical and emancipatory power in anger and pride:

> . . . anger becomes feminine anger when it is provoked by a perception that the harrassment endured by this woman is part of a pattern of systematic sexual harrassment, and pride becomes feminine pride when it is evoked by the realization that this person's achievement was possible only because the individual concerned overcame specifically gendered obstacles to success.[284]

Feelings and emotions are not necessarily 'confused ideas', as Descartes and Spinoza wanted to prove. Indeed, they are often a deep source of wisdom, of a knowledge much sharper and clearer than the ideas we can produce by the exercise of pure intellection.

In 'The Child's Relations with Others', Merleau-Ponty argues for a connection between perceptual rigidity and the authoritarian personality. Analogously, I wish to argue that perception, like knowledge, can be channeled in a 'feminine' mode as well as in a 'masculine'. (These terms – 'feminine' and 'masculine' – must be understood, of course, in their historical relativity. They describe the character of our perceptual channels as they have manifested in relation to the sex-gender system of our tradition. I take it as axiomatic that these channels appear as they do because of historical conditions which can be, and should be, changed.) Since morality concerns social relationships, and these are inseparable from forms of perception, the recognition and understanding of these two archetypal modes, ways of channeling perception, must be considered of the greatest importance.

Under the logocentric rule of the patriarchy, the ideal of *knowledge* is absolute certainty, a clear and distinct representation. Thus, the ideal to which our *vision* conforms is a vision of straight lines, fixed points and centres, and narrow, piercing rays.

Perception occurring under the influence of the 'feminine archetype' has traditionally valued and maintained a character which, in comparison with perception under the influence of the 'masculine archetype', has been, I think, relatively more global, more holistic, more syncretic, softer, more diffuse, gentler, more relaxed; more receptive, more responsive, more feeling, more open to being moved by what is being perceived, less intensely and less exclusively motivated and focused by personal desire. It is also a mode of perception which has been much more attuned to the presencing of the background: the field as a whole. This is why I find Mary Daly's asseveration, which I quoted at the beginning of this section, to be so very provocative. What she says is: 'the Background is the realm of the wild reality of Women's Selves. Objectification and alienation take place when we are locked into the male-centered, monodimensional foreground'. When our visionary being is controlled by the prevailing social mode of channeling, namely, by the 'masculine', it tends to be acutely focused, concerned with questions of certainty and mastery: with clarity, sharpness, fixation, frontality. Correlatively, what the traditionally 'feminine' channeling has tended to value and maintain with relatively greater consistency is a visionary being concerned with questions of contextuality, relationships, background, inclusion, and wholeness.

These characteristics are essential for moral development and should be more highly regarded by our society. At the present time, it is mostly through women that these characteristics are maintained in the channeling of perception. Men need to *develop* the traditionally 'feminine' mode of channeling; women need to be *released* from their historical burden of maintaining it; and women also need to be free to develop, in themselves, but presumably in a distinctively new way, the mode of channeling which has been traditionally occupied by men.

For me, our work in this book begins with an understanding of the task which Jaggar states so well that I should like to repeat it here, giving it pride of place as an ending for this analysis of the patriarchal paradigm:

> We can only start from where we are, beings who have been created in a cruelly racist, capitalist and male-dominated society which has shaped our bodies and our minds, our perceptions and our emotions, our language and our systems of knowledge.[285]

The Self and Its Moral Development: Another Paradigm Critique

In her recent book, *In a Different Voice,* Carol Gilligan attempts, as she puts it,

> to expand the understanding of human development by using the group left out in the construction of theory to call attention to what is missing in its account.[286]

It is Gilligan's contention that all the major theories of human development, theories both past and current, are systematically distorted by assumptions which reflect the fact that they have been formulated by men; and that, in consequence,

> The elusive mystery of women's development lies in its recognition of the continuing importance of attachment in the human life cycle. Woman's place in man's life cycle is to protect this recognition while the developmental litany [of the male-dominated theoretical consensus] intones the celebration of separation, autonomy, individuation, and natural rights. . . . Only when life-cycle theorists divide their attention and begin to live with women as they have lived with men will their vision encompass the experience of both sexes and their theories become correspondingly more fertile.[287]

Since the men who have been making the theories tend to take male development as their paradigm for human development without any self-critical awareness of their limited perspective,

> development itself comes to be identified with separation, and attachments appear to be developmental impediments [or regressions], as is repeatedly the case in the assessment of women.[288]

Whereas male development entails a 'more emphatic individuation and a [simultaneously] more defensive [and more aggressive] firming of experienced ego boundaries',[289] female development involves a growing awareness of the importance of human relationships, and a deepening responsiveness in motivation to the interconnections and interdependencies which constitute our human world. Since 'masculinity is defined through separation while femininity is defined through attachment',[290] developmental theories which are modelled on male experience and therefore take what is characteristic of the male as the norm for women as well as men, are not only oppressive for women, but fail to recognize a dimension of self-development which is essential to the moral life. The 'primacy of separation or connection [in theories of self-development] leads', as she observes, 'to very different images of self and of relationships'.[291] For the maturing girl, the world is not a field of action organized around isolated selves positioned according to hierarchical laws, but a 'world of relationships and psychological truths where an awareness of the connections among people gives rise to a recognition of our responsibility for one another, a perception of the need for response'.[292] What constitutes maturity in women – principally, a capacity for responsiveness and empathy, and a willingness to assume responsibility where care-taking is needed – are attributes of character essential to moral life in general.

Gilligan touches on a matter of great significance, therefore, when she points out that the 'images of hierarchy and web, drawn from the texts of men's and women's fantasies and thoughts, convey different ways of structuring relationships and are associated with different views of morality and self.'[293] Thus, as she says,

> The reinterpretation of women's experience in terms of their own imagery of relationships clarifies that experience and also provides a non-hierarchical vision of human connection.[294]

This non-hierarchical vision enables us to look beyond the horizon of our patriarchal tradition, whose conception of the self and its moral life has been unremittingly committed to defending the necessity of hierarchical order. But, whereas hierarchy is compatible with an ethics of rights and duties, it is inimical to an ethics of justice and care:

> the vision that self and other will be treated as of equal worth, that despite differences in power, things will be fair; the vision that everyone will be responded to and included, that no one will be left alone or hurt.[295]

For too long, our society has been ruled exclusively by a monadological vision of ourselves and an ethics of rights and duties. The time has indeed come when we need to develop, as a society, as a 'volonté générale', a vision of ourselves as essentially connected with others in a web of interactions and interdependencies. We need both visions; and we need to integrate both ethics. Only if we succeed in this task will we achieve a society in which the feminine spirit, embodied not only in women, but in men, too, is truly released from its conditions of oppression.

I see women taking a critical historical role in this process, for they, much more than men, have understood, as Gilligan says, that,

> The truths of relationship . . . return to the rediscovery of connection, in the realization that self and other are interdependent and that life, however valuable in itself, can only be sustained by care in relationships.[296]

In arguing for a conception of morality concerned with responsiveness and the activity of care, Gilligan centres moral development around the understanding of responsibility, rather than around the recognition of abstract moral principles and obedience to formal rules. Here, once again, is the contrast Jung noted between *Logos* and *Eros*: the paradigm of moral development suggested by Gilligan is a challenge to the logocentric paradigm of the patriarchy, which can see moral life only in terms of rules and rule-governed behavior. Moreover, it is becoming clear that, by setting itself up in the most dogmatic opposition to *Eros*, rather than working in co-operation with it, the logocentric tradition is compelled to define the goal of moral development exclusively in terms of autonomy and independence, and fails to understand the importance of relationship and interaction even for the achievement of mature individuation. It conceives of moral development in terms of the formation of a monadic, essentially isolated self: not at all a *social* individual, but an essentially *atomic* individual, inevitably standing alone, and in opposition to all other individuals. Thus, too, it must regard even the *ideal* society with a calculative eye, seeing it as a summed totality of individuals, rather than as an organic whole of interdependencies.

I see women taking the lead in returning our moral life to its concrete grounding in the cultivation of sensibility. When the body's felt sense of situational rightness and wrongness is properly cultivated, finely tuned, the self can often act by consulting this living, situationally very specific sense, instead of finding itself with no alternative to the following of general rules.

In this regard, I think it significant that, in his most recently published work, *The Nature of the Child*, Harvard psychologist Jerome Kagan writes:

> I believe . . . that, beneath the extraordinary variety in surface behavior and consciously articulated ideals, there is a set of emotional states that form the bases for a limited number of universal moral categories that transcend time and locality.[297]

And he points out that,

> The human competence to experience a small number of distinctive emotional states can be likened to the preservation of basic morphological structures in evolution – the eye is an example – each of which is expressed in varied phenotypes but descended from an original, fundamental form.[298]

Later in this section, we shall attempt to interpret this point phenomenologically, following Merleau-Ponty into the depths of what he calls 'the flesh'.

Continuing his argument, Kagan invites us to consider:

> One reason many scholars have preferred to base morality on logic, rather than on feeling, is that most Western philosophers have assumed human nature to be basically selfish, cruel, and deceitful. As a result, they could not trust a person's emotion as a basis for ethical choice and had to insert the idea of will between a person's strong desire and his behavior. Will is a planful, reflective executive amenable to reason. However, because the Chinese regarded human nature as more benevolent, they could be more trusting of human instincts and did not need to rely on a rational will to ensure that children and adults would behave in a civilized manner.[299]

Kagan oversimplifies, sounds naive. But I will say this: I do not see 'human nature' as 'essentially' or even 'basically' selfish, cruel and deceitful. I am inclined to believe that, both as individuals and as societies, we are potentially capable of rising above such comportment. But I also observe that we have so far failed to achieve social conditions which would enable our tremendous potential to be realized. What Kagan overlooks, though, when he reflects on the ancient civilization of China, is the fact that their 'trust' in the 'human instincts' was not a trust which simply let them grow wild, but a trust which informed their system of education, and which their institutional rationalization of society constantly and pervasively reinforced.

I see our present society rapidly disintegrating because, among other things, it has neglected the cultivation of sensibility, this crucial medium of the development of our moral, and therefore most essentially human capabilities. When we finally, as a society, recognize the moral significance of responsiveness and care, we shall begin to develop a deep respect for the traditional strengths of women; but this will also give women the opportunity to develop new kinds of strength. And who can doubt that this would improve the moral life of our society as a whole?

In *Being and Time*, Heidegger turns our thinking in the right direction when he returns human being to the safekeeping of 'Care': 'But since "Care" first shaped this creature, she shall possess it as long as it lives.'[300]

7 REVISIONING THE BODY POLITIC

(1) At its most materialistic, materialism comes to agree with theology. Its great desire would be the resurrection of the flesh, a desire utterly foreign to idealism, the realm of the absolute spirit.

Theodor Adorno, *Negative Dialectics*[301]

(2) . . . an ideal community of embodied subjects . . . intercorporealty.

Merleau-Ponty, 'The Concept of Nature'[302]

(3) . . . every form of the reality principle must be embodied in a system of societal institutions and relations, laws and values which transmit and enforce the required 'modification' of the instincts. This 'body' of the reality principle is different at different stages of civilization. . . .

Marcuse, *Eros and Civilization*[303]

(4) A class is defined as much by its *being-perceived* as by its *being*. . . .

Pierre Bourdieu, *Distinction: A Social Critique of the Judgment of Taste*[304]

(5) . . . we respond to gestures with an extreme alertness and, one might almost say, in accordance with an elaborate and secret code that is written nowhere, known by none and understood by all.

Edward Sapir, 'The Unconscious Patterning of Behavior in Society'[305]

(6) Everything takes place as if the social conditionings linked to a social condition tended to inscribe the relation to the social world in a lasting, generalized relation to one's own body, a way of bearing one's body, presenting it to others, moving it, making space for it, which gives the body its social physiognomy.

Bourdieu, *Distinction: A Social Critique of the Judgment of Taste*[306]

(7) The 'trivia' of everyday life – touching others, moving closer or farther away, dropping the eyes,

smiling, interrupting – are commonly interpreted as facilitating social intercourse, but not recognized in their position as micropolitical gestures, defenders of the status quo – of the state, of the wealthy, of authority, of all those whose power may be challenged. Nevertheless these minutiae find their place on a continuum of social control which extends from internalized socialization at one end to sheer physical force at the other. . . . In front of, and defending, the political-economic structure that determines our lives and defines the context of human relationships, there is the micropolitical structure that helps to maintain it. This micropolitical structure is the substance of our everyday experience'.

Nancy M. Henley, *Body Politics: Power, Sex, and Nonverbal Communication*[307]

(8) Let us ask . . . how things work at the level of on-going subjugation, at the level of those continuous and uninterrupted processes which subject our bodies, govern our gestures, dictate our behaviours.

Foucault, *Power/Knowledge*[308]

(9) The conservation of the social order is decisively reinforced by what Durkheim called 'logical conformity', i.e., the orchestration of categories of perception of the social world, which, being adjusted to the divisions of the established order (and thereby to the interests of those who dominate it) and common to all minds structured in accordance with those structures, present every appearance of objective necessity.

Bourdieu, *Distinction: A Social Critique of the Judgment of Taste*[309]

(10) The signs constituting the perceived body, cultural products which differentiate groups by their degree of culture, that is, their distance from nature, seem [as if they are] grounded in nature. The 'legitimate use' of the body [for example] is spontaneously perceived as an index of moral uprightness. . . .

Ibid.[310]

(11) . . . a norm exists in our society which prescribes different amounts of visual attention (in normal discourse) for persons having different degrees of social power: the higher one's standing, the less looking one has to give to others, and vice versa. This notion is supported by three experiments in which low-power members of dyads gave more visual attention to high-power members than vice versa.

Nancy Henley,, *Body Politics*[311]

(12) So much of women's interaction style focuses on kinesic prescription that special attention must be given to these restrictions in posture, gesture, and motion.

Ibid.[312]

(13) Thus one can begin to map out a universe of class [and sex-gendered] bodies, which (biological accidents apart) tends to reproduce in its specific logic the universe of the social structure.

Bourdieu, *Distinction: A Social Critique of the Judgment of Taste*[313]

(14) Another . . . [experiment]: subjects were divided by a one-way screen so that only one could see the other. Those who were able to see their partners looked significantly more at them than partners did in the mutual visibility experiment, suggesting that the possibility of being seen has an inhibitory effect on looking. And females were more inhibited than males by mutual gaze.

Henley, *Body Politics*[314]

(15) . . . superiors use two forms to communicate their superiority: *staring* is used to assert dominance – to establish, to maintain, and to regain it. On the other hand, superior position in itself, especially a secure one, is communicated by *visually ignoring* the other person – not looking while listening, but looking into space as if the other isn't there. This is the reciprocal of the visual attentiveness that the insecure subordinate must show. In other words, both looking and not looking may communicate

power, and their use is not hard to figure out. Women's tendencies both to look more at the other, and to avert the gaze, do not contradict each other, but are understandable in the power/status interpretation.

Ibid.[315]

(16) . . . when asked to assess the height of familiar persons from memory, [subjects] tended to overestimate most of all the height of those who had most authority or prestige in their eyes.
Bourdieu, *Distinction: A Social Critique of The Judgment of Taste*[316]

(17); When nursing students were asked to estimate the heights of known faculty and student members (female) of their school, the heights of the two faculty members were overestimated, and [those] of the two students, underestimated. Similarly, [in another experiment, testing for 'ascribed status' correlations,] . . . the estimated height increased as the [presumption of] ascribed status increased.
Henley, *Body Politics*[317]

(18) If one year in school has meant so much in lower-class children's development of gestural understanding compared to middle-class children's, might it not be because there is something they have encountered there for the first time, namely, middle-class cultural habits?

Ibid.[318]

(19) Taste, a class culture turned into [second] nature, that is, *embodied*, helps to shape the class body.
Bourdieu, *Distinction: A Social Critique of the Judgment of Taste*[319]

A New Subjectivity

'We have to promote new forms of subjectivity.'[320] This familiar position, echoing earlier positions in the Frankfurt School, is one of the last thoughts we have from Foucault. It recalls Marcuse, who asserted in *An Essay on Liberation* (1969) that 'radical change in consciousness is the beginning, the first step, in changing social existence: the emergence of the new Subject.'[321]

I agree. Here, then, I would like briefly to concentrate on one of the dualisms of our tradition: the dualism which splits the modern form of subjectivity, the modern form of the self, into an inside and an outside, inner life and outer life. This is not only a metaphysical dualism; it is an ideology with political consequences of the utmost importance. But even very astute thinkers like Adorno and Foucault get into difficulties which they could have avoided – and certainly should have. Consider, for example, what Adorno says in *Minima Moralia*:

> Not only is the self entwined in society; it owes society its existence in the most liberal sense. All its content comes from society, or at any rate from its relation to the object.[322]

In just a moment, we will continue our reading of this text. I interrupt it in order to indicate the point where Adorno's thinking goes astray and becomes self-defeating. The *first* sentence is not, for me, problematic. I see the self as inherently social, as emergent only within social interactions. But the question is whether we should concur with the sentence which immediately follows. Is it true of the self that 'all its content comes from society?' Is it true that the self enters the world entirely empty, and that our 'inner life' consists of nothing but externally, i.e., socially derived 'contents'? If Adorno were actually right, artists like Goya and thinkers like Adorno himself would not be possible. In order to emphasize a valid point, Adorno writes himself, here, into an untenable corner: his position is not only wrong; it is also self-defeating.

The passage continues:

> It [i.e., the self] grows richer the more freely it develops and reflects this relation, while it is limited, impoverished and reduced by the separation and hardening that it lays claim to as an origin.

Here he begins to escape from the corner, for he moves very close to recognizing that we must understand the problem of the socially entwined, socially constituted self much more radically: that it is necessary to overcome the traditional dualism; that the split between inner and outer has been concealing 'the monadological form which social oppression imposes on man'; that the split is one of the ways by which society is oppressive to the self at the same time that it is responsible for bringing it forth (*Ibid.*).

The self is not self-created, as the narratives of modern idealism tell us. It is not, in this sense, an 'origin'. Adorno is therefore right to criticize conceptions of 'authenticity' which assume the possibility of absolute self-creation, 'originality' in the sense of modern idealism. But he should have taken care not to repudiate the spirit which went into these ideologically corrupted endeavours. Instead of rejecting 'authenticity' altogether, we need to rethink it as a self-experiencing process without the dualism of inner and outer; and we need to rethink it without avoiding the problem of social domination.

Much of our so-called 'inner life' is really nothing but internalized social control. Freud saw this very clearly in the tragic vision of *Civilization and Its Discontents*. Moreover, as we have noted briefly already, the very notion of an 'inner life' can function to reinforce oppressive social control, since – as many women, for example, have recently come to understand – it pushes our experience of the world back into a monadic subjectivity without any reality of its own. But we must not abandon the touchstone of resistance. Naomi Scheman has argued well for this point in a paper on 'Anger and the Politics of Naming', where she writes:

> If there is the connection I have been suggesting between politics and . . . [women's discovery of their] anger, then emotions become much more threatening than they would be were they simply inner states.[323]

Gendlin has shown how, in 'focusing', we can use even the language of 'inner' and 'outer' to point to experiencing which goes beyond this language, and even works against it in a critical and deconstructive way.

Foucault did not see any of this. Thus Habermas, lecturing on Foucault, wants to ask whether we are only 'individual copies that are mechanically punched out'.[324] In his Howison Lecture on 'Truth and Subjectivity' at the

University of California in Berkeley (October 20, 1980), Foucault asserted that,

> If one wants to analyze the genealogy of the subject in Western societies, one has to take into account not only the techniques of domination but also techniques of the self.[325]

Foucault is making a valid point. However, when we consider in more detail just what he means by 'techniques of the self', we discover that the only practices he acknowledges, the only processes he includes in the denotation of his concept, are techniques through which the power of socially dominant groups and institutions is *applied* by individual selves to themselves. We become the instruments and accomplices of power.

As Reiner Schürmann says, 'Given Foucault's method, this much seems clear: the subject is constituted from without. This excludes any [self-constitution, any] practical constitution from within, be it transcendental or collective.'[326] The word Foucault uses – 'techniques' – is revealing: he seems unable to imagine any relationship I might have to myself, e.g., to my experiencing of myself and my experiencing of the world, other than one modelled on the mechanical application of a technological procedure. But I submit that, even when we grant him the pervasiveness of technology and concede the possibility that much of what we think and experience, even in regard to ourselves, is more dictated by ideology, by false consciousness, than we have been willing to believe, Foucault's position, here, is clearly too extreme. It is also self-defeating.

The problem of domination, of social control, is of course very complex. And it is certainly essential to see that not all forces of subjection are simply and directly imposed on the self by sources 'outside' it. The self can easily become, and often is, an accomplice in its own oppression. After all, if Freud's theory of ego-formation is basically true, then the ego, the self as an ego, never comes into being except in a process of adaptation to social power. The ego-logical self is from the very beginning a product of the channeling of power – of what he calls adjustments to the 'reality principle'. So Kant's correlation of the 'inner' and the 'free' does not work: the differentiation of 'inner' and 'outer' is itself an 'effect' of power, and does not in the least ensure that the self has escaped social domination: our inner life may still be controlled by 'heteronomous voices that tell us our identity'.[327] Schürmann accordingly declares – and I agree with him – that 'new forms of inner, although heteronomous modes of subjection have appeared as urgent targets in today's struggles.'[328]

Beyond this, however, I cannot agree with his project. From my point of view, his concept of 'anarchistic self-constitution' is fatally flawed, and in the last analysis, his position is no less self-defeating than Foucault's. Let me quote the step which I find problematic. In commenting on 'Des supplices aux cellules', an interview with Foucault published in *Le Monde* (21 February, 1975), he writes:

> For a culture obsessed with what is deep inside the self – hidden, unconscious, profoundly and unfathomably my own – anarchistic self-constitution means the dispersal of inward-directed reflection into as many outward-directed reflexes as there are 'systems of power to short-circuit, disqualify, disrupt'.[329]

Some of the problems I see here could perhaps be worked out in the direction I will soon suggest. But there is at least one problem which is more fundamental, and I do not see how it could be resolved without a more comprehensive change in theory: Schürmann unwittingly perpetuates the Nietzschean belief that the only alternative to action under Apollonian control, action ordered by oppressive power, must be the chaotic, disorganized action of Dionysiac frenzy.

It is telling that Schürmann speaks of 'dispersal' and of 'reflexes'. This is a romantic conception of action which is not at all helpful. And the assumption on which it implicitly depends, namely, that turbulent gestures of resistance are the only alternative to Apollonian control – an assumption unquestionably at work in Foucault as well – is simply not one that we are compelled to make or acknowledge. Since this assumption is clearly self-defeating – this much Foucault himself more or less admits when he concedes that a sense of futility and despair haunts his conception of resistance – we must ask ourselves why it has drawn their reluctant, desperate consent.

Briefly stated, I think the problem lies in the fact that

neither Foucault nor Schürmann have connected the discourse of philosophy, which concerns the being of the subject, and the discourse of critical theory, which concerns the historical vicissitudes of the will to power, with a self-reflective discourse on psychology, in which the character of different self-experiencing processes would be questioned and evaluated from the standpoint of constructive social action. If Shürmann had taken the time to reflect on the different experiencing processes in which subjectivity may be engaged; if he had put them to rigorous phenomenological examination and given special attention to the ways in which they happen, perhaps he would have seen an alternative to his concept of 'anarchistic self-constitution'. If 'anarchy' is turbulent and chaotic, there is not much to recommend it.

The Body: Organ of Political Thought and Action

I would like to open our reflections on the body of vision as an organ of thought and action with a poem by Michèle Najlis, a Nicaraguan poet. The poem is called 'Las viejas tribus':

> The old tribes are here
> with their malarial myths,
> their warriors with weapons ready,
> the elders secretly vigilant.
> The legend – remote voice of dreams –
> returns as a spell,
> and profound drums resound in the veins,
> summoning distant rebellions.
> The old tribes return,
> not as a call to the past,
> but as the throbbing weapon,
> as the chief alive in the rebel,
> as the united voice that comes to us from
> within.[330]

This voice is a voice of intelligent resistance; it is the body speaking; purposefully, coherently and creatively. The poem comes from what the poet sees through her visionary body of feeling.

Foucault returns the body, which centuries of Jewish and

Christian asceticism had cast to the margins of the texts our philosophical tradition recognized, to a position of major importance. The body, he asserts, is a primary site for the application of power. Thus he gives it fundamental importance as a locus of resistance to power. And yet, paradoxically, his understanding of the body deprives it of the capacity to resist in any organized, coherent, intelligent, creative, and purposive way.

His genealogy, for example, is assigned the task of exposing 'a body totally imprinted by history and the process of history's destruction of the body.'[331] 'What I want to show', he says, 'is how power relations can materially penetrate the body in depth, without depending even on the mediation of the subject's own representations.'[332] Here we see Foucault explicitly avoiding any reference to the subject's body of experience.

When analyzing the economic changes of the eighteenth century, Foucault takes pains to show how these changes 'made it necessary to ensure the circulation of effects of power through progressively finer channels, gaining access to individuals themselves, to their bodies, their gestures and all their daily actions.'[333] This attention to the body is welcome, after so many centuries of philosophical indifference and neglect. But unfortunately, the 'body' in terms of which Foucault is thinking is not the 'lived body' which appears in the phenomenology of Merleau-Ponty; it is the 'objective body' of the metaphysical tradition, the body represented as a 'material substratum' for the imprinting of social forms, or as a 'machine' of drives and desires. Consequently, at the same time that he introduces the body into critical theory as a potential locus of resistance, he also contributes to the 'progressive subsumption of bodies under technologies of power'.[334]

Nevertheless, it is possible to find passages where, despite himself, Foucault unwittingly seems to imply a very different conception of the body. In *The Order of Things*, for instance, there is a passage which sets in motion a contrast between an order which is given and an order which is created. In this instance, it is an order created by a glance:

> Order is, at one and the same time, that which is given in things as their inner law, the hidden network that determines the way they confront one another, and also that which has no existence except

> in the grid created by a glance, an examination, a language.[335]

To be sure, Foucault does begin, here, to adumbrate a very different role for the human body; but this is not a problem which he sufficiently explored. For the most part, he falls into line with the prevailing metaphysical ideology that has surrounded the body of modern experience. Consider what Heidegger says in *What is a Thing?*. He contends that, in the modern age,

> nature is no longer an inner capacity of a body. . . . Nature is now the real of . . . uniform space-time. . . . Bodies have no concealed qualities, powers and capacities. Natural bodies are now only what they *show* themselves as, within this projected realm [of science]. Things now show themselves only in relations of places and time-points, and in measures of mass and working forces.[336]

The 'natural body' of science and science-saturated 'common sense' is a well-ordered body; but it is not accurate to call it 'natural'. Nor is it a body capable of resisting political exploitation and domination in a purposive and creative way – a body capable, that is, of creating a new social order from out of its own distinctive sense, its ownmost vision, of what is needed and called for. Seyla Benhabib is right: 'While men humanize outer nature through labor, inner nature remains non-historical, dark, and obscure.'[337]

In *The Constitution of Society*, Anthony Giddens criticizes the sociology of Parsons and Goffman on the ground that 'Each tends to emphasize the "given" character of roles, thereby serving to express the dualism of action and structure characteristic of so many areas of social theory.'[338] The same charge could be made, however, in regard to the human body: most of our leading thinkers in social and political theory can see only the *given* character of the body. They see only two kinds, or two sources, of order: the order of our biology and the order imposed by society. They cannot see the role of the body in the creation of new order, or new kinds of order. And even the 'givenness' of the body is seen only as facticity, not potentiality; as a state, not a process; as ontic, not ontologically dimensional.

Pierre Bourdieu is a perceptive sociologist, and he has

given to the body an unusual amount of attention. Let us take some time to consider his observations and reflections. According to him, there are 'Strictly biological differences [among individual people, which] are underlined and symbolically accentuated by differences in bearing, differences in gesture, posture and behaviour which express a whole relationship to the social world.'[339] These differences are inscribed into the body 'at its deepest and most unconscious level, i.e., [as] the body schema, which is the depository of a whole world view and a whole philosophy of the person and the body'.[340] The social imposition of meaning penetrates into the innermost recesses of the body: for him, there is no bodily felt sense of a dimensionality transcending the imposition of social interpretations. For Bourdieu, the 'principle' which structures our activity is not 'a system of universal forms and categories but a system of internalized, embodied schemes'.[341] And these are schemes which, 'having been constituted in the course of collective history, are acquired in the course of individual history. . . .'[342] This analysis of embodied schemes makes a promising start. But he also holds the view that,

> The social representation of his own body which each agent has to reckon with, from the very beginning, in order to build up his subjective image of his body and his bodily hexis, is thus obtained by applying a social system of classification based on the same principle as the social products to which it is applied.[343]

Here we find him moving towards a conception of embodiment which sees the imposition as total and sees the functioning of the body schema in mechanistic terms. Indeed, Bourdieu does not hesitate to say, without feeling a need to comment on the political implications, that 'the "capacity to discern aesthetic values" is . . . turned into muscular patterns and bodily automatisms'.[344] Now, I do not altogether disagree with this observation: I see its truth. But I cannot refrain from pointing out that, if perception is nothing more than socially imposed interpretations translated into automatisms organizing the body of experience, then we are destined to a sorry political fate.

Bourdieu insists that,

> The schemes of the habitus, the primary forms of classification, owe their specific efficacy to the fact that they function below the level of consciousness and language, beyond the reach of introspective scrutiny or control by the will. Orientating practices practically, they embed what some would mistakenly call *values* in the most automatic gestures or the apparently most insignificant techniques of the body – ways of walking or blowing one's nose, ways of eating or talking. . . .[345]

These observations, these facts, are not in dispute here. What I want to question is rather their generalization in a theory which seems to leave no space for the creativity of the individual body.

Bourdieu, like Foucault, occasionally says things, however, which imply some recognition of the body's capacity to generate new images, new visions, new order. For example, he notes that

> Charm and charisma . . . designate the power, which certain people have, to impose their own self-image as the objective and collective image of their body and being, to persuade others, as in love or faith, to abdicate their generic power of objectification and delegate it to the person who should be its object, who thereby becomes an absolute subject, without an exterior (being his own Other), fully justified in existing, legitimated.[346]

Thus it would seem that Bourdieu is theoretically committed to the position that there exist human bodies which are not totally determined, totally schematized, by socially imposed forms and norms. Another passage supports this position. Bourdieu writes that 'The sign-bearing, sign-wearing body is also a producer of signs which are physically marked by the relationship to the body. . . .'[347] Presumably he would see the difference, here, between production and reproduction: the difference between a gesture which initiates something new, something not completely formed and given in the existing corporeal schematism, and a gesture which only repeats, only manifests or expresses, what was already fully determinate. But where in his theory would he fit what Merleau-Ponty calls 'consecratory gestures'?[348]

Bourdieu makes some intriguing remarks about the 'alienated body', which he characterizes in terms of class identity:

> . . . the experience *par excellence* of the 'alienated body', embarrassment, and the opposite experience, ease, are clearly unequally probable for members of the petite bourgeoisie and the bourgeoisie, who grant the same recognition to the same representation of the legitimate body and legitimate deportment, but are unequally able to achieve it. The chances of experiencing one's own body as a vessel of grace, a continuous miracle, are that much greater when bodily capacity is commensurate with recognition; and, conversely, the probability of experiencing the body with unease, embarrassment, timidity grows with the disparity between the ideal body and the real body, the dream body and the 'looking-glass self' reflected in the reactions of others. . . .[349]

One of the reasons I value this analysis is that, by clearly articulating the difference between bodily capacity and social 'recognition', and the difference between the ideal body and the real body, Bourdieu protects the body from its reduction to social 'recognition' in a theoretical position which would conceive all forms of embodiment as being subject to the most extreme, because total, alienation. There are times when Foucault comes dangerously close to this helpless position.

Since Bourdieu refers to the dreambody, I should like to call attention to some reflections recently published by a Jungian psychologist with an extensive clinical practice. According to Arnold Mindell,

> Integrating the dreambody into your life, bringing your symptoms into your personality and living your dreams, is bound to take you to the edge of your own personality and bring you into conflict with the world around you.[350]

There is, then, a dreambody which can function as a source of creative resistance to prevailing social practices and institutions. Thus, Mindell argues that there is an important sense in which we are bodily responsible for the collective in which we live, although we must always bear in mind that

> The inner world and the outer world dreambodies are two-way streets, and it's impossible to place blame, for we all contribute to the body as a whole. Our dreambody is part of the entire world's dreambody, yet the world's dreambody is also found within us.[351]

Because of these transactions, Mindell envisions the possibility of a new relationship between the individual and the collective. But he also sees this possibility emerging from a situation of tremendous conflict, pain and suffering:

> I believe that the individual of the future, like the individual today, faces the lonely task of transforming himself, with or without the agreement and understanding of those around him. He needs only to know that transforming himself means coming up against interiorized cultural edges. If this transformation is to occur, he will have to disturb the status quo of the world around him as well. The person in the midst of an individuation process must know that when his symptoms disappear, a new kind of pain is likely to arise: conflict with the history of the world, of which he has been an integral part. How he deals with this conflict is a creative task which no-one can predict. But one thing is certain. Becoming an individual means stepping over cultural edges and therefore, paradoxically, also freeing the public to communicate more freely. This means that the collective could [begin to] integrate double signals, diseases and madnesses, which otherwise only operate in the sick, dying or insane.[352]

In *Civilization and Its Discontents,* Freud suggests that human beings shape their individual identity by learning to repress their 'chaotic' inner nature for the sake of dominating nature and the world 'outside' them. I am beginning to wonder to what extent this repression, creating a split between the 'inner' and the 'outer', is responsible for conditions which make us feel so often that our 'inner life' is empty or chaotic. There is a passage in his essay 'On the Uses and Disadvantages of History for Life' where Nietzsche – who is himself responsible for promoting the view that our 'inner life' is inherently and inevitably chaotic, a cauldron of unmastered drives and irrational, violent emotions – seems to move in the direction of such a diagnosis. He writes:

> Knowledge, taken in excess without hunger, even contrary to need, no longer acts as a transforming motive impelling to action, and [it consequently] remains hidden in a certain chaotic inner world . . . and so the whole of modern culture is essentially internal. . . .[353]

If this interpretation is sound, and I think it is, then I think we should turn to Merleau-Ponty, who makes a suggestion which could be useful in overcoming this widespread cultural theory of inner chaos. The suggestion is based on his intuition that the (sense of) inner chaos is a function of the metaphysical split between inner and outer, oneself and others; and he therefore proposes self-transforming processes, self-constituting activities, which involve working with the body schema. In a major essay, 'The Child's Relations with Others', he endeavours to formulate this provocative suggestion:

> To the extent that I can elaborate and extend my corporeal scheme, to the extent that I acquire a better organized experience of my own body, to that very extent will my consciousness of my own body cease being a chaos in which I am submerged and lend itself to a transfer to others.[354]

Thus he sees self-defeating self-interpretations, such as the sense of oneself as 'internally' nothing but chaos, as directly related to the self's entanglement in metaphysical dualisms, and in an ideology which encourages a separation of our inner experience (of ourselves) from the world of action – precisely the withdrawal of knowledge from practice against which Nietzsche struggles in the text we just read.

What needs to be understood, here, is the capacity of the embodied self, as embodied, i.e., as capable of being bodily motivated, to take part in the life of the world through socially constructive, and consequently self-constituting, self-individuating action. (Schürmann's concern.) More specifically, what needs to be understood is the relationship between actions and the body: the process whereby actions are based on motivations which come from the body's felt sense of what its situations call for. Perhaps we could derive some benefit from a fragment attributed to Herakleitos, and to which Heidegger calls our attention. In the seminar he gave with Eugen Fink, Heidegger quotes Herakleitos and comments as follows:

> When he speaks of father and ruler, Herakleitos grasps in an almost poetic speech the sense of the *archē* [the primordial principle] of movement: *prōton hōthen hē archē tēs kinēsēos*. 'The origin of movement is also the origin of ruling and directing.'[355]

Even if we suppose, as Herakleitos probably did, that movement does not originate in the body but only takes hold of it or passes into and through it, there still would be a phenomenological sense in which it would be true that the body is the 'origin' of movement: 'origin' not in any ultimate metaphysical sense, a sense which is patriarchal, but 'origin', rather, in the sense of 'locus' or 'proximate cause': the point where a succession of events of particular importance to embodied mortals always begins. However, the reason for introducing Herakleitos, here, is not to make sense of his words by adjusting them to accommodate present-day common sense; it is, on the contrary, to compel us to adjust our understanding of the body to accommodate the thought of a deeper, more radical 'origin' (*archē*) than our prevailing patriarchal consensus has acknowledged. (This is 'anarchism'. But it is different from Schürmann's.) Merleau-Ponty's late work, in which he articulates a conception of the flesh, will take us into the *depths* of movement: closer, perhaps, to the 'origin' Herakleitos may have had in mind.

As I would like to read this fragment, it provocatively adumbrates the role of the body – the body which in the first instance is a body in movement – in the activities of the *polis*. In other words, it suggests the possibility of grounding political action in the depths of (our sense of) the moving, self-motivated body. This suggestion raises a question about the motivation of action – a question which again returns us to the Nietzsche text we just read: 'where' in the body of experience does motivation arise and how can we bring it forth? We shall consider two answers to this question. The first, coming from Gendlin, will be outlined in the remaining pages of this section. The second, coming from Merleau-Ponty, will be taken up in the next section, the section with which this chapter concludes.

Before we move into Gendlin's answer, however, I would like to quote from Edward Bellamy's once very popular book, *Looking Backward*. It may help us to begin with a very concrete sense of this question; so I will quote a passage in which the author gives voice to his question about the 'origin' of action in the movement and response-ability of the body of felt experience:

> When I saw that things which were to me so intolerable moved them not at all, . . . I was at first stunned and then overcome with a desperate sickness and faintness at the heart. What hope was there for the wretched, for the world, if thoughtful men and tender women were not moved by things like these![356]

What moves our vision and how – how, in other words, the body of experience can question given social reality and initiate actions of constructive political resistance in response to oppressive conditions – will now be a question which takes us from Schürmann's unsatisfying 'anarchism' into the process Gendlin calls 'felt sensing' and 'focusing'.

Since every interaction between the body and society involves a certain degree of social control, Gilles Deleuze and Felix Guattari attempted to visualize a 'body without organs'.[357] I can see no point in this exercise of the imagination. Instead of making a surrealist gesture, poetic but futile, Gendlin confronts it. In a recent paper called 'Process Ethics and the Political Question', he continues the project of Dewey and Rogers, and demonstrates that more (order) can arise from the individual body than whatever order society programmes into it; that there can be a morality which comes from the sensibility of the body, rather than from the imperatives of the superego; and that it is not inevitable that our perception of social and political reality be channeled exclusively through the socially constituted ego.[358]

Nietzsche envisioned a morality beyond good and evil: a morality of real autonomy, independent of the authority of principles and precepts; a morality which would be rooted in the nature of our physiology. But his vision was doomed to failure, because he could not see his way beyond a body of drives, inwardly chaotic. For Nietzsche, the body knows no order of its own; its only order is socially imposed. John Dewey and Carl Rogers were able to go farther than Nietzsche in establishing the body as a radically different basis for morality. In *Freedom to Learn*, for example, Carl Rogers argued that 'There is an organismic base for an organized valuing process within the individual.'[359] Unlike Nietzsche, Foucault, and Schürmann, Rogers could see in the body the subtle operation of organized, and self-organizing processes: an order which is constitutive of our embodiment, but one which does not come from the social imposition of socially established order. Gendlin continues the work of Dewey and Rogers by defining the steps and phases of these organizational processes – very clearly, and very precisely. Gendlin also relates their work, as they do not, to the question of political control. He notes that,

> Adorno and Foucault find it impossible [to believe] that a freeing process could arise within an organism and go counter to the socially imposed forms. This has the effect of reinforcing the status quo. . . . Social change has its own supra-individual laws and developments, such as the evolution from agriculture to industry. Individual bodies live and are programmed by these social developments. But theories after Marx have assumed that there can be no feedback from the human organism, except perhaps disorder and disorganized resistance. These theories deny the body an order of its own and imply that social change can come only from engineering on the social level. It must be imposed on individuals, and in particular, on their bodies, since it cannot come *from* them. But recent history shows that social change is not necessarily freeing. What Adorno and Foucault see is that, even though social patterns have changed, they are still being imposed by people in some positions on people in other positions: change can only be different imposed forms, or different people in power. As

> they see it, real change seems virtually impossible.[360]

According to Gendlin's diagnosis, the political theories in terms of which we have been accustomed to thinking create a false conflict between the individual and society, because they implicitly identify authenticity, i.e., genuine experiencing processes, with a monadic, non-social individual, and see social existence only in terms of control. But, since individuals are always social, he argues that genuine individual processes are also inherently social. Critical thinking is social: inherently social. But we must be careful not to miss differences: as a process, thinking is different in kind from obedience. Even the intricate texture of personal feelings is social; yet these, too, can be newly elaborated and more complex than the socially imposed forms.[361]

It is Gendlin's contention that,

> Most political theories hold that the direction can only be one way: social control provides order for individual experience. There can be no orderly feedback; certainly not the more intricate feedback we actually observe. Foucault thinks that creative feedback assumes a nonsocial individual and a body capable of existing outside social interactions, and he rightly denies that possibility. But 'subjectivity' is *not* the separated unsocial source whose possibility he is denying. If we see that subjectivity is always social, and see also that it has its own order, then we should not let the word 'social' mean only 'imposed forms'. Imposed form is not the only kind of order.[362]

Important though this argument is, I think that Gendlin's most original, most revolutionary contribution to the discourse of political theory is his detailed phenomenological analysis of different (kinds of) experiencing processes. 'Ethics', he says, 'is best cared for as distinctions among kinds of processes':

> After all, it is the process which determines the content. Thoughts, feelings, desires and other experiences are not just given things. They are *generated* by processes. A certain kind of [self-

> reflective, self-constituting] process, for example, created the ancient virtues. It is not the case that just anything at all can be the content of just any process. Far from it.[363]

The example Gendlin gives at this point is instructive:

> Suppose your good friend has decided to marry someone, and you like the person. Marrying that intended spouse seems (in general) a good thing. Is that enough for you to call the decision right? Would you not need to know more about how your friend decided? What if the decision was made on a drunken afternoon and called for getting married later that very day? Suppose your friend badly wants money and the intended spouse is rich? Suppose the wedding was announced and your friend now wants to back out but is going ahead only because he is scared of disappointing the relatives? What if your friend talks mostly of not wanting to live alone?[364]

These questions call attention to the importance of taking into account the character of the decision-making process. Of course, as he observes,

> It is very difficult to delineate or define 'the right kind of process', even though it is already familiar to us all. People therefore tend to describe its character indirectly. For example, they describe the process in terms of time. You might ask how long your friend has thought about his decision. Or you might describe a 'right process' in terms of 'depth': how far down inside has your friend examined his decision? Or you might put it in more factual terms: how much is known about the person, the family, where they will be living?
>
> These questions of length, depth, and knowledge show that we know of a 'right' process, but time, space and facts are accidental parameters. We do not mean mere length of time. We hope, for example, that the time was not wasted going round in circles. We mean the process which is *opposite* to going round in circles. We mean a process of steps which

> can *correct* what one thought, felt, or was before. . . . Similarly, depth does not help, if nothing changes while digging deep. Sheer depth does not make something right. If the friend comes upon a deep wrong motive, we trust that the decision will change, somehow. Nor are the most relevant facts those that now exist as facts. Rather, the process should raise new questions, newly needed facts.[365]

Gendlin's phenomenological analysis of different processes does not stop, however, at this level. Because he appreciates the importance of embodiment – the role of the body – in human living and experiencing, Gendlin takes his analysis into processes of the lived body, processes of the body of experiencing. This enables him to demonstrate very concretely a way out of the impasse into which political thinkers such as Adorno and Foucault have erred:

> Social patterns are not imposed on mere chaotic drive energy. They are imposed on a more intricate experiential texture, a greater order. But this shows itself only when we give thought to process-questions. There is not a second, natural person inside or under the socially formed person. But not all order comes from society. . . . Society develops the human being; but the body-as-experienced also contributes to the further development of the forms we live in. The body is not just a copy of society.[366]

He warns that 'We cannot say what is from the organism and what is imposed. We can only differentiate kinds of process, kinds of steps.'[367] In particular, therefore, Gendlin helps us to recognize and appreciate *the difference* between steps which simply follow the body's social preprogramming and steps in which the body feeds back, speaks back, with some new form. As an example, he examines the transcript of a 'focusing' session with a patient who initially was disturbed by a conflict between her feelings of superiority and her commitment to an ethics of equality. What distressed her was that she considered herself to be deeply committed to this ethics, yet also felt that what she took to be her feelings of superiority were also somehow justified. However, after she had been working in a focusing way with her body's felt sense of the dilemma, her body finally brought forth a surprising resolution of the conflict: what her body-sense told her was that the feeling she took to be 'superiority' was 'really' something very different: something not in conflict with a respect for equality. Thus, she had *not* betrayed her democratic values after all; what changed everything was her bodily realization that what she was feeling was not really 'superiority' but rather a need to be *included* somehow in the group of people in relation to whom she had been thinking herself superior.[368]

Gendlin draws an important lesson from this example:

> Conceptual thinking alone [e.g., an ethics of universal principles] does not usually change us, but it has an essential role in the process. . . . This role differs, however, from the usual. In our example, her bodystep . . . might not have come without her conceptually formulated ethics of equality, which initially made a conflict with her feeling of superiority.
>
> Theories, concepts and values do not float on an independent conceptual level. They certainly may enable a step of process. But when they do, what comes is more intricate than the concepts were. We have seen that concepts and old forms do not 'determine' such a step, do not impose their form on it. In fact, the process can lead to rejection or modification of the very concept that helped the bodystep to come.[369]

Gendlin is not arguing that the body is all-wise, so that we do not also need to think and reason. He is not arguing that ethics can dispense with moral principles and arguments. What he is arguing, rather, is that the body is not a primordial chaos and that it is a mistake to think that the body's only socially relevant organization is either biologically preprogrammed or imposed by social domination. And this lays the ground for an argument that the body of felt experience can sometimes contribute a very articulate speech to the political discourse on power and its effects. In the experiential process Gendlin calls 'focusing', the body's creative, self-constituting, and critical capacity has been empirically demonstrated again and again.[370] The human body is inherently social; it is also inherently capable of rendering moral and political discriminations which might otherwise not have formed and been visible.

In our logocentric, patriarchal society, the body is not appreciatively understood, and often not authentically experienced, as an organ of visionary being. Thus, the body's natural gift of political vision remains for the most part unrealized – even today. And yet, because the body we live and experience is historical, it can tell us, sometimes very precisely and with the utmost specificity, just what it is we need from, or need to see changed in, the present lived moment of our historical situation.

This book is concerned with the development of our capacity for vision. This capacity is a question of our self-development as human beings. But self-development is a question that is neither only individual nor only psychological. In Chapter One, I set in motion a critique that would move back and forth between individual experience and society. I want to clarify, here, how I conceive this critique.

We began with reflections that diagnosed some of our common experience with vision. This diagnosis attempted to make contact with our pain and suffering and to let it form in speech. I hold that there are compelling moral claims which arise from the human capacity to feel pain, experience suffering and sense unfulfilled needs, and that to neglect these claims is to neglect and betray this capacity. Thus, the speech of suffering, the pathology, that comes from our experience with vision motivates a corresponding critical analysis of society. Who we are, and how and why we are suffering, is a reflection of prevailing social conditions. But it is equally true that what is wrong with society is a reflection of who we are – and an indictment of the *character* of our vision. We suffer; we suffer, sometimes, because of what we see; but our limited vision, a capacity whose potential we have only begun to develop, contributes to social conditions that perpetuate, even intensify, this suffering. The forms of our suffering are distinctive manifestations of our historical situation. In Chapter One, we reflected on what our sufferings as visionary beings could tell us about ourselves and the age in which we live. Since this suffering can be suppressed, concealed, unconscious, and misunderstood, our reflections were intended to raise our consciousness and focus awareness.

The suffering we experience through our vision, and the

ills of society that are visible to us, urgently call for a transformation of our vision through our capacity for further self-development. In order to change the social ills we see, we need to change our vision; we need to change ourselves. But society itself needs to be changed. It is *not enough* simply to give voice to the pain, the suffering and the need – to let it be seen. If this is all we do, nothing really changes. The experience of the individual must be *connected* to a critical theoretical analysis of society and culture – and to an appropriate social praxis. 'Inner' changes are no substitute for necessary changes in our social reality. The pain, the suffering, and the genuine unmet needs, are not only 'inner', and they cannot be overcome, or transformed, by only 'inner' changes. They are, and need to be understood as, essentially connected to social, political, economic and cultural conditions. We need to *see* our pain and suffering, and needs, as the 'effects' of changeable social arrangements – although they are not only 'effects'. (To suppose that they are is to fall prey to a skepticism or determinism that is irremediably self-defeating.)

The pain and suffering, and the needs we can sense, imply, and can inform us about, what social conditions we need to change and how to proceed. We can *work* with our sense of affliction to conceive, beyond mere 'strategies' of localized and immediate resistance, some appropriate modes of constructive engagement with society. We must expect, however, that as we commit ourselves to self-transformative work – to a 'caring for the self' which is also a social practice, we will be confronted, at certain points of pressure, by powerful obstacles to our further development: limits that are manifestly social in origin. Moreover, we will inevitably encounter social conditions that distort and confuse this process of self-development, and even operate, sometimes, to suppress what cannot be assimilated by the existing institutions. Self-development will always reach a point where it should become clear that social conditions must be changed before further individuation is possible.

It is at this point that we may begin to appreciate the fact that the processes of transformation in our vision are inherently social. Self and society are not separate systems. Needful changes in the one call for, and are responsive to, corresponding changes in the other. Both the social changes that our 'internal' processes call for, and these processes themselves, which concern the *social character* of vision, depend on, and need, specific conditions of communication, reciprocity, and solidarity. In brief, they demand the building of new kinds of community.

The Body of Depths

(1) Prior to all ethical behaviour in accordance with specific social standards, prior to all ideological expression, morality is a 'disposition' of the organism, perhaps rooted in the erotic drive . . . to create and preserve 'ever greater unities' of life.

Marcuse, *Essay on Liberation*[371]

(2) I live in the facial expressions of the other, as I feel him living in mine.

Merleau-Ponty, 'The Child's Relations with Others'[372]

(3) The behaviourists' explanation of a new action relies on the psychological consequences of that action. But this force for change, although operative on many occasions, does not seem helpful in accounting for the smile following mastery or the distress after task failure. It is unlikely that the three-year-old's upset after being unable to finish a puzzle is a generalized response traceable to parental punishment for failing to finish breakfast or to complete a sentence correctly. Rather, as nineteenth century theorists suspected, the child is biologically prepared to acquire standards.

Jerome Kagan, *The Nature of the Child*[373]

(4) To understand and judge a society, one has to penetrate its basic structure to the human bond upon which it is built. . . .

Merleau-Ponty, *Humanism and Terror*[374]

(5) The social-spiritual sphere . . . is said by Hegel to repeat at a higher level *the interplay of forces* dealt with by the Scientific Understanding. There we had *an underworld of distinct forces*, mutually releasing and inhibiting one another; here, on the other hand, we have *an upper world of mutually* acknowledging

conscious persons, who are also conscious of their mutual acknowledgement, and who are thereby raised to the fullest self-consciousness.

John Findlay, *Hegel: A Re-examination*[375]

(6) The communication or comprehension of gestures comes about through the reciprocity of my intentions and the gestures of others, of my gestures and the intentions discernible in the conduct of other people. It is as if the other persons' intention inhabited my body and mine his.

Merleau-Ponty, *Phenomenology of Perception*[376]

(7) My body . . . discovers in that other body a miraculous prolongation of my own intentions, a familiar way of dealing with the world. Henceforth, as the parts of my body together comprise a system, so my body and the other person's are one whole, two sides of one and the same phenomenon, and the anonymous existence of which my body is the ever-renewed trace henceforth inhabits both bodies simultaneously.

Merleau-Ponty, *Phenomenology of Perception*[377]

(8) . . . the body is a unity of unformed and formed existence, and is the reality of the individual permeated by his reference to self.

Hegel, *Phenomenology of Mind*[378]

(9) . . . the original primordial being, the connate body. . . .

Ibid.

(10) Our political system is placed in a just correspondence and symmetry with the order of the world and with the mode of existence decreed to a permanent body composed of transitory parts, wherein, by the disposition of a stupendous wisdom, moulding together the great mysterious incorporation of the human race, the whole, . . . in a condition of unchangeable constance, moves on

through the varied tenor of perpetual decay, fall, renovation, and progression.

Edmund Burke, *Reflections on the Revolution of France*[379]

(11) [T]he problem is not so much that of defining a political 'position' (which is to choose from a pre-existing set of possibilities), but to imagine and to bring into being *new schemas* of politicization.

Foucault, *Power/Knowledge*[380]

(12) The roots of the term *order* are from the Latin *ordiri*, which means 'to lay the warp, begin to weave'. Since the prevailing order is warped, dis-ordered, we unweave it as we begin to weave. Since it is a source of our known dis-ease, we unweave it with increasing ease, uncovering its previously unknown causes. . . . As Spinsters whirl, we continue to unweave the prevailing dis-order, weaving our way deeper into the labyrinth.

Mary Daly, *Gyn/Ecology: The Metaethics of Radical Feminism*[381]

(13) . . . the intertwining of my life with the lives of others, of my body with the visible things, by the intersection of my perceptual field with that of the others. . . .

Merleau-Ponty, 'Reflection and Interrogation'[382]

(14) The chiasm, reversibility, is the idea that every perception is doubled with a counter-perception, . . . one no longer knows who speaks and who listens. . . .

Merleau-Ponty, 'The Intertwining – The Chiasm'[383]

(15) . . . postural impregnation . . . initial sympathy . . . initial community. . . .

Merleau-Ponty, 'The Child's Relations with Others'[384]

(16) The logocentrism and intellectualism of intellectuals, combined with the prejudice inherent in the science which takes as its object the psyche, the soul, the mind, consciousness, representations,

. . . have prevented us from seeing that, as Leibniz put it, 'we are automatons in three-quarters of what we do', and that the ultimate values as they are called, are never anything other than the primary, primitive dispositions of the body, . . . in which the group's most vital interests are embedded. . . .
Bourdieu, *Distinction: A Social Critique of the Judgment of Taste*[385]

(17) Hume regarded moral sentiments as 'so rooted in our constitution and temper, that without entirely confounding the human mind by disease or madness, 'tis impossible to extirpate and destroy them.' Why should an appreciation of standards emerge so early in development, long before language and motor coordination are mature, and a decade before reproductive fertility? One possibility is that a 'sense' of right and wrong, an awareness of one's ability to hurt another, and the capacity to empathize with a victim are accidental accompaniments to the more fundamental competence of retrieval memory, symbolism, playfulness and language. But it is also reasonable that these distinctly human characteristics appear early in development because they are necessary for the socialization of aggressive and destructive behavior. . . . Without this fundamental human capacity, which nineteenth century observers called a moral *sense*, the child could not be socialized.
Jerome Kagan, *The Nature of the Child*[386]

(18) I am convinced, from [my] daily-life observations of infants and young children, and [also] from psychoanalytic and psychotherapeutic work with neurotic and psychotic adults, that lovingness is the basic stuff of human personality, [and] that it is with a wholehearted openness to loving relatedness that the newborn infant responds to the outside world, with an inevitable admixture of cruelty and destructiveness ensuing only later – being deposited on top of the basic bedrock of lovingness – as a

> result of hurtful and anxiety-arousing interpersonal experience.
>
> Harold Searles, *Collected Papers on Schizophrenia and Other Related Subjects*[387]

It is not clear to me, from what Kagan says, in text (17) above, whether or not he is taking moral sense and aggression to be equiprimordial. Nor is it clear what equiprimordiality would mean for him. I certainly do not want to deny the fact that there is much aggressive and destructive behavior in the world – nor, *a fortiori*, that there must be a basic capability, a biologically ingrained predisposition, to such forms of behavior. But I do want to deny that aggression in human beings is instinctual, i.e., spontaneous and unprovoked. I believe, with Searles, that human nature is more primordially oriented by a rudimentary moral sense; that, however, a certain amount of aggressive and destructive behavior is an *inevitable* reaction-formation, given the fact that 'hurtful and anxiety-arousing interpersonal experience' cannot be avoided altogether in the normal course of social life; and that it is on the presence of this universal 'moral sense', however, rudimentary it may be, that we must rely in order to socialize aggressive and destructive behavior when and as it appears. Thus I am also disturbed by what Bourdieu says, in text (16) above, for his language is exceedingly reductionistic: 'the ultimate values', he writes, 'are never anything other than the primary, primitive dispositions of the body'. This makes moral comportment essentially automatic. It overlooks the fact that the organismic values, the process-values inherent in our bodily dispositions, will not develop on their own: they need to be cultivated, further socialized, before they can ever become habitual, automatic, second nature. And it overlooks the fact that automatically moral behavior is still heterogeneous, and therefore falls short of the ideal.

Gendlin has shown how a certain way of working with the body of felt experience can bring forth really new understanding, specifically appropriate understandings, of our lives and situations. He has shown that the (healthy) body is so finely attuned to its specific situation that it bears within itself a felt, preconceptual sense, a situationally specific and potentially individuating sense, of its *Befindlichkeit*: how one is faring, how things are going. If, for example, a woman should involve herself in the focusing process, she would first of all get in touch with some experience that constitutes her individuality: her own body's felt sense, perhaps, of what it is to be living as a woman in these times and what needs to change in order for this living to be distinctly better; and she would then attempt to let this gradually forming sense, at first unclear, give her the words, the concepts, to speak about her situation from an individually derived and authentically processed experience.

Now we shall go deeper into the nature of this body of experience. We shall attempt to understand how the focusing process Gendlin shows us is made possible by the deeper nature of the body. We shall see, I think, that if focusing can bring forth organismic values and motivations which are appropriate to the specific needs of our existential situations, that is because the body has already been inscribed with the intricate texture of this potentiality.

How shall we get to know this body of depths? Our access to its nature will be, first of all, through *The Survivor: An Anatomy of Life in the Death Camps*, a book in which the author, Terrence Des Pres, makes use of numerous testimonies and documents to support his argument that 'Something innate – let us think of it as a sort of biological gyroscope – keeps men and women steady in their humanness, despite the most inhuman pressure'.[388] Having researched the stories of the survivors, Des Pres noticed that they all referred, in one way or another,

> to something other and greater than the personal ego, a reservoir of strength and resource which in extremity becomes active and is felt as the deeper foundation of selfhood. This is as much as survivors can say of their experience, but in coming to this limit we touch upon a further implication – a view reached precisely at the limit of personal experience. Survivors act as if they were prepared for extremity; as if anterior to learning and acculturation there were a deeper knowledge, an elder wisdom, a substratum of vital information biologically instilled and biologically effective.[389]

This is important empirical evidence for a 'body of depths', a body deeper and more primordial than the body engaged by focusing, and from out of whose flesh our moral sense of political order is originally derived.

The so-called 'state of nature', which Hobbes defined as 'that condition which is called war, and such a war as is of every man against every man',[390] is not, according to Des Pres, at all 'natural':

> A war of all against all must be imposed by force, and no sooner has it started than those who suffer it begin, spontaneously and without plan, to transcend it.[391]

Much of this book is concerned with the documentation of this thesis. What the survivors reported enabled him to see the deeper history of the death camps: what they manifested of human nature and the more concealed basis of political order. At first:

> Civility disintegrates and disorder prevails. Then slowly, in sorrow and a realism never before faced up to, the mass of flailing people grow quiet and neighbourly, and in the end rest almost peaceful in primitive communion. . . . Order [eventually] emerges, people turn to one another in 'neighbourly help'. This pattern was everywhere apparent in the world of the camps. Giving and receiving were perpetual, and we can only imagine the intensity of such transactions.[392]

According to Des Pres, who took a very keen interest in the question of underground resistance within the camps,

> through time, most survivors developed a degree of political consciousness, an awareness of the common predicament and of the need to act collectively. But what came first was spontaneous involvement in each other's lives on the immediate level of giving and receiving. And like the need to bear witness, which might also be viewed rationally, there was yet an instinctive depth to the emergence of social order through help and sharing. . . . Judging from the experience of survivors, 'gift morality' and a will to communion are constitutive elements of humanness. In extremity, behaviour of this kind emerges without plan or instruction, simply as the means to life.[393]

The next step in the argument is very important. Des Pres argues that,

> with the fall from civilization into extremity, with that fierce descent to nakedness, significant changes should have appeared in the behaviour of survivors – *if*, that is, the basic components of humanness are [exclusively] culturally determined and [exclusively] culturally upheld. But apart from the period of initial collapse, no dramatic change took place. The elementary forms of social being remained active; dignity and care did not disappear. These facts argue an agency stronger than will or conscious decision, stronger even than the kind of practical intelligence which made the need for moral order and collective action obvious.[394]

The point is, the camps did nothing to reinforce the conditions necessary for co-operative social behavior; on the contrary, they existed for the sole purpose of dehumanizing and destroying their inmates. To be sure, it could be argued that the survivors knew that, to survive, they needed each other: knew that they needed to co-operate with one another and maintain at least a minimum level of reliable reciprocity. But utilitarian rationality cannot account for what Jorge Semprun, a survivor, described as the 'hidden structure of available wills'.[395] As Des Pres says,

> The depth and durability of man's social nature may be gauged by the fact that conditions in the concentration camps were designed to turn prisoners against each other; but that in a multitude of ways, men and women persisted in social acts. Fear and privation increased irritability but did not keep inmates from joining in common cause.[396]

Something other than a calculative rationality, something more powerful than volition, kept inmate behavior as close as camp conditions would permit to the essential ground-rules of moral, social and political order. How shall we explain this?

In his study *On Aggression*, the result of years of research into the behaviour of primates, Konrad Lorenz defends his conclusion that,

> If it were not for a rich endowment of social instincts, man could never have risen above the animal world. All specifically human faculties, the power of speech, cultural tradition, moral responsibility, could have evolved only in a being which, before the very dawn of conceptual thinking, already lived in well-organized communities.[397]

Taking sides with Lorenz against thinkers such as Hobbes, Rousseau and Freud, who were apparently convinced that only force and fear could socialize an inherently selfish and aggressive 'human nature', Des Pres argues that the evidence at hand – not only survivor testimony but also other sources of documentation – does not support the view that a 'state of nature' prevailed in the death camps, 'or that a war of all against all necessarily erupts as soon as constraints are removed'.[398] On the contrary, it seems, rather, that

> primary aspects of the camp experience – group formation, 'organizing', sharing and the giving of gifts – are evidence amounting to proof that in man social instincts operate with the authority and momentum of life itself, and never more forcefully than when survival is the issue.[399]

The survivors survived because camp conditions *failed to cut them off* from the normativity, the inherent sociality of the 'body of depths'. The camps could more easily extinguish the *life* of the 'normal' body than it could cut off the identity of those who survived from the basic sanity, the *archē tēs kinēsēos*, of the 'body of depths'.

When Habermas takes up the question of moral development in his book on *Communication and the Evolution of Society*, he looks, as we should expect, into the conditions of society. 'How', he asks, 'do the . . . basic institutions of a society interfere with an ontogenetic developmental pattern?'[400] But this question raises, in turn, another, which he defines in terms of 'ego identity'. Thus he ends up reflecting on how the organization of a society can be destructive, inimical, to the identity, and that equally means the moral life, of the ego. The death camps are, of course, our most extreme example. What Habermas says on this point is worth noting:

> as long as the ego is cut off from its internal nature and disavows the dependency on needs that still await suitable interpretations, freedom, no matter how much it is guided by principles, remains in truth unfree in relation to existing systems of norms.[401]

I think this analysis is true. I should add, however, that this being 'cut off' is today a very widespread epidemiological condition: we might say that it is a psychopathology from which very few of us, if any, can escape. Now, I do not want to underestimate the effectiveness of our social practices and institutions in maintaining this condition. But I am convinced that the project to which this book is devoted constitutes a task which, to the extent that it can achieve its purpose, may contribute in some way to an emancipatory reconnection with our body's 'internal nature'.

I see this reconnection as an historical task which requires, among other things, the integration of a deeper sense of embodiment. Habermas speaks of needs we are disavowing. I submit that our vision *needs* to be rooted more deeply: needs to be rooted in the visionary being of the 'body of depths'. Our progress towards freedom depends on the recognition of this need. And as long as we disavow this need, we shall remain in a state of bondage in relation to the prevailing systems of norms. I am convinced that the 'body of depths' is a body whose radical nature holds within it an inherent sociality which is *different* from the alienated sociality that presently holds sway, cut off from (the needs of) our deeper human nature. If this conviction continues to be supported by empirical research, then, to the extent that our practices, institutions and social theory can be grounded in the visionary life of this deeper sense of the body, we may find new resources for questioning and changing our prevailing norms and standards. This way of seeing 'human nature' also has implications for education, because, if pro-social behavioural tendencies are already present in the biological programming of the child, then education can indeed become a process of *alētheuein*, of bringing out and educing, whereas, if human beings are innately brutish, aggressive and antisocial, then education must essentially involve a process of imposing restraints and inhibitions.

In the remaining pages of this chapter, we will turn to the late work of Merleau-Ponty and consider how his hermeneutical phenomenology of the flesh could provide a radical grounding for critical social theory. I will argue that his concept of a transhistorical 'flesh' enables us to bring out the deeply concealed potential, always carried by our body of experience, for a radical critique of the existing body politic. I will also argue that this concept brings to light the essentially social nature of our bodily being at the deepest level of our 'immediately' felt experience. Lorenz and Des Pres speak in objective, biological terms of a 'social instinct'. I want to use Merleau-Ponty's concept of 'flesh' to articulate, in terms of an experientially accessible dimension of our bodily being that is deeper than the dimension where Gendlin's 'focusing' process takes place, an *experiential equivalent* of this 'social instinct'.

'Flesh' is a phenomenological concept; it is not biological. It is to be realized 'directly', through reflective experience. And as we use it hermeneutically, to open up our experience of being embodied, I think we shall be able to flesh out a phenomenological understanding of 'human nature' which is suggested, but very darkly, by the biological concept of 'social instinct'. This should make it possible for us to see in the flesh a very radical schematism for the future of our body politic.

But what is this – the 'flesh'? In 'The Intertwining – The Chiasm', Merleau-Ponty explains it as follows:

> The flesh is not matter, is not mind, is not substance. To designate it, we should need the old term 'element', in the sense it was used to speak of water, air, earth, and fire, that is, in the sense of a *general thing*. . . . The flesh in this sense is an 'element' of Being.[402]

(Compare this with Hegel's formulation of the concept of 'matter', in *The Phenomenology of Mind*.[403] According to Hegel, 'matter' is 'neither a body nor a property of a body'. 'Matter . . . is not a thing that exists; it is being in the sense of universal being, or being in the way the concept is being'.)

Now, in a statement the importance of which Merleau-Ponty did not live long enough to think out, he maintains that 'We will have to recognize an ideality that is not alien to the flesh, that gives it its axes, its depth, its dimensions.'[404] It is this 'ideality' which we shall be looking into here, because I want to show how the flesh schematizes the structure of an ideal political body.

The concept of 'flesh' lets us see that the body is not inherently disorganized, turbulent, chaotic; it is not even anarchic, though its *archē*, its principle of order, is primordial, radical, inevitably subversive, and inevitably opposed to any order, any *archē*, which lays claim to exclusive and totalitarian authority. Relative to all orders which claim such authority, the flesh is unremittingly anarchic, because it is nothing if not the incessant interplay between sameness and otherness, identity, or conformity, and difference. The flesh, as we shall see, is a unique kind of order, organically, organismically organized. It *is* an order, but its order does not come from 'the outside', from society; its order is not imposed. Nevertheless, the inherent organization of the flesh is rudimentary, undeveloped; it is the gift of a potential which needs to be developed – developed, of course, through the care-taking of society. There is no guarantee, however, that, in the process of development, societies will take care not to impose on the body of experience meanings somehow felt to be inimical to the fulfillment of its deepest ideality.

For Merleau-Ponty, 'flesh' designates a dimension of our embodied being in which all individual lives are inseparably intertwined. The flesh consists in processes of interaction manifesting various degrees of symbiotic interpenetration and interdependency: an essential sociality or gathering (*legein*) already present at the inception of all individuated existence. In what Merleau-Ponty describes as the 'intercorporeality' of the flesh, individual beings continuously mirror and reflect one another, setting in motion a process of reversibility in which sameness and difference turn into one another: the seer is seen; the seer sees herself *being* seen; the seer sees herself *as* seeing, as a seeing being. And the same is true for the other, the one whom she sees. It is in the medium of this reversibility, this interplay, that rudimentary forms of reciprocity first emerge.

In 'The Child's Relations with Others', Merleau-Ponty makes visible a mode of interaction among 'subjectivities' which operates, by grace of a universal embodiment, 'beneath' cognition, 'beneath' an epistemological relationship.[405] Taking his thinking 'down' into the lived body, he brings back, 'up' into the light, a 'pre-personal existence', a dimension of our being through which we participate in 'an anonymous collectivity, an undifferentiated group life'.[406] There is an unconscious, involuntary 'postural impregnation' which takes place in our bodily presence to one another.[407] In her review of the sociological literature, Nancy Henley points out, for example, that

> Participants in an interaction often show synchronized nonverbal behavior with such postures as legs crossed in the same manner, similar standing positions, hand to chin or hip, or arms similarly folded (and maybe mirror-image style). Several researchers have studied this interactional synchrony, and have noted that even fast-passing motions of which most of us are unaware exhibit this synchrony. Persons in agreement will show high synchrony, and those in disagreement, dissynchrony. The effect of power may be seen in a group by considering whose motions the other interactants are in harmony with – it is the high-power person whose behaviors will be unconsciously reflected by others.[408]

To be sure, hierarchical power-relations and deep-seated prejudices are often unconscious; but the synchrony to which Merleau-Ponty is pointing operates at an even deeper level of social interaction. What his concept of 'flesh' lets us see is a *more fundamental dimension* of our bodily being: a dimension which is always already organized, already oriented, towards social synchrony and attunement precisely because its structure *precedes* the process of socialization which takes place in the years following earliest infancy. According to Merleau-Ponty, there is always an organismic bonding among infants, and even among children: an 'initial sympathy'. It is by this bonding that we always find ourselves already governed.[409]

This natural attunement is a transhistorical and pre-civil synchrony, and its primordiality makes all the difference in the world. But we need to *see* that it will, in fact, be decisive in making possible the kind of political consensus whose rational groundrules thinkers like Rawls and Habermas have been attempting to formulate in a consensus-making way. In 'Moral Development and Ego Identity', Habermas asserts that 'competent agents will – independently of accidental commonalities of social origin, tradition, basic attitude, and so on – be in agreement about such a fundamental point of view only if it arises from the very structures of possible interaction. The reciprocity between acting subjects is such a point of view'.[410] I want to argue that such structures of interaction do exist; that in rudimentary form they are already inscribed and preprogrammed into our flesh; and that, although they certainly cannot guarantee social harmony, the primordial inscription of these structures does make interactions of reciprocity possible – and even, in the absence of countervailing conditions, more likely than not. Furthermore, I want to argue that the 'corporeal schema' of which Merleau-Ponty speaks in *The Phenomenology of Perception* and 'The Child's Relations with Others' operates at a level much deeper than even he was disposed to think, for it is constitutive of the flesh. Thus I suggest that what Merleau-Ponty disclosed through the concept of 'flesh' outlines with remarkable clarity the already socially attuned character of this concealed schematism.

Earlier in this chapter, it was shown how, in beholding, we are ourselves beheld, held in a beholdenness that calls for a response. Radical reflection, a rigorous application of hermeneutical phenomenology, brings to light an interpenetration of glances, of gazes, of beings whose visionary capacity is inseparable from their visibility to others. It brings to light a seeing which is always, regardless of conscious desire and the effects of socialization, touched and moved, even at a distance, by what is seen. It brings to light an intertwining of gazes: interdependencies in every interaction. It brings to light the phenomenon, the fact, of reversibility: the seer is seen and sees himself as seen, seen through what he sees. The seer can feel his seeing as it is felt, or received, by the other, the one whom he sees. The seer and the other as seen belong to the same flesh. Contrary to Sartre's claims in *Being and Nothingness*, two seers, seeing one another, cannot avoid an involuntary, organismic acknowledgement of their primordial kinship. Reciprocity at the level of moral choice, moral autonomy, is possible because of the fact that, as soon as there are two visionary beings within the same field of visibility, there is *already* a primordial, inaugurative layout (*legein*) of reversibilities: a rudimentary structure of interactions on the basis of which more developed forms of reciprocity become impossible.

Reversibility is inherent in the very structure of visual

perception. But, as Kant understood, reversibility is a necessary precondition for the possibility of genuine reciprocity.[411] Now we can see that reciprocity is an ideal which appears, not only as the end, but also as the beginning, of our moral and political life. Thus it may be said that the universalism of the generalized other is already schematized – schematized concretely, in every real encounter – by the 'intercorporeality' of our embodiment: schematized, in Merleau-Ponty's words, cited earlier, as 'an ideal community of embodied subjects'.

Of course, genuine universality, i.e., a society in which the ideal of reciprocity is fully incorporated, can never be more than an ideal to which we approximate.[412] Nevertheless, widespread recognition of the fact that this universalism is already schematized in the body ought to make a significant difference. And this is why I argue that we need to see and feel this schematism. The more we see and feel it, the more, that is, we can sense and recognize it, the more we can develop its potential, drawing on our bodily felt sense of its' image' for the building of a new body politic: a body politic more responsive to the deepest needs of the individual body as a body of depths, a body of flesh.

Now, in her paper on 'The Generalized and the Concrete Other', Seyla Benhabib proposed a 'feminist critique' of what she calls 'substitutionalist universalism'. I want to consider it here because it bears directly on the problem we have been addressing. The question we need to think about is this: How does our retrieval of a reversibility and reciprocity schematized in the depths of the flesh help us to break the hold of our logocentric moral tradition? The critique Benhabib formulates may suggest an answer:

> In this tradition, the moral self is viewed as a *disembedded* and *disembodied* being. This conception of the self reflects aspects of male experience; the 'relevant other' in this theory is never the sister but always the brother. This vision of the self, I want to claim, is incompatible with the very criteria of reversibility and universalizability advocated by defenders of universalism. A universalistic moral theory restricted to the standpoint of the 'generalized other' falls into epistemic incoherencies which jeopardize its claim adequately to fulfill [the goals of] reversibility and universalizability.[413]

Benhabib contends that 'universalistic moral theories in the Western tradition from Hobbes to Rawls are substitutionalist'.[414] in the sense that they make *their own experiences*, which are really only the experiences of some specific subject (or group of subjects), into their paradigm of the human as such, and interpret the experiences of *all other* subjects (or groups of subjects) in terms of this paradigm. In other words, they test for universalizability by a method of substitution in which one category of experience (say, that of men) is taken as paradigmatic and substituted for other categories. In effect, then, these theories achieve 'universality' only by domination, suppressing the differentness of the other. Their 'universality' requires that we abstract what is common from all individual differences, and that we teach ourselves not to be responsive to the concrete identities these differences constitute.

Instead of basing the ideal of moral and political life on the perspective of the disembedded and disembodied generalized other, an approach which, despite its high aspirations, unknowingly imposes its own paradigm of the moral self on those who are different and tends to see moral dignity only in a human nature which appears in conformity to this paradigm, Benhabib proposes a theory she calls 'interactive universalism':

> Interactive universalism acknowledges the plurality of modes of being human, and differences among humans, without endorsing all these pluralities and differences as morally and politically valid. . . . [It] regards difference as a starting point for reflection and action. In this sense, 'universality' is a regulative ideal that does not deny our embodied and embedded identity, but aims at developing moral attitudes and encouraging political transformations that can yield a point of view acceptable to all. Universality is not an ideal consensus of fictitiously defined selves, but the concrete process in politics and morals of the struggle of concrete, embodied selves, striving for autonomy.[415]

Her interactive universalism does not deny the need for a formal consideration of the *generalized* other in abstraction from all particularities; but it insists that we must acknowledge every *generalized* other as also a *concrete* other: as an embodied being: a being of needs and feelings; a situated, historical being; and a being with whom we must interact. This acknowledgement makes a decisive difference, because it understands that universality must not be imposed, as it is when we generalize from our own needs, desires, and affects, and because it sees universality as an ideal *to be achieved* through interaction, through communication. Moreover, whereas the test for universality in substitutionalist theories is strictly cognitive, the interaction theory sees the test in a moral consensus achieved through a process of communication in which embodied feelings are respected and given consideration.

I appreciate Benhabib's argument for a universalism grounded in experiences with the concrete other, and I ask accordingly whether the social and political theory I am eliciting from Merleau-Ponty's conception of the flesh continues the traditional representation of universality as a question of reversibility by substitution, or whether the concept of 'flesh' contributes a strong grounding in deeply thought-through experiences for the kind of theory she is advocating. It should be evident by now, I hope, that I want to affirm the second possibility.

In fact, the major reason that I find Merleau-Ponty's conception of the flesh so appealing is that it encourages us to think in terms of a moral and political theory of the kind for which she is arguing. When the Kantian test of universalization is grounded in the nature of the flesh, it is made concrete and situational; its traditional formalism and abstractness are overcome. And since it stays close to lived experience, it is deeply felt: it is not purely cognitive, purely epistemic. We begin with the unquestionable fact of a lived acknowledgement between myself and others. Clearly, the reversibility in question, here, is inherently interactional; it does not take place by theoretical, and merely contemplative substitutions. The substitutions are not based on theoretical reversibility, but on actual, concretely lived interactions. Universality is therefore not a paradigm introduced by thought, but that *archē* which appears through the communions of the flesh. The ideal does not originate in a pure monadological Reason, but in the intercorporeality of the body's rudimentary schematism. The sense of universality comes from a depth which is beyond the metaphysical dualisms of inner and outer, self and other: it comes from a depth 'within' ourselves where our identity is always already intertwined with the identities of all others. Consequently, it does not require any formal generalization to the other by a method of substitution. But of course, the most important point of all is simply the fact that the concept of 'flesh' focuses moral and political theory on the experiencing life of the embodied, concretely situated self.

Adorno partially anticipated the critique Benhabib is proposing when he argued, in *Minima Moralia*, that 'An emancipated society . . . would not be a unitary state, but the realization of universality in reconciliation of differences'.[416] He repudiated the idea of the 'abstract equality of men', warning that we need to achieve the ideal of 'universal reason' in such a way that different people 'could be different without fear'.[417] In 'Poetically Man Dwells', Heidegger defined this problem very sharply. He declared:

> We can only say 'the same' [in contrast to the 'equal' or the 'identical'] if we think difference. It is in the carrying out and settling of differences that the gathering nature of sameness comes to light. . . . The same gathers what is different into an original [and therefore opening] being-at-one. The equal, on the contrary, disperses them into the dull unity of mere uniformity.[418]

We shall return to this passage in the fourth chapter, when we take up the relation between truth and power. In the context of our present concerns, what makes this passage useful is its language of dynamism: sameness and difference are understood dynamically, as they should be, as they require. And this makes it easier for us to connect the political problem of sameness and difference with the nature of the flesh in the original 'body politic'. Critical social theory needs to be grounded in a hermeneutical phenomenology of the flesh, for it is in the radicality of the flesh, first of all, that an emancipatory interplay between sameness and difference is schematized. There can be no grounding more radical than the intertwining of visionary subjects in the chiasmic flesh.

Our political tradition, now manifesting ever more visibly a nihilism already at work in its modern hegemony, has not let us see the body politic radically schematized in the very flesh of our being. It has cast out the body of experience and turned us away from a very powerful, and in truth essential 'touchstone' for the articulation of concrete political needs, legitimate political consensus, realistic political dreams, and accurate political perceptions; and it has denied us a body from out of whose transhistorical nature we could have made more intelligent demands on history.

Today, when there is a crying need for new political visions, new visions of the body politic, I see a need to retrieve the political wisdom already schematized in the intertwining flesh of our visionary bodies. The wisdom in our bodies is the truth of intercorporeality, made visible, if we are at all attentive, through the phenomenon of reversibility. My conviction is that, if each individual were to make some contact with this latent wisdom, and were to allow that wisdom to schematize appropriate political comportment, new opportunities for historical initiatives would increasingly become apparent.

Agonizing over the historical significance of the Nazi death camps, Des Pres touches on a sense of our present historical situation, and on a vision of the future, which is very close to my own, and I should like, therefore, to let him speak. The time has come, he says, to recognize the narcissism in modernity: to see that 'Man's kinship with the gods is over'.[419] The time has come, he thinks, to see that 'Our Promethean moment was a moment only, and in the wreckage of its aftermath a world far humbler, far less grand and self-assured, begins to emerge.'[420]

In a sense, all of us who are alive today are 'survivors'. Like the inmates who survived those camps, we too are living in their shadow, haunted by their ghostly presence. We are compelled by the still incomprehensible facticity of the camps to interrogate the very foundations of our civilization. Freud tried to escape the task of diagnosis, though in the end, he could hardly avoid shaking the foundations of what we call 'our civilization'. But the task he avoided, we must, I'm afraid, accept. For this reason, I like the direction of the thoughts Des Pres records:

> We look into the darkness of our own hearts and behold every kind of destructiveness – the whole demonic surge of savagery erupting daily in war, torture, genocide, or spilling out imaginatively in much of modern art and literature. What we have not seen is that our rage stems from nihilism, and

> that nihilism is the outcome of allegiance to a mind-body split which makes hateful the body and its functions, and storms in spiteful execration against the whole of existence as soon as life is no longer justified by firm belief in 'higher' values. And we have altogether missed the fact that beyond our lust for disaster there is another, far deeper stratum of the human psyche, one that is life-affirming and life-sustaining.[421]

I have tried to show that Merleau-Ponty has something important to say about that deeper stratum where we are already in touch with the 'values' of Being. But I have also found it necessary to say what he himself left for the most part unthought: that in the universal, transhistorical nature, the 'wild being', of this deeper, more ontologically oriented stratum, there is a utopian image of the body politic, a schematism of political life, already functioning, and that there is a way to retrieve its radical wisdom – the process of reversibility – for the future development of our visionary being.

Modernity, our present historical situation, is a time of crisis and danger. But it is also a time, because of this, in which, if we are to survive, we must somehow find within ourselves those life-affirming resources and capabilities with which we never thought we were endowed, and which we can now leave split off from present living only at a terrible price.

Nihilism rages; but it is *life-affirming* to realize that, as Merleau-Ponty has demonstrated by phenomenological analysis, 'It is characteristic of cultural gestures to awaken in others at least an echo, if not a consonance.'[422] For this fact, which I regard not as the end of the matter, but rather as only the beginning, *opens up* for us the possibility that a new political order, a new body politic, could perhaps be achieved on the basis of a consensus allowed to emerge from the depths of our visionary bodies: a consensus not imposed by alien powers, but allowed to happen in accordance with the intercorporeality of the body's own processes, the body's own felt sense, very concrete, of what the intertwining of beings calls for . . . and what needs, in the time of each particular circumstance, to take place.

Foucault, in the passage quoted at the beginning of this section, argued that we need to 'imagine and bring into being new schemas of politicization'. Taking to heart the words of Herakleitos, saying that 'the origin of movement is also the origin of ruling and directing', I hope to have shown that, in the body of depths, in the flesh of our visionary being, the call for new schemas, a crying need, may find an affirmative response.

In Plato's *Meno*, Sokrates declares that 'he who would answer the question "what is virtue?" would do well to have his gaze steadied'.[423] Such steadiness, the neutralization of inveterate dualistic tendencies that are metaphysical in character, comes from the rootedness of vision in the nondual dimension of our being. Since crying is the rooting of vision, we have in this chapter given thought to the root. If we have opened ourselves to this experience with thought, then we will have learned what can only be learned by crying for a vision. From out of this experience, it is perhaps possible to develop a felt sense of the body politic schematized in the intertwinings and reversibilities of the flesh, the body of depths. And if we let that sense determine the project that informs our vision, perhaps we can integrate our felt sense of the schematism into the discursive structures of everyday living. Today, we deeply need the vision this body holds within it.

CHAPTER 3

Lightning: The Transformative Moment of Insight

Opening Conversation

(1) In teaching philosophy in the Gymnasium, the abstract form is, in the first instance, straightaway the chief concern. The young must first die to sight and hearing, must be torn away from concrete representations, must be withdrawn into the night of the soul and so learn to see on this new [philosophical] level.

Hegel (letter dated October 23, 1812)[1]

(2) What do we mean by *world* when we speak of a darkening of the world? World is always world of the spirit. The animal has no world nor any environment. Darkening of the world means emasculation of the spirit, the disintegration, wasting away, repression, and misinterpretation of the spirit.

Heidegger, *Introduction to Metaphysics*[2]

(3) What if this present were the world's last night?

John Donne, *Devotions upon Emergent Occasions*[3]

(4) . . . Lightning steers beings as a whole.

Herakleitos[4]

(5) In this truth, therefore, the deed comes to the light . . .

Hegel, *The Phenomenology of Mind*[5]

(6) In a dark time, the eye begins to see.

Theodore Roethke, 'In a Dark Time'[6]

(7) . . . my glance . . . like lightning . . .

Rainer Maria Rilke, 'The Words of the Lord to John on Patmos'[7]

(8) [A divine presence hides] in lightning, thunder, storm and showers of rain.
Heidegger, 'Hölderlins Erde und Himmel'[8]

(9) [A]lthough the thunderclouds veil the heavens, they belong to it and show the joy of God.
Heidegger, 'Hölderlins Erde und Himmel'[9]

(10) This is the teaching of Brahman, with regard to the gods: It is that which now flashes forth suddenly in the lightning, and now vanishes again.
Talavakara Upanishad[10]

(11) Pure Light scatters its simplicity as an offering to self-existence, that the individual may take sustainment to itself from its substance.
Hegel, *The Phenomenology of Mind*[11]

(12) The longing for light is the longing for consciousness [i.e., individuationl].
Carl G. Jung, *Memories, Dreams, Reflections*[12]

(13) Understanding constitutes rather the being of the 'there' in such a way that, on the basis of such understanding, a Dasein can, in existing, develop the different possibilities of sight, of looking around [*Sichumsehens*], and of just looking.
Heidegger, *Being and Time*[13]

(14) Yet when a self stands in the gateway where past and future 'affront one another' and 'collide', existence ceases to be a spectator sport. In the 'flash of an eye' the thinker must look both fore and aft, 'turned in two ways', and must study the internecine strife of time. 'Whoever stands in the Moment [*Augenblick*, time measured by a glance or a wink of the eye] lets what runs counter to itself come to collision, though not to a standstill, by cultivating and sustaining the strife between what is assigned him as a task and what has been given him as his endowment.' It is the effrontery of time that in

it we collide against mortality and strive with it, closing in the glance of an eye and not in some remote infinity.
David Farrell Krell, 'Analysis' of Heidegger's *Nietzsche*, vol. II: *The Eternal Recurrence of the Same*[14]

(15) The unbearable white radiance of primordial light is broken up by the prism of consciousness into a multicolored rainbow of images and symbols.
Erich Neumann, *The Origins and History of Consciousness*[15]

(16) The uroboric goddess of the beginning is the great Goddess of the Night, although she is seldom worshipped directly as such.
Neumann, *The Great Mother: An Analysis of the Archetype*[16]

(17) Night sky, earth, underworld, and primordial ocean are correlated with this feminine principle [of wisdom], which originally appears as dark and darkly embracing.
Neumann, *The Great Mother*[17]

(18) In contrast to the feminine mysteries, the transformation mysteries of the Archetypal Masculine have the character of a surprise attack, and sudden irruptions are the decisive factor. Consequently, lightning is the characteristic symbol for them.
Ibid.[18]

(19) The [ego-logical] *subject* is conceived as *a fixed point of arrest* in cognition, as *unchangeable*, existing once and for all time, and thus all movement is mistakenly accorded only to the *object*. Should then contradictions become evident in the course of the cognitive process, because the subject itself is *entwined* in that process, is itself a moment in the movement, is itself also *moved*, then all panic breaks out.
Theodor Adorno, *Über Mannheims Wissenssoziologie*[19]

(20) . . . the Glance takes on the role of saboteur,

trickster, for the Glance is . . . repressed, and as it is repressed, is also constructed as the hidden term on whose disavowal the whole system depends. The flickering, ungovernable mobility of the Glance [in contrast to the rational, absolutely fixed Gaze, driven to see things as either ready-to-hand or present-at-hand] strikes at the very roots of rationalism, for what it can never apprehend is the geometric order which is rationalism's true ensign: . . . unable to participate in the unitary mysteries of reason, the Glance is relegated to the category of the profane, of that which is outside the temple. Before the geometric order [of the imposing Gaze] the Glance finds itself marginalized and declared legally absent, for . . . all it knows is dispersal – the disjointed rhythm of the retinal field. . . . Against the [well-disciplined] Gaze, the Glance proposes [the play of] desire, proposes the body, in the duree of its practical activity: in the freezing [and reifying] of syntagmatic motion, desire, and the body, the desire of the body, are exactly the terms which the tradition seeks to suppress.

Norman Bryson, *Vision and Painting: The Logic of the Gaze*[20]

(21) But in the glow of the newly kindled light, with which she [Psyche] illumines the unconscious darkness of her previous existence, she recognized Eros. She *loves*.

Neumann, *Amor and Psyche*[21]

(22) Then cometh the god, . . . and by a flash of his eye burns up the veil which shrouded all things, and the meaning of the very furniture, of cup and saucer, of chair and clock and tester, is manifest.

Emerson, 'Circles'[22]

(23) . . . something that passes quickly by, glimpses, brief flashing revelations that last a second in you under the influence of some occurrence; all the unimportant that often becomes significance through a passing intensity of our vision or because it happens at a place where it is perfected in all its fortuity and perpetually valid and of deep import for

some personal insight which, appearing in us at the same moment, coincides meaningfully with that image. Gazing is such a wonderful thing, of which we still know so little; with it we are turned completely outward, but just when we are most so, things seem to be going on in us *without us* . . .
Rilke (letter to Clara Rilke, Capri, March 8, 1907)[23]

(24) ['Illumination'] may not be just a metaphor. [It may be correlated with] an actual sensory [i.e., neurological] experience occurring when, in the cognitive act of unification, a liberation of [nervous electrical] energy takes place, or when a resolution of unconscious conflict occurs, permitting the experience of 'peace', 'presence', and the like. Liberated [charges of physical] energy *experienced as light* may be the core sensory experience of mysticism.
Arthur Deikman, 'Deautomatization and the Mystic Experience'[24]

(25) A small patch of light, falling within the central excitatory region of a receptive field, will cause brisk firing of a retinal cell.
Lloyd Kaufman, *Perception: The World Transformed*[25]

Part I

The Night of the Soul: Time for Dreambody Vision

(1) Becoming yourself can be understood as knowing your dreambody, becoming whole or round, developing your full experience through awareness of each of your different channels [of perceptiveness].
Arnold Mindell, *Working with the Dreaming Body*[26]

(2) Archetypally, luminous bodies are always symbols of consciousness, of the spiritual side of the human psyche.
Neumann, *The Great Mother*[27]

(3) The night sky and the daytime sky were

apprehended earlier than the heavenly bodies, because the whole was seen as a unitary being and the religious awareness of the heavenly bodies was often submerged in fusion with a vision of the sky as a whole.

Konrad Preuss, *Die geistige Kulture der Naturvölker*[28]

(4) The Great Round embraces in itself light and dark, day and night, but priority is given to the night. . . . Throughout the world, lunar mythology [Feminine Archetypes] seems to have preceded solar mythology [Masculine Archetypes]. But we also know that in the human psyche, the experience of totality [in the sense of wholeness] always precedes the experience of particulars.

Neumann, *The Great Mother*[29]

(5) The plane of images and symbols is closer to the unconscious than it is to the plane of consciousness, but often the process that takes place on the symbolic plane has an anticipatory character and makes possible the later conscious process.

Neumann, *The Great Mother*[30]

(6) The mystery of transformation, in which the 'spirit' comes into being, is . . . a product of the Great Round; it is its luminous essence, its fruit, its son. . . [T]he favored spiritual symbol of the matriarchal sphere [of influence] is the moon in its relation to the night and the Great Mother of the night sky.

Neumann, *The Great Mother*[31]

(7) . . . the night is the Great Round, a unity of underworld, night sea, and night sky encompassing all living things.

Neumann, *The Great Mother*[32]

(8) The night sky is a reflection of the earth. . .

Neumann, *The Great Mother*[33]

(9) After all that has been said, it is not hard to see that space [the Open] is one of the most important projections of the Feminine as a totality. . . But the Feminine is also the goddess of time, and thus of

fate. The symbol in which space and time are achetypally connected is the starry firmament, which since the primordial era has been filled with human projections.

Neumann, *The Great Mother*[34]

(10) The dependency of all the luminous bodies . . . on the Great Mother, their rise and fall, their birth and death, their transformation and renewal, are among the most profound experiences of mankind.

Neumann, *The Great Mother*[35]

(11) . . . modern consciousness is threatening the existence of Western mankind, for the one-sidedness of masculine development has led to a hypertrophy of [rational] consciousness at the expense of the whole human being.

Neumann, *The Great Mother*[36]

(12) Supposing truth is a woman – what then? Are there not grounds for the suspicion that all philosophers, insofar as they were dogmatists, have been very inexpert about women? . . . What is certain is that she has not allowed herself to be won – and today every kind of dogmatism is left standing dispirited and discouraged.

Nietzsche, *Beyond Good and Evil*[37]

(13) We all recognize that dreaming is somehow different from waking. As soon as we begin to define that difference, we find ourselves groping in the dark.

Medard Boss, *Last Night I Dreamt*[38]

The teachings of wisdom we have taken to heart in thinking about our experience with crying brought us into creative contact with our historical need and bring us now to a new phase in the development of our capacity for vision. Further transformations in our visionary being – individuating processes of many different kinds – are possible but seem to require, sooner or later, that we be open to the radical and essentially subversive teachings of the night. For it is during the night, the night of the soul, that we open our eyes, the organs of our visionary being, to other dimensions of reality, of Being: dimensions we desperately need to remember and integrate into the prevailing, consensually validated reality of daily life, since everyday life in our present epoch is controlled by the ego-logical subject and its metaphysics of presence.

In this chapter, our concern is to understand some extremely important connections and to challenge and bestir ourselves – our ego-logical selves, that is – as we enter into a thoughtful relationship to the experiences they propose. The connections will appear, somewhat obscurely no doubt, as we unravel the various hints, the various threads provided by the texts we have just been reading, and collect them to spin our story, weaving the elements of our interpretation loosely together. Our task is a recollection, and in undertaking this task, we are gathered into a process of deep experiential change. This process is insight, vision 'outside' of the ego's control, vision rooted in the soul, the spirit, our deeper self, the 'inner light' of our visionary being. Insight betokens deep changes in the character of our visionary being. These changes take place 'within' the individual; yet they are inevitably amplified 'within' society and culture as a whole: they are, in fact, fraught with historical significance.

Throughout the first two chapters, our thinking about vision has depended on a fundamental metaphysical assumption, an *epistēmē* (as Foucault would say) which, in conformity to the long-standing historical domination of philosophy by the Masculine Principle and its 'organs' of amplification within individuals, society and culture, we have not at all, until this very moment, recognized and acknowledged. The assumption, which from this moment of vision on we should be radically calling into question, is that our vision always takes place in the clarity of daylight – in the constant, continuing, and utterly full presence, the *parousia*, of the lighting. We take for granted the constancy and omnipresence of that by grace of which we are enabled to see, and we assume conditions that ensure total visibility, total illumination, total clarity, and total distinctness.

Western metaphysics emerges from a worldly vision which takes the *gift* of daylight for granted and assumes, deeply unconscious of itself and its projections, the permanent presence – *parousia* – of our source of illumination: conditions of total unconcealment, making possible a vision of total lucidity in perfect possession of its (transparent) object. Western metaphysics reflects a worldly vision of truth which sees only sharp boundaries and divisions, the oppositions permanently fixed in duality: inner/outer, self/other, subject/object, reality/delusion, presence/absence. But this is a vision of truth which *occludes* our experience with shadows and shades (of meaning); the enchantment of the sunset hour, the uncanny lighting of the twilight, when the owls of Minerva, wise, like all the animals, in their own way of vision, are suddenly appearing in the trees, connected somehow with the spirit of the hour; and, most dangerously, it forgets the time when our vision is abruptly plunged into darkness, into a night which eventually closes our eyes and makes us heavy with its sleep, into a night of visionary dreaming, a night where our vision subsides into a different logic, the intertwining of opposites, a 'confusion' of identities and difference, a matrix of Being where it is temporarily released from its ego-logical and logocentric obsession with visibility.

It is not accidental, considering our patriarchal culture, that Western metaphysics cannot see truth as alētheia. In order to *see* truth as *alētheia*, we should have to see truth as unconcealment, and metaphysics would be compelled to consider the vision of the night, the vision attuned by darkness and concealment, the vision whose character is rooted in an awareness informed by authentic encounters with primordial darkness – and with the vision of beings, and of Being as a whole, which takes place through the organs of our dreaming unconscious. Western metaphysics has forgotten, has suppressed, this *other* vision, this vision without the presence, the *parousia*, of the light of the day: a vision which understands (the ontological significance of) the *absence* of light and is open to learning from the greatness – even the terror – of the night.

Western metaphysics traditionally sees things – sees beings and the very Being of beings – with a vision of the head, the disembodied mind, the pure intellect, reason and rationality: a vision for which the values of clarity and distinctness are supreme and exclusive; a vision of instrumentality, calculation, analysis, and power of domination; a vision constituted as an organ of the ego, the ego of the Cartesian *cogito*, the ego of adaptation and the logic of common reality: a vision which happens only during the day.

There is a wisdom in our experience with the night that we desperately need to learn: an experience with absence, with fusion and indistinctness, with ambiguity, shifting boundaries, elusive and transitory presences, insubstantial apparitions, concealments, a sense of wholeness and integration, encounters with the night which disturb our settled sense of reality and penetrate our culturally established egological defenses. Under the spell of the night, our vision goes *down* into the body: goes down within the individual, down into the collective unconscious, takes leave of the (consensually validated) world and returns for a while to the underworld, the realm of the dead, rejoins the experience of nature it shares with the animals, and enters into the world of the dreambody.[39] Under the spell of the night, vision takes root in the intertwining, the elemental, the flesh of Being, and practices ways of seeing it otherwise could not know: synchronicity, for example, and vision *sub specie aeternitatis*. . .

Or imagine, for example, this: your head is inexplicably turned towards a china cup on display in the cabinet, something you have lived with for many years and to which you have not given any conscious attention in almost as many years, and you suddenly notice, for the first time – so far as you can tell – and without any hovering or hestitation of the glance, a tiny gossamer crack, almost invisible from the distance at which you had been at the time of the perception, and certainly much too small to have, by itself, so strongly seized your attention in the normal way; and the crack brings before the mind, in an involuntary flash, a different crack, a crack of equally small dimension, one which happened many years ago to a lens of your binoculars – the ones your mother gave you years and years ago – in fact, as you suddenly realize with a jolt, on *this* very day of the year; in fact, on the very *same* day of the year, many years ago, when, on an anniversary day celebrating her wedding, she had passed on to you the old binoculars, which had belonged, once upon a time, to your great grandfather. This kind of seeing, this channeling of vision happens through the dreambody: it is a visionary capacity of the dreambodyself.

Marcel Proust, a denizen of the night, wrote with great insight into the nature of 'perception involontaire'. The recollection of his past through the medium of the little madelaine is an example of his exceptionally developed capacity for dreambodyvision.

Writing about 'The Question of the Self in Nietzsche

during the Axial Period (1882-1888),' Stanley Corngold notes that,

> if Nietzsche attacks a superficial, merely instrumental, 'herd' ego [a shallow self], . . . he affirms [in opposition] a deep, creative, authentic self or *being* which founds (and eludes) the ego – a being linked to 'the body' and to 'the soul'.[40]

In the mythology of many cultures, a mythology whose wisdom still haunts our own culture despite its apparent marginality, the night is a time when the influences of the Feminine Principle are strong. Orpheus and Dionysos, and Hermes, too, are figures of the night. The night is dark; it denies our eyes the possession of clarity; it compels us to see through the body. But the body's vision is not ego-logical; nor is it raylike, 'phallic' – as is our vision during the day. The body's vision, during the night, is differently mooded, is global, synthetic, round: it is a mooded vision attuned by a sense of wholeness; it is vision aware of the encompassing, vision entrusted to the matrix of Being. The vision of the night draws its strength, not from duality, but from the *integration* of subject and object, the *dissolution* of the ego's defensive boundaries, the *questionableness* of daytime certainties. The night questions our 'presence', our self-mastery, our self-possession: we lose our footing, we go astray; and if we depend on 'normal' channels of vision, we get lost, our eyes play tricks on us, we see things that 'are not there,' and we fail to see 'the things that are'. We encounter our own projections: we *could* go mad. In the night, it is the 'soul' which sees; and it sees through the body. (The 'self', says Nietzsche, 'is your body'.[41] And because of this understanding, he can argue, in *Beyond Good and Evil*,[42] that 'it is not at all necessary to get rid of "the soul" . . . and thus to renounce one of the most ancient and venerable hypotheses – But the way is open for new versions and refinements of the soul-hypothesis. . . .') In the night, our vision belongs to the soulbody, not the ego. In the night, our vision comes under the influence of the Feminine Archetypes, the mystery of the Feminine. It is no longer regulated by the fact of patriarchal culture and the ego's domination of the *archē* during the rule, the ascendancy, of the Masculine Principle. In the night, vision is connected to the visible by way of a dimension of Being more 'archaic' than what is admitted into the visionary consciousness of our social and cultural life during the rule of the Masculine Principle. It is vision moved by the invisible, vision born of the invisible, vision nurtured and instructed by the hermeneutical darkness of the night.

In the night, our vision belongs to the unconscious; it can no longer repress what the ego has denied in the process of self-mastery, where the ego *stands in* for the vision of social and cultural conformity: as controls are rendered ineffective or useless, we see projections, phantasms; we see externalizations of our irrepressible desires; we see things we could only see through eyes that are organs of the soul, organs of the flesh of Being.

In the night, our vision happens through a dreambody. Writing about the dreaming we experience when asleep, Medard Boss touches on an important question for us:

> The question now is, which body is the 'real' one: the body that others see lying in bed, though the dreamer is unaware of it, or the body that the dreamer himself feels so intensely but that no waking observer can perceive?[43]

I think we should take this question to heart. But, whereas Boss frames his question only in terms of the dreambody 'taken over' by sleep, I want to question our understanding of 'reality' in terms of the dreambody which is active even when we are 'awake' – and which is particularly influential during the night, when the channels for vision used by the ego can no longer be trusted. In the night, *other* channels are opened up and it is the 'soul' which sees: the soul, the dreambody, a visionary being in touch with its more 'feminine' wisdom, and with long-concealed needs of mortal life.

Our story is a story of transformations and individuation, death and rebirth, history and its 'overcoming'.

This part of our story concerns the thunderbolt: that which, in the words of Herakleitos, 'steers beings as a whole.' The violent thunderbolt penetrating the night is an archetypal image of masculine power; but it is from out of the (womb of the) night that the thunderbolt suddenly bursts forth; thus it is the *night* which gives birth to the flash of lightning. Now, the thunderbolt is symbolically connected with the flash of insight. Insight is an 'inner' vision which comes to us during the dark night of the soul. In the course of the night, when 'outer' things are taken into concealment, we learn that the light of consciousness – of the ego – is not self-originating, that it comes from the darkness of the unconscious, and that the 'masculine' form of vision, vision motivated by the will, the will to power as the will to master and dominate, vision bound to the ego's subjectivity – this form of vision desperately needs to get its bearings, reconnecting itself to the wisdom of the night.

The thunderbolt makes visible an energy which speaks to us in terms of the Masculine Archetype; but its relationship to the night lets us know that it inaugurates and celebrates a moment of change – a shift, a transition. It announces a difference; it even lights up the sky with the intensity of its pointed decisiveness; and as it amplifies the concealment of the night, it portends a vision in which the channels of awareness formed during the night in relation to darkness and concealment, the absence of light, vitally contribute to decisive changes in its character.

The darkness of the night is today an experience haunted by the death of God. (See Heidegger, 'Opening Conversation', no. 2.) Since the death of God is a story in our history, the darkness of the night is now heavy with this historical significance for us. The death of God plunges us into the night of the soul. If insight is to come, it will be an inwardly generated vision born of our encounter with death, with mortality, with the absence of light. In this story, the thunderbolt also indicates a time of healing vision, a healing *of* vision, the possibility of redemption, the turning point; our release, perhaps, from the present nihilism of our history.

As we shall see, the nature of the thunderbolt's visibility, its rending of the sky, its flashing, searing irruption, its bursting forth, its violent penetration of the night, can be understood as pointing to a moment in nature which corresponds, or is, rather, attuned, to a decisive moment of transition in our visionary being, and to what is, for Heidegger in particular, a decisive moment in our history: a moment which makes all the difference in the world, because it concerns the visibility of the ontological difference between beings and the Being of beings – visible beings and the ground of their visibility – in our historical time. In the moment of the thunderbolt, the lightning flash of insight, our character is being tested. We are turned inward, turned to question ourselves: as beings belonging to the invisible surround; and as beings who are gifted, because of this connection, with the wondrous capacity to illuminate beings and shed light on the Being of all these beings as a whole. We are turned to reflect on our *capacity* to make visible.

Lightning is a rending of the night, an irruption, a rupture. As such, it points to the most primordial of all scissions: the ontological difference itself, as a decisive event (*Ereignis*) in the visionary field of Being. In the field of our vision, the ontological difference presences in – and as – a primordial differentiation between figure and ground, beings and ground, awareness and field, beings and Being, the visible and the not-visible. Lightning gathers into itself the meaning, for our vision, of this primordial difference, the *Riss*, the *écart*, the decisive moment when vision suddenly appears, an ek-static opening in the clearing of Being. Lightning is nature's gift of a trace: a trace of the primordial decision in relation to which all beings are steered. Thus the lightning ultimately points into ourselves.

Our experience with the night deepens our understanding of the historical need in which our vision presently suffers; and this understanding sets us a difficult task, laying claim to our capacity for existential resolve, authentic decision. The lightning amplifies our judgment, points to the need for decisive historical changes in our visionary being; and before our very eyes it lights up for us, in a primordial repetition, the *taking place* of the ontological difference.

In his book on dreams, Boss comments that,

> Of course, the western mind has never devoted much thought to the way we exist in deep, dreamless sleep. In fact, many an eastern thinker has pointed to this as the reason why so little is known in the Occident concerning the true nature of waking human life.[44]

Boss is right. But the same could be said of dreaming itself. In the discourse of Western philosophy, in Descartes, for example, dreaming becomes a matter of concern only because its subversive vision of reality must be decisively refuted and excluded. Few indeed are the philosophers who have turned to the dreambody to learn a vision steered by its insight and wisdom.

Nietzsche once wrote this:

> Not only the reason of millennia – their insanity,
> too, breaks out in us.[45]

When I think about the darkening of our historical world, the night of the soul through which we are now passing, the madness of our lives and the fury of Being, I wonder: perhaps we now need to *reconnect* our vision, our vision of reason and daylight saving, with that *other* vision, the vision of our double, the vision of our dreambody, the vision whose 'crazy' wisdom belongs to the night. Perhaps our insanity is too much sunlight. *Alētheia,* truth as unconcealment, requires a vision which understands the hermeneutical teachings of the night: a vision in which these teachings have made a decisive difference.

As we move deeper into the night of the soul and begin to see by its truth, I would like to recall some words from Goethe's *Italian Journey,* words to which I am unavoidably giving, because of the context surrounding this quotation, a different sense:

> My purpose in making this wonderful journey is not
> to delude myself but to discover myself in the
> objects I see.[46]

Goethe's 'self' is self-possessed and self-assured, ego-logical. He will look at objects in order to discover himself. He says 'discover', not 'create', because the 'self' he means when saying 'myself' is already there – so to speak, 'within' him, and does not exist outside the metaphysics of an 'encapsulated' ego, the metaphysics of *Zuhandensein* (being ready-to-hand) and *Vorhandensein* (being extant, being present-at-hand, being in reserve). However, the 'purpose' we have before us in making this visionary journey through the night is more complicated, for the night into which we are going is a night *following* the death of God and threatening the very survival of a culture dominated by the technologies of an ego-logical self. Goethe takes the self's being-there for granted; seeing himself in the objects he sees, he assumes he can simply take (in) what he has seen, 'increasing' himself in that way. There is no questioning the 'substantial' being of the self – nothing like an episode in which, seeing himself in the mirroring of objects, he suddenly finds himself lost in them, or lost among them, radically decentred, dispossessed. No question of any radical shattering of the ego's logical boundaries of defense against 'delusion,' no questioning of the *structure* which divides vision into self and other, subject and object, seer and seen.

What is delusion? *Where* is this 'self' which finds itself over there, in the objects it sees? If the 'self' is over there as well as here, the 'self' cannot be completely identified with the ego-logical subject which inhabits the structure of subject and object. In this night of the soul, the 'soul' I am calling, here, the dreambodyself, we will find ourselves compelled to question the consensual boundaries which separate the historical self's sense of 'reality' from its sense of 'delusion'. If our experience with the night teaches us to see ourselves in the objects around us and catch sight of the sense in which these objects are mirrors of our own projections, mirrors of our character, and of the prevailing sensibility in which we all participate, the sight of their violence, their terror, their demonic rule in our political economy, their spiritual ugliness will perhaps turn us back upon ourselves, turn us into ourselves, to question the individual and collective 'self' whose vision – visionary being under the spell of its ego – brought forth such disturbing objects. But seeing all this could also *open* this 'self' to a different vision. . . .

As Nietzsche says:

> . . . this ego: . . . the more it learns, the more words and honors it finds for body and earth.[47]

The more it learns, the more it opens to the wisdom of the dreambody and the wisdom which comes (*es gibt*) from the night.

Part II

Pointed Turnings

By what is our vision moved? – by what our gaze? our glance? our look? By what kind of *understanding* is it moved? What is the character in its turnings?[48] We should recall Herakleitos: *prōton hōthen hē archē tēs kinēseōs,* which translates, following Heidegger, as a statement about movement: *The origin of movement is also the origin of ruling and directing*.[49] Is our vision to be *steered* by the thunderbolt? What could this possibly mean? Perhaps our own vision, though bereft of the 'superstition' and 'mythology' that moved Herakleitos, can nevertheless derive some guidance and direction from its lived, and now 'proprioceptive' experience with lightning. We could, for example, see what happens when we let our vision be steered by this experience: steered by the way this experience, together with its amplifications through culture, is carried by the body of vision as a deeply felt sense; steered by the way our body feels when we sit 'proprioceptively' with our sense of the lightning's symbolic presence – an impression perhaps indelibly inscribed into the primordial recesses of our visionary being. We could let our vision be steered by the body's felt resonance to the thunderbolt and its cultural stories. Such steering would arise from the depths of our visionary being.

Herakleitos spoke of *tropai*, turnings of fire.[50] We shall speak of changes, turning points, bodily felt shifts, individual turnings pointed towards a redeeming of history. We shall work with a vision returning to bodily felt experience, going into the night of the soul, into the darkness of the unconscious, into the madness of recent history. And we shall return to symbols that appeal to an elementary visual awareness.

What Heidegger calls a 'moment of vision' is such a turning point. It may be symbolized by a flash of lightning. And, as we learned through the element of water our need for continuity and contact, so now, through the element of fire, we develop insight into possibilities for transformation, and into the channels and processes through which they could be realized. The moment of crying is followed by a moment of insight. This is our narrative: a story of possible change, a story whose truth is in the making.

Heidegger's most thorough interpretation of this moment, understood ontologically as an event (*Ereignis*) of Being, will be found in 'The Turning': *Die Kehre*. Before we undertake the amplification of our own interpretive responses to Heidegger, we will let him tell us, in his own words, just what this turning, as an event taking place within the world-field of vision, means to him.

Near the beginning of the essay in question, Heidegger looks into the shadow that has cast its danger across his field of vision and speculates in the following terms:

> Yet probably *this* turning – the turning of the oblivion of Being into the safekeeping belonging to the coming to presence of Being – will finally come to pass only when the danger [of Enframing], which is in its concealed essence ever susceptible of turning, first comes expressly to light as the danger that it is. Perhaps we stand already in the shadow cast ahead by the advent of *this* turning. When and how it will come to pass after the manner of a destining no one knows. Nor is it necessary that we know. A knowledge of this kind would even be most ruinous for man, because *his essence is to be the one who waits*, the one who *attends upon* the coming to presence of Being in that, in thinking, he guards it.[51]

One of the most pressing questions this statement raises for us is, I think, this: Granted that we must, as befits our nature, wait and attend upon the coming to pass of the turning, does this mean that there is 'nothing at all for us to do'? 'Does this mean,' Heidegger asks himself, 'that man is helplessly delivered over' to blind fate?[52] As an event of Being, the turning will certainly not come to pass *merely* by our doing; the turning, after all, is not just another event among the various events of our day; it is not, for instance, like the event of an autumn fair, which obviously will not come to pass unless we ourselves consciously make it happen. And yet, I think it is also true – and also what Heidegger himself believes – that the turning cannot happen *without* us. The turning calls for, and we might say it needs, the response of our thought: it needs the gift of our waiting and attending; it needs the reciprocating effort and resolve of our vision, organ of thought, organ of Being. The 'waiting' and 'attending' Heidegger has in mind are difficult existential tasks for us; they are not ways of being into which we could easily fall. We must attempt to think 'waiting' and 'attending' as 'something we do' in a sense of 'doing' that is not completely controlled by the metaphysical dualism enshrined in our everyday language.[53]

Thinking, he tells us, 'is genuine activity, genuine taking a hand, if to take a hand means to lend a hand to . . . the coming of presence of Being.'[54] If it is true that, within what

Heidegger terms 'the field of vision of the self-willing will to power,'[55] our seeing is:

> that representing which since Leibniz has been grasped more explicitly in terms of its fundamental characteristic of striving (*appetitus*)[56]

then it would seem that a different seeing, a kind of seeing which strives not to strive, and gradually learns how to take the truth of Being more under its care, must somehow make a new kind of *place* for the presencing of Being, doing so in a distinctively different moodedness. If our seeing could *lend its field* to the visible presencing of the invisible, would it not contribute to a turning?

Heidegger writes:

> When the turning comes to pass in the danger, *this can happen only without mediation*. For Being has no equal whatever. *It is not brought about by anything else nor does it itself bring anything about*. Being never at any time runs its course within a cause-effect coherence. . . Sheerly, out of its own essence of concealedness, Being brings itself to pass into its epoch.[57]

But then, in the very next sentence, he says: 'Therefore we must pay heed.' Since giving heed *is*, however, a doing, it can hardly be argued that Heidegger is urging us to go to sleep, while we *wait* for the epoch of Being to turn. The point of this passage is, rather, *to remove the event of Being from the technological realm of causality*. Nothing we do, or could ever do, will *cause* Being to presence, or to turn.[58] The *concept* doesn't fit.

So let us consider how Heidegger begins to characterize this turning in terms of its visible presence within our visual field:

> The turning of the danger comes to pass *suddenly*. In this turning, the clearing belonging to the essence of Being *suddenly clears itself and lights up. This sudden self-lighting is the lightning-flash*. It brings itself into its own brightness. . . When, in the turning of the danger, *the truth of Being flashes*, the essence of Being clears and lights itself up.[59]

The turning is called, here, a flash of lightning. Since the turning is an event in which the *truth* of Being presences in its unconcealment, the turning is phenomenally visible as lightning. And since Being presences as the lightning of our field of vision, the *suddenness* of the turning is appropriately manifest as a *flash* of light.

In the next gesture of thought, Heidegger sees this flash of lightning as the flashing glance of Being:

> To flash [*blitzen*], in terms both of its derivation and of what it designates, is 'to glance' [*blicken*]. In the flashing glance and as that glance, the essence, the coming to presence, of Being enters into its own emitting of light. Moving through the element of its own shining, the flashing glance *retrieves* that which it catches sight of *and brings it back* into the brightness of its own looking. And yet *that glancing, in its giving of light, simultaneously keeps safe the concealed darkness of its origin as the unlighted*. The inturning [*Einkehr*] that is the lightning-flash of the truth of Being is the entering, flashing glance – the insight [*Einblick*]. . . When oblivion turns about, . . . then there comes to pass the in-flashing [*Einblitz*] of world into the injurious neglect of the thing. That neglect comes to pass in the mode of the rule of Enframing. In-flashing of world into Enframing is in-flashing of the [luminous] truth of Being into [the darkness, the diminished light, of] truthless Being. In-flashing is the disclosing coming-to-pass within Being itself. Disclosing coming-to-pass [*Ereignis*] is a *bringing to sight* that brings into its own [*eignende Eräugnis*].[60]

I think it could be said that this is a 'repetition' of 'that flash of one trembling glance' by the grace of which, according to St. Augustine, his soul was turned around, converted, and finally 'arrived at that which is'.[61] But Heidegger's naming of the glance of Being is entirely different from Augustine's naming of the glance of God.

Heidegger's words seem to suggest an anthropomorphic projection. What could be the point? Perhaps he wants to suggest a comparison, a 'proportion' which will set in motion some deeper process of self-examination. By projecting the glance from the human world, where glances naturally occur, into the dimensionality of Being, he lets our sense of the significance of Being reflect back to us the ontological destination of the human glance and make visible the great claim on our eyes, the organs of Being in the field of its visibility.

The apparent anthropomorphism is a matter for reflection: if 'God' reflects back upon us the condition of our consciousness, the light of our present awareness, then the *return* of the reflection here proposes a comparison, a *homology*, which challenges us to contemplate the glance of our *own* being and ask ourselves how it measures up to the ideal of a human glance whose depth and reach, and whose felt source of movement, would inspire us to call it a glance of Being. Such a glance, though human, would in a special sense 'belong' to Being itself. The projection is therefore not anthropomorphism, but the recognition that, in the very nature of our projection, the ontological possibility of a *homologein* is proposed for our reflection, our thought.

The flashing glance is called 'insight': 'insight into that which is'.[62] But Heidegger must warn us against taking what he says to be nothing but a report of our own experience in the *ontic dimension* of the visual field:

> From the first and almost to the last it has seemed as though 'insight into that which is' means only a glance such as we men throw out *from ourselves* into what is. . . But now everything has turned about. Insight does not name any discerning examination [*Einsicht*] into what is in being that we conduct [only] *for ourselves*; insight [*Einblick*] as in-flashing [*Einblitz*] is the disclosing coming-to-pass of the constellation of the turning within the coming to presence of Being itself, and that within the epoch of Enframing.[63]

Now, when we read this passage with care, it should become clear that Heidegger is not at all denying that the glance of insight *is* a human glance, an event which comes to pass in and through the realm of human vision. On the contrary. It is of decisive, even fateful importance that it be understood *in relation to* the human capacity for vision. What he wants to deny, rather, is an understanding of the glance which restricts it and its significance for our lives to the ontic dimension of the visual field: the human dimension in its consensually determined, inauthentic, ontologically forgetful everydayness. We need to think and see the human glance in the light of our sense of what that glance would be like, if it were an organ of Being – a glance returned to Being, belonging to Being, attuned by Being, turned by Being, arising from our felt sense, our wholebodied sense, of Being as a whole. We would therefore entirely miss the revolutionary meaning at work in Heidegger's thinking if we did not relate what he has to say about the 'glance of Being' to the character of our own glance, the glance that is our own being, who we are.

What does our glance show about the character of our own way of being, and about the character of our historical world? There is forgetfulness of Being in the character of our glance. If we can see the lightning-flash *as* the glance of Being; if we can see *in* the flash the presencing of Being, then we may be reminded of the potential within our own capacity and opened up to the opportunities which the field grants us, and through which our own glance could experience a profound turning. (Recalling our work in Chapter Two, we can now say that this turning is a 'reversal', and that, in this reversal, our glance and gaze are returned to their rootedness, their grounding, in the *reversibility* of the flesh.) By a transfer of sense, a reflection of difference, Heidegger *opens* our vision to the in-flashing presence of its ontological setting, its primordial *Legein*, its primordial topology.

Thus, in the next movement of his thought, Heidegger articulates the significance of the glance of insight for the modern epoch whose vision has been determined by enframing. And he emphasizes that the glance in question is not a completely ordained fate to which we can only acquiesce with passive resignation:

> And yet – in all the disguising belonging to Enframing, the bright open-space of world lights up, the truth of Being flashes. At the instant, that is, when Enframing lights up, [when it comes to light] in its coming to presence, *as* the danger, i.e., as the saving-power. In Enframing, moreover, as a destining of the coming to presence of Being, there comes to presence a light from the flashing of Being. *Enframing is, though veiled, still glance, and no blind destiny in the sense of a completely ordained fate.*[64]

'Enframing' is not just a word to describe our present historical epoch; it is also a word we can use to describe the character of *our vision* at this time. Enframing may not be readily apparent in the glance, but it is 'at work' wherever we glance: at work in its medium, its element, its channel and organ. It *hides* within the glance, and yet makes itself visible, through this glance, in the field of vision. It is *through the glance* that enframing determines the being of the visible, determines visible beings, determines the horizon, the clearing, the character of the field as a whole, the openness of Being in our world.

When our normal thinking is upset, is reversed, and we suddenly realize that the glance of Being *is* our glance – our 'own' glance, however, in a nonreductionistic sense, i.e., by virtue of its relation to a deeper, ontological dimension – and understand, further, that the future of our glance, and of the human realm of vision in general, is *not* a preordained fate, we naturally find ourselves deeply shaken, jolted as if struck by lightning:

> When insight comes disclosingly to pass, then men are the ones who are struck in their essence by the flashing of Being. In insight, men are the ones who are caught sight of.[65]

The next passage to consider moves our thinking still further:

> Only when man, in the disclosing coming-to-pass of the insight by which he himself is beheld, *renounces human self-will* and projects himself towards that insight, away from himself, does he correspond in his essence [i.e., respond by attunement and in reciprocity, in *homologein*] to the claim of that insight. In thus corresponding [and manifesting the *homologein*], man is gathered into his own, that he, within the safeguarded element of world, may, as the mortal, look out toward the divine.[66]

(The 'divine', here, does not have to be thought in onto-theological terms. It can be understood phenomenologically, i.e., as expressing an experience of enchantment. Old words can be used in new ways.) The 'corresponding in his essence' is, as I have indicated, a *homologein*: it is an 'attunement' into which we are gathered, for example, when our gaze is appropriate to its ontological potential – to that which lays claim to our vision and calls upon us to look, to gaze, to see with deeper insight, and to lend the clearing of Being, without which we couldn't see, the inner light of our own mortal vision.

In the primordial, non-dual dimension of the field of vision, Being presences holistically, as the interweaving unity of the flashing glance and the element of light, whose prolonged, sustained 'flashing' is the field of energy which encompasses that glance. The human's glance responds to the claim in the layout (*Legein*) of that deeper ontological dimension when it is appropriately turned and correspondingly transformed. When the character of our gaze, our glance, 'corresponds' to, lives up to, our ontological insight into the presencing of Being in and as our field of vision, and we see how enframing prevails in this presencing, and therefore 'even' in our own gaze, then there is a *homologein*, and we are gathered into a new vision of our historical situation. The 'glance of Being' refers to our *own* glance *insofar as* it moves – is moved – with an awareness of, and an attunement by, the whole of Being. (In *The Body's Recollection of Being*, I have attempted to amplify Heidegger's interpretation of the *homologein* by working through the hermeneutical character of our gestures.) The 'moment of vision' discussed in *Being and Time* concerns precisely this existential metamorphosis in our vision and its field.

Heidegger returns to this moment of vision, the moment when insight flashes, in his *Nietzsche* work. The second volume (English translation) is devoted to the eternal recurrence of the same. Here, in a chapter on 'The Convalescent', Heidegger comments on the insight which flashes through the text at the point where Zarathustra encounters the 'wisdom' of the animals:

> The two animals [eagle and serpent] open the conversation. They inform Zarathustra that the world outside is like a garden that awaits him. They sense somehow that a new insight has come to him, an insight concerning the world as a whole. It must therefore be a pleasure to proceed to this newly constituted world, since all things are bathed in the light of the new insight and want to be integrated into the new dispensation. Insofar as they are so

> illuminated and integrated, things corroborate the insight in a profound way; they heal the one who up to now has been a seeker, they cure him of the disease of inquiry. That is what the animals mean when they say to Zarathustra, 'All things yearn for you. . . . All things want to be doctors to you!' And Zarathustra? He listens to the animals' talk, indeed gladly, although he knows that they are only jabbering. But after such solitude, the world *is* like a garden, even when it is invoked by mere empty talk, in the sheer play of words and phrases.[67]

Despite their 'silly' talk, the animals do teach him something after all. We may read the above passage as pointing out that the animals *helped* Zarathustra to see the world as a garden. Even if the 'new insight' came to him without the animals' influence (a supposition that is questionable, since the animals had been accompanying him for some time), it seems clear that the 'consummatory moment' of the insight – a sense of 'pleasure' if nothing else – has been facilitated in some crucial way by the 'jabbering' of the animals.

What does the companionship, the presence of the animals signify? What is their wisdom? Recalling the texts quilted together at the beginning of this chapter, I would like to suggest that the animals contribute to the insight an essential dimension of wisdom. The peoples upon whom our Western 'civilization' has always looked down and sought to destroy have, however, understood this point. Perhaps our 'civilization' is founded upon the 'rational' denial of their wisdom: founded politically, therefore, upon the 'rational' annihilation of the peoples with whom this wisdom was historically entrusted. We are also destroying the animals, and with them, the wisdom and ways of seeing of which they have been, since time immemorial, the endowed guardians. In the nihilism of this epoch, it seems that we may have been appointed to be the instruments of their destruction, bringing to an end all 'primitive' cultures and all animals, and with them – the visionary capacities they kept alive. As regards the animals, we might note their amazing instinctual attunement; their primordial integration into the environment; their global awareness; their 'unpolluted' perceptiveness; their 'precision' and 'clarity'; their sense of balance, their timing. Every animal makes visible a prowess, a perfection, of visionary wisdom: the eyes of the bat, the owl, the peregrine falcon, the frog, the mouse. Every animal illuminates and amplifies the presencing of Being within our world. Like the gods, they reflect back to us a vision of Being. Nietzsche says, in *The Will to Power*:

> [I]f consciousness .. . . [is] at first at the furthest distance from the biological center of the individual, . . . it deepens and intensifies itself, and continually draws nearer to that center.[68]

We have lost touch with the animal wisdom that sleeps and dreams in the recesses of our visionary being. We need to retrieve, and develop in our *own* ways, this animal wisdom. We need to learn how to see with *their* sense of awareness, with *our sense* of their awareness. To learn this is *not* regression; it is a *new phase* in the (historical) development of the visionary self, because we must return to retrieve our implicate biological awareness of Being in order to move from a highly rationalized vision to a vision changed by its awareness of the interconnectedness of all beings and its sense of the whole.

In 'The Turning', Heidegger says:

> Only when insight brings itself disclosingly to pass, only when the coming to presence of technology lights up as Enframing, do we discern how . . . the truth of Being remains denied as world. Only then do we notice that all mere willing and doing in the mode of ordering steadfastly persist in injurious neglect.[69]

The last two passages are gathered together, I feel, in a passage whose final question opens our vision at the same time that it brings 'The Turning' to a close:

> So long as we do not, through thinking, *experience* what is, we can never belong to what will be.
>
> Will insight into that which is bring itself disclosingly to pass?
>
> Will we, as the ones caught sight of, be so *brought home* into the essential glance of Being that we will no longer elude it? . . .
>
> Will we correspond to that insight, through a

> looking that looks into the essence of technology
> and becomes aware of Being itself within it?
> Will we *see* the lightning-flash of Being?[70]

The answers to these questions *could* come from the heart; they *could* arise from the depths of our body of vision: we *could*, for example, sit quietly with our body's felt sense of what a vision 'brought home' would be, or be like; and we could let *that* felt sense tell us how to work with that question and with the task it proposes. Here is where the kind of thinking which Gendlin calls 'focusing process' could be extremely helpful.[71] Whether and how we might 'correspond' to the insight which comes to us is a question we cannot answer. But I am sure that we can 'correspond' only by giving thought to our capacity to respond to the clearing and lighting of Being.

What looks at first like an anthropomorphic projection makes sense in terms of this question of correspondence. For the 'glance of Being' is Heidegger's metaphorical projection, his way of seeing and saying the ontological potential inherent in the human glance. The words turn our glance in a way that opens it. We suddenly *see* the human glance in its character as the organ for disclosing the clearing and lighting of the field. The 'glance of Being' is that visual truth of which we are capable; it is that depth of Being into which our vision is summoned. Insight is the gaze turning back upon itself, the gaze so turned that it can look into its own ontological depths, and which, by virtue of a gesture which 'corresponds' to the 'glance of Being,' fits Heidegger's description of a glance whose insight 'retrieves that which it catches sight of and brings it back into the brightness of its own looking.' What this gaze of insight catches sight of, what it retrieves and brings back, back into the world where it makes it radiantly visible, is the depth of its invisible nature as an organ of Being, deeply beholden to the elemental clearing and lighting.[72]

We are gifted with vision; and our insight into this giftedness is what brings forth within us the 'glance of Being'. *This* is the glance which is turned into the openness. . . .

What is the significance of the references to lightning? Heidegger's text remains abstractly philosophical. Consequently the lightning becomes a 'mere' metaphor, a trope. It loses its reality as an experience, as an amplification of visionary processes. I wish to *retrieve* its psychological truth. The philosophical significance thereby shows up more clearly.

In his essay, 'The Oversoul', Emerson speaks with great understanding of the experiential process through which insight is bodied forth:

> Every distinct apprehension . . . agitates men with awe and delight. A thrill passes through all men at the reception of new truth, or at the performance of a great action, which comes out of the heart of nature. In these communications, the power to see is not separated from the will to do, but the insight proceeds from obedience, and the obedience proceeds from a joyful reception. Every moment when the individual feels himself invaded by it, is memorable. Always, I believe, by the necessity of our constitution, a certain enthusiasm attends the individual's consciousness. . . The character and duration of this enthusiasm varies with the state of the individual, from an ecstasy and trance and prophetic inspiration – which is its rarer appearance, to the faintest glow of virtuous emotion, in which form it warms, like our household fires, all the families and associations of men, and makes society possible.[73]

(I suggest a comparison, here, between Emerson's sense of 'obedience' and Heidegger's notion of *Zugehörigkeit*. Vision whose channels of seeing are opened through such 'obedience' takes place in an 'ecstasy' beyond the will.) The vision which comes from this obedience needs, however, to be channeled into the world of daily life – into the world, where it can make a difference. When this vision is not channeled into the world, not authentically integrated, the ego becomes inflated. The ego which cannot cope with the sudden shift, the sudden opening, is threatened by madness:

> A certain tendency to insanity has always attended the opening of the religious sense in men, as if 'blasted with excess of light'.

Emerson understood the danger which comes with insight;

but he also understood something about the way out.

The bolt of lightning, seen as a glance sent into the sky of the night from the darkness beyond, lights up the sky and discloses its depth. Analogously, the flash of ontological insight make visible the fact that we draw our experience from a deeper spring of inwardness than we had thought: an inwardness which, as we go down into it, discloses to us the ultimate contingency of the subject-object structure and lets us experience its utter openness to the field of Being as a whole. 'Affective space' has greater depth, as Welwood points out, when it is vividly textured with many dimensions, or levels, of meaning:

> Whereas expansion is marked by a sense of moving beyond boundaries, depth is characterized by a sense of richness and untapped potentiality.[74]

As lightning deepens and expands our sense of nature, so insight deepens and enriches our sense of personal existence. It lights up the being of the Self, and the beings who we are. The lightning shows the immeasurable; we are *measured* by our relationship to *its* truth, its *alētheia*. This disclosure of the deep, this violent rending open of the flesh of the night sky, shows in a flash that the field of our vision, the restricted field to which we belong insofar as we tenaciously identify with the ego that presides over it, is not closed, but radically open. The bolt of lightning shatters our ego's comfortably narrowed vision, destroys the theoretical definition of its empire, invades its defenses against suddenness, against changes in standpoint and viewpoint, against the uncontrollable, against nature's claim to our mortality. The thunderbolt brings with it, as its 'gift' to our vision, a deep insight into the nature of our mortality. Its appearance strikes us with dread and marks, in the furrows of amazement on our brow, the assignment of our inevitable death.

But I am reminded of Wittgenstein:

> If we take eternity to mean not infinite temporal duration but timelessness, then eternal life belongs to those who live in the present [the clearing of Being]. Our life has no end in just the same way in which our visual field has no limits.[75]

And also of Rilke:

> . . . *in* us who with a part of our natures partake of the invisible, . . . *in* us alone can be consummated this ultimate and lasting conversion of the visible into an invisible no longer dependent upon being visible and tangible, as our own destiny continually *grows at the same time* MORE PRESENT AND INVISIBLE in us.[76]

Perhaps our existence, as visionary beings, is to be keepers of the invisible. Perhaps our death could be a vision of our assumption of this service. But the ego must pass through its death to be reborn in this other vision.

The 'ego' in question here is not only an individual ego; it is also social and historical. The thunderbolt of nature bursts into our social and historical world; and it awakens our collective unconscious. The thunderbolt is an archetypal symbol. Erich Neumann writes that,

> The order that we find in the unconscious as well as in consciousness – the spiritual order of the instincts, for example – long before the rise of consciousness as a determinant of organic life and its development, lies in a plane of experience to which the normal experience of our polarizing consciousness does not attain. On this plane, the human community lives with the relative solidity and security of a world supported by the cultural canon, and only seldom does an earthquake, a subterranean encounter with the repressed power of chaos . . . disturb the security of the human collectivity.[77]

Great insight is always deeply disturbing, for we tend to *cling* to habitual patterns of vision. Rilke alludes to the *clinging* which happens in the wake of attention to the invisible when he describes the 'angel':

> The angel . . . is that being who vouches for the recognition in the invisible of a higher order of reality. – Hence 'terrible' to us, because we, its lovers and transformers, do still cling to the visible.[78]

But where does this clinging come from, if not from the (historical) ego, attached to its boundaries, its defensive skin, needing to master and be in control, living sometimes in mortal dread of any deeper opening up?

Wrathful visions correspond to the *struggle* around this clinging. 'Malevolent demons' may appear at this time. We may find ourselves reflected in their eyes. And what we see may look 'evil'. But we must look into it for the healing of understanding. In his 'Letter on Humanism', Heidegger says that,

> With healing, evil appears all the more in the lighting of Being. The essence of evil does not consist in the mere baseness of human action, but rather in the malice of rage.[79]

I should like to close this phase of our thinking with an observation made by Jung in his 'Phenomenology of the Self':

> it is quite within the boundaries of possibility for a man to recognize the relative evil of his nature, but it is a rare and shattering experience for him to gaze into the face of absolute evil.[80]

If we should have the courage to gaze into this face, we would *learn* about the 'malice of rage', the 'fury of Being', which works its way through the channels of our vision, making it a vision full of pain and hate, a vision of hells.[81]

Part III

Borderline Awareness: Flashes from the Heidegger/Fink Seminar on Herakleitos

(1) In the dark I see nothing, and nevertheless I see.
Heidegger, *Seminar on Herakleĩtos 1966/1967*[82]

(2) A human kindles [touches on, approximates] a light in the night, when his eyesight is extinguished. Living, he touches on [approximates] death in sleep; in waking, he touches on [approximates] sleeping.
Herakleitos[83]

(3) The little light [we humans can kindle in the night] stands in opposition to the rhythmic, great light of day that befalls us and that has nothing dark about it. The human is the light-related being who, it is true, can kindle light; but never such as would be able to completely annihilate the night.

Fink (129)

(4) But a human is not only a cleared being [i.e., a being with awareness, knowledge, understanding, a capacity to see and make visible]; he is also a natural being and as such he is implanted in a dark manner in nature.

Fink (144)

(5) Is *lēthē* [the sleep of forgetting] to be identified with night?

Heidegger (147)

(6) Is sleep the genuine understanding of the dark ground?

Heidegger (148)

(7) . . . the experience of the dark ground of life is the experience of . . . *the coincidence of all distinctions* . . . the relationship of the human, who stands in individuation, to the nonindividuated but individuating ground.

Fink (147)

(8) But if a human exists *between* light and night, he relates himself to night differently than to light and the open. He relates himself to night or to the nightly ground insofar as he belongs bodily [through the visionary channels of the dreambody] to the earth and to the flowing of life. The dark understanding rests as it were on *the other principle of understanding*, according to which *like is [re]cognized through like*.

Fink (145)

(9) But how do you understand the night?

Heidegger (53)

(10) If I have spoken of another, more original night, of the nightly abyss in explication of the sun fragment, I did so in preview of the death-life fragments. From there I have viewed the deeper sense of the phenomenon of closedness of the earth. . . . Only when we first consider the relation of life and death will we see how the realm of life is the sun's domain and how a new dimension breaks open with the reference to death. The new dimension is neither the domain of openness nor only the closedness of the earth, although the earth is an excellent symbol for the dimension of the more original night. Hegel speaks of the earth as the elemental *individuum* into which the dead return. The dimension of the more original night is denoted by death. That dimension, however, is the realm of death, which is no land and has no extension, the no-man's land.

Fink (54)

(11) Because a human being does not dwell in the great light, he resembles the night owl, that is, he finds himself *on the border* of day and night.

Fink (129)

(12) If a human is *the in-between being*, between night and light, then he is also the in-between-being between life and death, the being who is already near to death in life.

Fink (131)

(13) Gods and humans [humans and demons] in their *intertwining relationship* have a mirroring function. . .

Heidegger (117)

(14) The gods live the death of mortal humans.

Fink (100)

In the night of the soul, the ego confronts the demons of hate which continue to influence its vision, and it *clings* to the visible as it meets its own death. Heidegger sees possibilities for individuation in encounters with death, for death points into the invisible; and it offers a wisdom which annihilates the historical egobody's *attachment* to the visible:

> In the clear night of the nothing of anxiety, the original openness of beings as such arises: that they *are* beings – and *not* nothing. (*What Is Metaphysics*?, *Basic Writings*, 105)

Part IV
Bardo Vision: The Fury of Being

We are in-between. We continue the process of thinking and letting our vision open. But we are beginning a new developmental phase.

There is no text I can think of more provocatively concerned with the ways in which the energies of light enter into the pain and healing of mortals than *The Great Liberation Through Hearing in the Bardo*, more commonly referred to as *The Tibetan Book of the Dead*. *The Great Liberation* is an ancient liturgical text; but it is also a text of healing, a text used by Tibetan physicians administering to the needs of the dying. It is, finally, a text in the phenomenological psychology of the ego: for use in psychotherapy with people concerned about their deadness and aliveness, and about their capacity for growth . . . and a different vision.

According to the Introduction by Chogyam Trungpa, 'The fundamental teaching of this book is the recognition of one's projections and the dissolution of the [ego-logically determined] sense of the self in the light of reality.'[84]

The Great Liberation is a story, a discourse told in the symbolical terms that would speak most directly and most effectively to the culture of its origin. It is also a *hermeneutical* discourse: a text which displays its meaning in the light of an unending exchange of interpretations; a text which plays in the texture that is wrought by its innumerable shades of meaning. In the Tibetan culture, the symbols speak; they are disclosures of meaning. For us, here and now, the symbols are more concealing. But the concealing is also a protecting, a preserving of the text's capacity to speak, to teach, to transmit its precious wisdom. We must work with the text, bringing it our experience, our questions, our needs, our dreams. In order to participate in *its* vision, we must be willing to *share* with it the vision by which *we* are living.

The Great Liberation is a discourse, then, which reminds us of our inherent capacity – call it the gift of our inborn sanity

– to let ourselves see things in their true light: in the *truth* of their light and in the *light* of their truth. And it is a discourse which recognizes the predominance of our inveterate *incapacity* – call it, if you like, our vulnerability, or our madness – to *accept* things as they are, to *open* to things in the true dimensionality of their being: to *see*, in other words, with an understanding that everything we encounter, whether visible or invisible, is nothing but the cosmological 'caprice' of Being, manifesting through the organs of our eyes (an organizing principle of Being) in, and of course also as, the field of vision. What the text tells us is that nothing we see is other than the play of Being, manifesting, through the organizing principle that is embodied in our eye-organs, as local disclosures of its invisible presencing within the luminance of a field of vision. But if we do not understand this in the living of our lives, then what we are vulnerable to seeing, what we will tend to see, are configurations of this presencing that cause us to suffer, take fright, fall ill, turn destructive, or lose our minds – perhaps in different forms of cultural delusion regarding the truth of reality. So corresponding pain and suffering are already 'in' the eye of the beholder as well as 'out there'. And some of the ugliness and malevolence we see in the world is a product of the corresponding character deeply motivating our vision. The world mirrors, it reflects our visual presence, the way we are present, the way we are with things, the way our awareness is present in gaze and glance, the way we are emotionally open or closed to things as they are and appear. A wrathful manifestation is always a reflection of some consciousness that has been raging unrecognized within us, in the very depth of our visionary being. What happens when we see *ourselves* in the violence, the Fury of Being, which surrounds us?

The concern of the *Bardo* text is for our way of experiencing what Jaspers called 'extreme situations': situations of crisis that test, that call into question, the very limits of our existential capabilities. Such situations are 'spiritual emergencies' – for a new spirit, a new self, is struggling to emerge, and there are many dangers, but also unimagined opportunities, for insight and self-development.

The *Bardo* is said to be that moment between the ego's perceptual and conceptual fixes, readings, takes, and takings-for-true (*Wahr-nehmungen*), when it becomes *possible* to experience the presencing of beings in, as, and with a radically different light: a different level, gradient, intensity, and quality of illumination, a luminance which is felt to be deeper, richer, clearer, and more enhancing than the lighting which normally appears through our representations, which gets diminished thereby, and to which we are accustomed by our cultural training. This different light is a 'true' light, in the sense that it is a light in which the ontological difference becomes manifest much more clearly within the structural dynamics of subject and object, self and other, figure and ground, visible and invisible. This light may also be visible as a very *beautiful* lighting, since it is a clearing which is good to things, good in particular to their truth, letting them be, and be seen, 'as they are'. (We all know, I believe, the sense in these words; and I think we all know, also, that they do not have to be construed in any strangely metaphysical way – for example, as designating fixed essences, mere substances, or matter.) And it is, finally, a lighting (in) which it could feel very good to see, and see by.

And yet, this different lighting – it could be the light of the day under the rule of the Sun, or it could be the diffuse light of the night – can appear very frightening, very wrathful, to the degree that our being human is not attuned by openness to Being, but rather by the worldly reference points, the representational schemes, through which the human ego projects an identity and clings to it. In the *Bardo* experience, we exist *between* our representational schemes; we are thrown into an unfamiliar situation which calls for creativity and openness. We are seeing things, and must even see ourselves, in a new, and therefore still undefined light. The truth of our ego-logical identity is very much at stake. The claims of the lighting are testing our limits – even testing the boundary between us and our death, us and our madness.

As a text of ancient provenance speaking very directly about the different ways, actual and possible, in which we are called upon to experience the great lighting of Being, *The Great Liberation* is, I think, very much worth our attention. We read:

> O son of noble family, when your body and mind separate, the *dharmata* [the forms of existential awareness] will appear, pure and clear and yet hard to discern, luminous and brilliant, with terrifying

> brightness, shimmering like a mirage on a plain in spring. Do not be afraid of it, do not be bewildered. This is the natural radiance of your own *dharmata*, therefore recognize it.

(It might prove helpful to compare the *Bardo* experience, as described here, to Heidegger's interpretation of light and color in his 'Moira' essay on Parmenides.)[85] The locution 'separation of mind and body' refers in the (superstitious) language of the people to the time of death. However, since 'Bardo' is the Tibetan word for an interval or interlude, any in-between or transitional situation, Trungpa argues that the text does not necessarily refer to the sort of event we would call 'death'. It may also refer to any moment when we confront the world of our habits and routines, and see it concealing an abyss, a void, nothingness. *The Bardo* is a text, a story about our experience at the moment of a 'turning'. The *Bardo* is always a test of our basic, inborn sanity: that wisdom, that insight we need to rely on whenever the everyday and consensual reality is no longer convincing, no longer to be taken for granted. But if the necessary wisdom is lacking, the betweenness, the splitting open of the *Bardo* space, can be terrifying. Medard Boss has described the panic we may experience as we recognize the metaphysical groundlessness of our existence and are thrown back on our own inborn resources:

> In any attunement to anxiety, e.g., in panic, the perceptive open world-realm as which this human being basically exists is narrowed down to such an extent that the only threatening traits of all that is encountered can now enter into the existence's 'field of vision'.[86]

The narrowness of which he speaks is of course also somewhat characteristic of our vision in its ontical everydayness; and, for the most part, most people go through life more or less comfortably settled into the 'normal' patterns of experiencing. But if they are suddenly confronted with apparently random flashings of light, with spontaneous materializations of light, or simply with the reality of the utter openness of Being, as a field of vision, to the formations and suggestions of the lighting, it often happens that the threat to the boundaries of their (ego-logical) security is overwhelming. The night holds us with a special power; in its uncanny light, neither daylight nor total darkness, we see the forms of our projection: what we fear, what we dread, what we do not know because we cannot see. A primordial anxiety, because vision is 'stopped', taken over, dispossessed: it belongs to the enchantments of the night. In the *Bardo,* we see what happens when this kind of confrontation takes place. If what we see is initially horrifying or terrifying, our horror or terror will be amplified, since it is reflected back to us, made visible by the light. And if we do not see the wrathful forms which appear in the light as our own visionary projections, do not see that the configurations of light simply reflect back to us the spontaneous externalizations of our own state of mind, our *Gemüt,* the mere 'autonomy' of the forms – their apparently independent metaphysical reality – will increase their authority, their power, their wrathful character. As James Glass, summarizing his many years of work with schizophrenic patients, observes:

> Delusion functions as a desperate measure to protect the self from having to deal with unconscious feelings that carry with them an unbearable pain.[87]

As he points out, 'for the schizophrenic, delusion defends against a truth that threatens to annihilate the self.'[88] But the schizophrenic is really different from the rest of 'normal' humanity only in the sense that he feels himself to be in a more extreme situation of threat. Although we 'others' live for the most part in the visual world of anyone-and-everyone, we too are vulnerable to such anxiety, panic, and self-protective delusions.

The *Bardo* threatens the ego's identity, substantiality, and permanence; it threatens annihilation, nothingness. Recalling Heidegger's words, quoted in Part II: 'In the clear night of the nothing of anxiety, the original openness of beings as such arises: that they *are* beings, – and *not* nothing.' But it takes *being at risk* in order to learn and grow and develop. For Heidegger, this means: 'without the original revelation of the nothing, [there can be] no selfhood, no freedom.[89]

Who are we? As *Dasein,* we occur essentially as 'thrown': thrown into the openness or clearing of Being. The human being, 'in his essence, is *ek-sistent* into the openness of Being.'[90] 'The ecstatic essence of man consists in *ek-*

sistence. . .'[91] To be fully human, then, in the sense of measuring up to the human potential, is to *live* 'an ecstatic inherence in the truth of Being.'[92]

Only a few mortals (can) experience this ek-static inherence with ecstasy; most of us prefer, or manage, to live out our lives in a duller, more narrowly circumscribed dimension, protecting ourselves from the ontological claim upon our being by falling into line with the host of defenses established for anyone-and-everyone. It is too difficult to live with the thought that, as Heidegger says, '*Dasein* means: being held out into the nothing.'[93]

In his *Concluding Unscientific Postscript*, Kierkegaard tells us that, 'An existing individual is constantly in process of becoming.'[94] We are by nature process, becoming, a dynamic body of experiential shifts. We will therefore always 'find' ourselves already in process of becoming: already *in* the *Bardo* space of the in-between. Thus, what matters, what makes all the difference, is how we (are able to) relate to that consciousness of process. Some people can grow with it, enjoying the fulfillment, the sense of deep contingency, that come from going into that process as a way of living. Some people, more people than the first, must radically close themselves off, and withdraw into the annihilation of delusion: they are the mad. But most people shut their eyes to the space, the spaciousness, the silence and stillness, the ground of openness, which appears, in flashes, breaking into their consensually validated and settled world. Most people escape madness: somehow graced, they do not need such strong defenses; they can live comfortably in a world which supports their less extreme measures of defense.[95]

The possibility of growth, of achieving a new (grounding of) sanity, of what might be called, as we keep in mind Heidegger's text about 'the malice of rage', a process of 'healing', depends on the extent of our capacity to turn the opening into a process of growth, individuation, *Jemeinigkeit*. Thus, the *Bardo*, the visionary space of the ontological difference, is also a space and time for deep personal decisions, further phases of individuation, further learning in the *light* of that space, the light, that is, of the ontological difference.

The *text* of the *Bardo* provides instructions for the 'second day' of confrontation with the Fury of Being: a second chance, as it were, to understand the deeper nature of the 'authoritative light' which presences in wrathful forms during the time of the *Bardo* experience. We are given a second chance to respond appropriately; but we must achieve some radical insight into our habitual way of representing the configurations of light that disturb us. We are called upon to look into ourselves, to see in our visions the projections of our own states of mind:

> the white light of the *skandha* of form in its basic purity, the mirror-like wisdom, dazzling white, luminous and clear, will come towards you from the heart of [the deity] *Vajrasattva* [pristine awareness] and his consort and pierce you so that your eyes cannot bear to look at it. At the same time, together with the wisdom-light, the soft smoky light of hell-beings will also come towards you and pierce you. At that time, under the influence of aggression, you will be terrified and [will try to] escape from the brilliant white light, [and] you will [also] feel an emotion of pleasure towards the soft smoky light of the hell-beings. At that moment, do not be afraid of the sharp, brilliant, luminous and clear white light, but recognize it as wisdom. . . It is the blessed *Vajrasattva* coming to invite you in the terrors of the *Bardo*; it is the light-ray hook of *Vajrasattva*'s compassion, so feel longing for it.[96]

'Vajrasattva' refers to the 'fact', the 'truth', that radiant, luminous energy is manifesting in the field of vision (of Buddhist culture) as a deity, a mythopoetic being of adamantine wisdom and enlightenment. 'Vajrasattva' is a (cultural) vision of the embodiment of pristine awareness. As a 'visible' being, visible in the field of light, 'Vajrasattva' corresponds to a certain achievement of vision: not only a certain capacity for projective visualization, visionary ideals externalized and embodied in a disciplined projection of desire into light, but also, and much more importantly, a certain capacity to see with a very deep, metaphysically self-deconstructive understanding of the visionary process. In meditation, the lighting of the field may appear as a deity in interdependent co-existence with a seer whose gaze is a gesture of mirror-like wisdom: the lighting mirrors the gaze, fearless, without aggression, open to whatever passes in front of it, neutral, free of the *need* for projections.

We should not be taken in by the mythopoetic forms: in *whatever* configurations of light these forms manifest, they are but the reflections of our own emotional condition, the projections of our individual and collective unconscious. Insight into our visionary processes means confronting the ego, whose investments in projections, whether they be deities like *Vajrasattva* or beings in some more wrathful form, are extremely powerful and well-concealed.

The (peaceful and wrathful) deities of meditative practices in Buddhism reflect the *polarities* of our experiencing, as we struggle with our deepest anxieties, our defenses, our need for an absolute point of reference, something fixed and permanent to hold on to, clear boundaries of separation. The deities test, in their mirroring of our condition, the insight we have achieved into the nature of vision.

Of course, a great spectrum of shifts, transitions and turning points separates the vision of madness from the vision of enlightenment, the condition of basic sanity. Not all shifts happen suddenly, like lightning; nor do they necessarily come as flashes of deep insight; bodily felt shifts, shifts in our feelings, our moods, our perception, our way of seeing and understanding, may be very small, and yet important as openings and clearings. 'I have been feeling depressed; but now that I see why I have been feeling that way, I realize that what I was calling "depression" is really a deeply smouldering anger. And now that I see this, I see that there are ways for me to communicate the anger, and I can feel the mood lifting.' In his book on *Focusing*, Eugene Gendlin amplifies the process of such shifts.

The Heidegger texts we have been considering in this chapter, however, are concerned, as is the *Bardo* text, *The Great Liberation*, with more extreme shifts, transitions, in-between states, and turning points. Returning to Herakleitos, Heidegger speaks of the lightning flash, the thunderbolt, the flashing of insight. In *Being and Time*, he speaks of the 'moment of vision'; and in his work with Nietzsche's *Zarathustra*, he returns to the thought of that moment – an 'instant'. In 'The Turning', he speaks not only about a *human* glance, but also of the 'glance of Being'. The 'turning point', here, is an ontological moment: a moment when a pre-ontological understanding which had been implicate within our vision, our ontical way of seeing things, suddenly comes into awareness and our visionary being begins to become an organ of Being, its channels suddenly opened up to the depth and matrix of Being as a whole. When this happens, our eyes are turned and moved by their ontological understanding, their attunement by Being. In the *Herakleitos Seminar* he gave with Fink, Heidegger goes into the night, into sleep, which is the realm of our dreaming, and into the darkness of death. In many different texts, of course, he speaks of the clearing and lighting of Being – the clearing and lighting which Being *is*, and in which and as which Being presences, giving itself to our vision by giving way to the visionary capacity assigned to us through nature. In his 'Letter on Humanism', however, Heidegger touches on 'the closure of the dimension of the hale', the 'malignancy' distinctive of our age, 'the upsurgence of the healing', 'healing', 'raging', 'compulsion to malignancy', 'evil' and 'the essence of evil', 'the malice of rage', 'healing and raging', 'grace'. And in 'The Question Concerning Technology', he speaks of 'the danger' of 'enframing' and of 'the saving power' concealed within it – and the language he uses, here, is drawn from, and is intended to appeal to – address and lay claim to – our *visionary* being: 'an adequate look into what enframing is', 'the upsurgence of the saving power in appearance', the necessity that we 'look with yet clearer eyes into the danger', a conviction that, 'wherever man opens his eyes. . .'

Thus Heidegger speaks of the *betweenness* of our present historical situation: we, the postmoderns, living out our lives with a growing awareness of the legacy of modernity: its danger, the raging of nihilism and its pathologies, the night of the oblivion of Being, whose caretakers we are. Heidegger's words for this oblivion, this historical closure, often refer to visionary experience, as do his words for the 'malice of rage', the psychopathology of our time, the danger, the healing, the grace, and the insight. We of today live in a world *between* the danger of the modern and the hope we see for a future without nihilism. We exist in a space *between* pathology and healing, in a time of great pain and suffering, in a world of wrath and rage – in a world, however, where healing *does* take place, where grace does sometimes touch and move us, and where the very depths of madness can bring forth healing insights. With an understanding awareness of our suffering, we can struggle to separate ourselves from the metaphysics of the moderns; but our inseparability is a more powerful fact and holds us in the vice of the between: between the old vision of modernity, now steered by death, and an altogether different vision, already perhaps implicate, but which has in any case not yet unfolded.

We are living in a fateful time, a time for fateful commitments and decisions, of decisive turning points. The thunderbolt is *nature*'s decisiveness: it announces a change, and it does so decisively, cutting through the night. The lightning flash is a symbol for the instant of vision, when Being as a whole lights up, becomes visible. It is a symbol of the flash of our insight, the glance which suddenly sees with ontological insight. The thunderbolt is a symbol of the *Logos*: it amplifies and is articulate; it lights up the darkness; it turns our heads and points our eyes; it steers vision into a sense of Being as a whole. But lightning, as we know, is a form of the lighting. Herakleitos calls it *ho logos* and asks us to see, and also to see by the guidance of, its 'steering'. Into what does it steer our eyes? Into the darkness of the night? Into the lighting of Being? Into the opening and clearing? Heidegger speaks of the 'lighting of Being': this lighting, or clearing, is the *Logos*, it is that which steers, it is the thunderbolt.

The lightning flash is said to be a 'glance of Being'. We could perhaps say, then, that, in our present epoch, in the time for a decisive turning, this glance manifests the effects of nihilism, makes them visible: this glance is the Fury of Being, flashing with the rage of nihilism, bursting with violence into the night of our world. The fury of Being is unleashed in a glance of terrifying rage, pointing out the danger, pointing into ourselves, into our fear, into the heart of our capacity for insight. Our vision is steered by the 'grace' of the light into the night of a vanishing point, into the openness of Being which clears space for, and illuminates, beings as a whole.

The thunderbolt frightens us; its decisiveness puts *us* into a state, a time, of profound in-decision: night, uncertainty, indistinctness, ambiguity, anxiety, panic, and confusion. We are between a state of 'normal' composure and an experience of something different. We are struck by our vulnerability: our mortality, our death, is for an instant lit up. Our eyes, normally steered by the ego, are suddenly turned, decentred, steered for an instant by the lightning, by that through which the lighting of Being is manifest in the fury of its decisiveness. For an instant, perhaps, our vision is touched and gifted with appropriate responsiveness, and we may see the wholeness of beings, and the openness of Being as a whole, through the subliminal channels of dreambodyvision.

The thunderbolt, symbol of the *Logos*, of *Phusis* and the flashing glance of Being, is also to be thought – as Heidegger himself, to my knowledge, has not – as the symbolic, tropological presence of the ontological difference. The decisiveness of the thunderbolt, the character of the de-scission it makes, and makes visible, in the nightsky, is a 'repetition' by nature which gathers our dreambodyvision into a recollection of the ontological difference – that de-cision, that rending of the tissue, the flesh of the field, which takes place wherever there is an event of vision. The bolt bursts into presence, rends, tears, tears open, divides, differentiates, amplifies difference. Thus, the ontological difference is made visible in the field of our vision as that primordial de-cision, or opening in the matrix of Being, through which a figure-ground structuration comes into Being. The figure-ground, centre-periphery, focus-diffusion differentiation is the most primordial difference, the most primordial inscription and layout that the ontological difference makes in the field of our vision. In the 'instant' in which this most primordial process of differentiation takes place, an *ek-static Gestaltung* comes into being: a *Gestaltung* very different from the one which Heidegger calls 'das Gestell', enframing; a *Gestaltung* the recollection and development of which will be the concern of our final chapter: 'Das Geviert', the Fourfold, the splendour of radiant vision.

So we see that the Lighting of Being gives way to the ontological difference, which makes itself visible through the structural differentiations it makes and lays down (*legein*) in the clearing of the field of vision.

If our experience of the night can teach us the need for a vision in touch with its wholeness and the wholeness of Being, our experience of the thunderbolt should emphasize, rather, the need for individual and collective commitment and decision – *Entschlossenheit*, as Heidegger called it in *Being and Time* – 'recollection' of that more primordial decision, the ontological difference. In such 'recollection', we go down into ourselves – this is the phase of *Erinnerung* – and we attempt to retrieve, on this journey of return, that which has been forgotten: the ontological difference. This retrieval is a *Wiederholung*, a process of healing integration, of re-membering, and making whole and healthy. The retrieval is steered by our insight into that which has been forgotten: the invisible lighting which cleared an open space for our vision to be.

When deep insight occurs here, the lighting of Being itself may take over the steering of our vision, our glances, our gaze, our turnings. Our vision is no longer in the possession of the historically (in)formed egobody. Instead, it belongs to the emerging body of ontological understanding, an essentially hermeneutical visionary being whose fulfillment is in the character of its *homologein*: its capacity to take care of Being and make the meaningfulness of the ontological difference *visible* in the world of our indifference.

We have not as yet given thought to the matter, the *Sache* proposed for thought, but left very obscure, in the 'Letter on Humanism'. In his writings on the history of metaphysics, Heidegger takes care to give thought to pain and suffering, and to the crying need they bespeak. Here, in his text on humanism, he writes of evil, of grace, of the malice of rage, of the raging of nihilism, of healing. But his writing is cast in shadow; its shades of meaning require interpretation.

All I can offer is a story: a story about the pathology of vision and the possibility of a time of healing: a story about the raging of nihilism in our time and the malice of rage which is correspondingly consuming our visionary being, destroying it from within. (Cf. the essay I wrote on 'Psychopathology in the Epoch of Nihilism', contributed to my collection, *Pathologies of the Modern Self: Postmodern Studies on Narcissism, Schizophrenia, and Depression*.)

In the epoch of nihilism, the 'glance of Being' manifests what Guenther, in the *Matrix of Mystery*, has called 'the *fury* of Being.' The fury of this glance is related to the character of truth, of unconcealment, of *alētheia*, which prevails in the epoch of nihilism. *'Das Ge-stell'* designates this character. This character is 'evil', is 'malignant', in the sense that it is closed to the lighting of Being, the present, the gift, or grace, of the clearing and lighting: no receptiveness, no acceptance, no recognition, no acknowledgment, no guardian awareness. And this closure is destructive, is self-destructive. Our deeper ontological nature always cries out somehow against this world-historical closure; but all it can do, perhaps, is rage underneath, suffering an *inner* death, a death of the spirit, in the destructive malice of rage. Historical conditions conflict with the visionary spirit of our ontological nature. Nihilism is a cancer of the spirit: this cancer mirrors, is in some way attuned to, the cancers of the flesh. The fury of Being rages within us, consuming the flesh, consuming the spirit, taking over our visionary being. In the epoch of enframing, in a time when *das Ge-stell* rules all processes of *Gestaltung*, our eyes are so many organs for the fury of Being. *Das Ge-stell* manifests, and makes visible, the fury which rages within our own visionary being. The nihilism in *Das Ge-stell* destroys the seer even as it destroys that which is being seen.

But let us recall some words of Merleau-Ponty: 'every object is the mirror of all others'. What I see mirrors who I am as seer: the object reflects back to me the face of my vision. Subject and object are co-emergent; they arise together, simultaneously and in interdependence. The object reflects back to me the character of my vision. Nihilism is the fury of Being: it is a wrathfulness which is pervasive, intrusive, enveloping and haunting beings. Nihilism inhabits things, and in this form of visibility it is projected and reflected. It inhabits things and shows me their wrathfulness. It inhabits my glance, my gaze, and I project it unconsciously into the world, onto things. I am consumed by the nihilism of history raging within me, but the wrathfulness of things shows me my projections.

We need insight, here: insight into the wrathfulness of things. We need to see in this wrathfulness a mirroring of the malice of rage raging within us. We need to see into ourselves. We need to go into the raging body of visionary experience and open it up, so that we can sense, with our pre-ontological understanding, its – our – relationship to the fury of Being. The wrathfulness of things points into the rage at work in our own gaze. From where does this rage come? It is pervasive in the visionary life of our culture. Between the ontological event, the *Ereignis* that manifests the fury of Being, and the psychological process, the suffering of rage in our visionary being, there is the long history of nihilism.

In 'The Turning', Heidegger says (in words I am recalling from the end of Chapter One, Part V):

> Enframing is, though veiled, still glance, and no blind destiny in the sense of a completely ordained fate.[97]

Enframing is the 'glance of Being' turning all its fury into a human vision turned by the malice of rage. *Das Ge-stell* rules the character of vision consumed by this rage.

Enframing, Heidegger says, is a diminishing of light. The epoch of enframing is one of closure, *epochē*. In his 'Letter on Humanism', this closure is said to be a closure in the dimension of the hale. The oblivion of Being takes place with this closure. In the realm of vision, this oblivion is closure to the *lighting* of Being – to that element through which the presencing of Being is manifest. In our epoch, how do things stand with this lighting of Being? In the epoch of nihilism, it comes to a stand in *das Ge-stell*. Since the malignancy, the raging, the pathology is in the *closure* of the dimension of the hale, the pain and suffering we make and endure *as visionary beings* is related to the closure which comes *between* us and the lighting of Being.

The closure and the raging are related to enframing: enframing is the historical form of the subject-object structure which prevails today in the modern world. The pathology latent in the enframing of this structure is its forgetfulness of Being: its closure to the grace, the *Es gibt*, of the lighting; its closure to that elemental matrix of intertwining in which both seer and seen inhere. The pain and suffering in this ontologically closed historical condition are mirrored back and forth between seer and seen. In the time of this closure, the oblivion of Being, our visionary being suffers and rages. This raging takes possession of our visionary organs, and is mirrored by the wrathfulness of things. Our subjection to their wrath of course only intensifies this raging, sometimes consuming us with its nihilism.

The lighting of Being is that by grace of which we see; it is therefore *like* 'grace', a kind of 'saving power'. And we find it, as the poet says, not by shutting our eyes or turning away, but by looking fearlessly *into* the danger. The healing is in our capacity for heartfelt openness to the lighting, our capacity for a deeply felt, bodily felt *recollection* of Being.

We are visible beings; the objects we see are visible beings. We are luminous beings, luminous bodies; the objects we see are also luminous beings. Some luminous beings, however, and above all, those beings who are called 'human', have been given the capacity to *make visible*, to bring to light, to illuminate; other luminous bodies cannot:

they only *reflect* the light. What we see, the objects whose presencing we behold, are mirrors of our projections: shapings of the light we cast in making beings visible. What we see always makes visible who we are. The wrathfulness of things in the epoch when enframing steers the human glance makes visible our *historical character*, the character of our culture.

The *Bardo* text tells us that what we see are the projections of our *egobodyvision*. In terms of our analysis in Chapter One, I think this makes sense. Indeed, I think it is a *compelling* story. The *Bardo* text goes on to teach that, if we look with insight into the projections, into the *processes* of their formation, into their 'genealogy', we may penetrate so deeply into the *essence* of the subject-object structure that we will see the ultimate *emptiness*, *openness*, and *contingency* of its functioning. The tenure of this structure is like the hold of a magical spell. If we look into it, look fearlessly into its raging, its danger, we may see its ultimate metaphysical insubstantiality. If we look with insight into the processes involved in the simultaneous and interdependent emergence of subject and object, we may see them turn into configurations of light with no more metaphysical substantiality than a flashing of light.

The *Bardo* text helps us to understand that the rise to power of the ego and its egobody in the modern epoch is *not* a preordained fate, but an event into whose *contingency* we need to be looking. The egobody's vision will perhaps always tend to settle, however uncomfortably, into the subject-object structuration; but perhaps we can see now that it has presided for centuries over the enframing of the lighting of Being: an increasingly pathogenic twist or torsion in the structuring of the primordial figure-ground differentiation which the ontological difference brought forth.

How could our insight into this process steer human existence into historical decision? How could insight tear open, like a thunderbolt, the defenses that serve our indifference? 'My glance', as Rilke says, 'like lightning. . .'?

Part V

Vision: Our Capacity to Make Visible

(1) *Phuseos katagoria*: Hinting at, making nature visible.
Periander of Corinth.[98]

(2) Since the infusion of grace is very clearly illustrated through the multiplication of light, it is in every way expedient that through corporeal multiplication of light there should be manifested to us the properties of grace in the good, and the rejection of it in the wicked. For in the perfectly good the infusion of grace is compared to light incident directly and perpendicularly, since they do not reflect from them grace nor do they refract it from the straight course which extends along the road of perfection in life. . . But sinners, who are in mortal sin, [i.e., visual pathology,] reflect and repell from them the [light which is the] grace of God.
Roger Bacon, *Opus Majus*[99]

We need to read Bacon's words, 'corporeal multiplication of light', in a phenomenologically self-reflective and hermeneutical way. His words would accordingly refer to our visionary capacity, our capacity as visionary beings, to receive, to be receptive to, the manifestations of 'grace' in the visible world. And suppose that 'grace' refers to the 'Es gibt', the emitting of light which is the 'mittance' of light, the clearing and lighting of Being, which *can be seen as* a 'gift' to our capacity to see.

(3) When a ray strikes a glass of water, its reflection
leaps upward from the surface once again
at the same angle but opposite direction
from which it strikes, and in equal space
spreads equally from a plumb line to midpoint,
as trial and theory show to be the case.
Dante Alighieri, *Purgatorio*, XV, 16-21

Dante's *Purgatorio* describes the newly emerging vision of his time: an optical theory, of course, but also a new way of seeing the human, a new way of seeing, and seeing into, ourselves; a vision for which the rationality of vision, and the rationality, the ratio and geometry of the entire visible world, at last became visible as such.[100] In its *purest* theoretical formulation, it is a vision of extraordinary beauty. Spinoza represents, for me, the most sublime formulation of this historical vision. But I reproduce the passage from Dante to show up a contrast, a difference, between *this* vision, the one about which he speaks, and the – shall we say 'mystical'? – vision articulated by Roger Bacon. (Bacon was still living during the first twenty-nine years of Dante's life. They were nearly contemporary.)

As a *bursting* of Being, of *Phusis*, in the field of natural light (*phaos, phōs*), lightning is an event (*Ereignis*) which, in its own way, is disclosive of the presence of light as that which lets be, lets things appear, and be visible. As Heidegger writes in 'The End of Philosophy':

> Such appearance necessarily occurs in some [kind of] light. Only by virtue of light, i.e., through brightness, can what shines show itself, that is radiate. But brightness in its turn rests upon something open, something free, which might illuminate it here and there, now and then. Brightness [e.g., in the form of lightning] plays in the open and wars there with darkness.[101]

Lightning makes the light visible *as* that which makes visible – or *as* that by grace of which there *is* visibility. Commenting on the references, in Herakleitos, to *Keraunos*, Eugen Fink states:

> Lightning, regarded as a phenomenon of nature, means the outbreak of the shining lightning-flash in the dark of night. Just as lightning in the night momentarily flashes up and, in the brightness of the gleam, *shows things in their articulated outline*, so lightning in a deeper sense *brings to light* the multiple things in their articulated gathering.[102]

Another statement puts it this way:

> As lightning tears open light, and *gives visibility to things in its gleam*, so lightning in a deeper sense lets

> *panta* (many things, the many, all things] *come forth to appearance* in its clearing.[103]

But what is the connection between lightning and vision? We have heard Heidegger call the lightning-flash of Being a 'flashing glance.' This is not just a whimsical ornament of speech. Nor is it a concession to the primitive thought of the early Greeks.

Lightning is *like* a glance in that it gives visibility, gives articulation, to things. This likeness then makes it possible for lightning to instruct us; to become, as it were, an 'authoritative' light, by the grace of which our vision is shaken out of its complacency, and deepened by a flash of insight into its own essential nature. Thanks to the flash of lightning, we may enjoy a story with a flash of insight. Our vision finds itself thrown into a world: sent on a mission, as it were, to make things visible. (Is our vision capable of *submission* to this destiny, this claim which appears in a flash?)

Perhaps we can now hear a deeper, more ontological resonance in what Merleau-Ponty has to say:

> Here a light [*lumière*] bursts forth, here we are no longer concerned with a being that reposes in itself, but with *a being whose whole essence, like that of light, is to make visible . . .*, to open itself to an other and to go outside itself.[104]

Years later, in response to Heidegger's meditations on *Lichtung* (*Sein und Zeit*, 133) and *Gelichtetheit* (*Sein und Zeit*, 350), Merleau-Ponty inscribed their textual meaning into the flesh:

> My flesh and that of the world thus brings about clear areas, clearings [*comportent des zones claires, les jours*], around which pivot their opaque areas.[105]

In their primordial, ontological ecstasy (*ek-stasis*), my flesh and the lighting of the field intertwine: they are one and the same. Thus we say that, in the gaze, it is the *light* which alights and makes visible. The gaze is that organ, or medium, *through which* the lighting of Being makes beings visible in its sensuous field.

But our vision is a capacity; it has an implicit, or latent

potential. And our vision is deeply implicated in the unfolding destiny of Being. Let us therefore reflect on this implication. Have we made as explicit as possible, yet, what is still *implicit* in vision's mission of making visible? Consider this. Making visible is potentially a question of seeing in a way that lets shine, lets gleam and glisten, sparkle and glow. Accordingly, the gaze which not only makes visible – by casting, or granting, the necessary *minimum* of light – but distinguishes itself as capable, in its truth, of making shine and glow, may be appropriately seen as a flashing glance: like a glance of lightning.

The Levite priests in the temples of ancient Palestine were called 'they whose faces are radiant as lightning, and as glistening as gold garments.'[106] Mere metaphors? No, *true* metaphors: metaphors true to the coming-to-truth of lived-in experience. Their countenance, it was said, was like 'the brightness of the vaulted canopy of heaven', and –

> As lightnings flashing from the splendour of the
> Chayoth,
> As the celestial blue thread of the fringes,
> As the iridiscence of the rainbow in stormclouds,
> [And] as a halo of purity shining forth from the
> mitre of holiness.[107]

I should like to interpret this 'presence' of the priests by reference to a phrase I can still hear from my reading of Heidegger's *Erläuterungen zu Hölderlins Dichtung*, a phrase which sings of 'joy in the presence of the Joyous'.[108] In the gaze which participates with care in the world of light and looks about with joy to give, we see the *fulfillment* of a gaze that can make things glow and shine in some quite extraordinary way for them to be present. To see the world with joy *brings* joy into it. But in the world, this joy is mirrored. To see this joy mirrored in the world – mirrored not only by things but much more so through other people – is itself a great joy. Joy is always returned, *reflected* in the *vision* of the beautiful, the good, the true. Since it is not a question of some causal relationship, but rather of a correspondence, i.e., a co-emergent co-responding, a reciprocity, the glowing and shining of the things which are visible presents a vision of *beauty* that quite naturally heightens the visionary experience of joy. The radiance of things reflects, and is simultaneously reflected by, the 'equivalent' radiance of the gaze. As it alights and lights things up, the gaze itself lights up in its delight. The two, the seer and the seen, are thus gathered together in an *ecstasy* of light.

In an essay called 'On the Supersensible Element in Knowledge, and on the Immaterial in Nature',[109] Leibniz observes that 'there is a light born within us.' A 'natural light,' of course. What kind of light is *this*? The eyes of most disciplined readers will not be caught by this phrase, will not be tempted to linger over it, will not be provoked into thought by it. Mere metaphor again? Mere embellishment? No, again. We could perhaps try reading these words as attempting to get at some distinctive visionary experience, saying it, bringing it to words, speaking its truth.

Being and Time offers a surprisingly firm interpretation:

> This term [*Augenblick,* 'moment of vision'] must be understood in the active sense as an ecstasis [*Ekstase*]. It means the resolute rapture with which Dasein is carried away to whatever possibilities and circumstances are encountered in the situation as possible objects of concern, but a rapture which is held in resoluteness [*in der Erschlossenheit gehaltene Entruckung des Daseins*].[110]

According to the anthropologist, Gerardo Reichel-Dolmatoff, the Tukano Indians of the Colombian rainforest are familiar with this experience, or with something akin to it, and 'even' make some extremely perceptive discriminations. According to his report,

> The soul [of a shaman] is compared with a fire whose light penetrates obscurity and makes things visible. . . . Of a payē [shaman] who is not very active it is said: 'His soul is not seen; it does not burn, it does not shine.[111]

(The cross-cultural appearance of such testimony is unsettling, but for that reason, it calls for thought. The roots of vision go very deep: into the ancestral body, the collective unconscious, the transpersonal and transhistorical, the cross-cultural.) The Tukanos believe that the 'mind' of a shaman 'does not have strength without the knowledge that the light provides. He can only act in a field explored by the light.'[112] 'On a much higher level' than the shaman, however,

> the *kumū* [priest] is a luminous personage who has an interior light, a brilliant flame that shines and unveils the intimate thoughts of all people who speak to him. His power and his wisdom are always compared to an intense light that is invisible but perceptible through its effects. This is the manifestation of the *kumū*, his capacity of fathoming the psyche of a person and thus knowing his intimate motivations.[113]

We would do well to hold this experience with vision in phenomenological respect. (On this point, see Merleau-Ponty, *Phenomenology of Perception*, 285-92 and 351.) The temptation to dismiss this 'marginal' knowledge from the discourse of our 'civilization' should be resisted. Here is Heidegger speaking:

> The basic fallacy . . . consists in the belief that history begins with the primitive and backward. . . The opposite is true. The beginning is the strangest and mightiest. What comes afterward is not development, but the flattening that results from mere spreading out; it is the inability to keep the beginning. . .[114]

Heidegger, of course, always looks back to the ancient Greeks. In the modern world of science and technology,

> The original emergence and continuing of energies, the *phainesthai*, or appearance in the great sense of a world epiphany, becomes a visibility of things that are already there and can readily be pointed out. The eye, a vision which originally could perceive the dynamic coming-into-being, *phainesthai*, becomes a mere looking at or looking over or gaping at. Vision has degenerated into mere optics.[115]

Hegel, as we have already seen, evokes a visionary consciousness akin to 'the sunrise, which, in a flash and at a single stroke, brings to view the form and structure of the new world.'[116] And Jung brought back this from his journey up the Nile:

> The sunrise in these latitudes [i.e., near the sources of the Nile] was a phenomenon that overwhelmed me anew every day. The drama of it lay less in the splendor of the sun's shooting up over the horizon than in what happened afterward. . . At first, the contrasts between light and darkness would be extremely sharp. Then objects would assume contour and emerge into the light which seemed to fill the valley with a compact brightness. The horizon above was radiantly white. Gradually the swelling light seemed to penetrate into the very structure of objects, which became illuminated from within until at last they shone translucently, like bits of colored glass. Everything turned to flaming crystal. . . At such moments I felt as if I were inside a temple. It was the most sacred hour of the day. I drank in this glory with insatiable delight, or rather, in a timeless ecstasy.[117]

One might want to say that Jung saw, in the gift of the sunrise, a repetition of the story of the Creation: the emerging of beings into the light. Some might want to call this experience 'mystical,' but we should note the disciplined perceptual detail. Vision, for an instant, was transformed; but with no loss of contact with everyday reality.

Martensen, writing on Jacob Boehme, the German mystic, tells another story:

> Sitting one day in his room, his eyes fell upon a burnished pewter dish, which reflected the sunshine with such marvelous splendor that he fell into an inward ecstasy.[118]

Strikingly similar accounts, however, can be found in the Zen Buddhist tradition. In his commentary on Ummon's famous koan, 'Rice in the bowl, water in the pail,' Katsuki Sekida observes:

> You must once have this experience [*samadhi*] and you will discover what a splendid thing the boiled

> rice in the bowl is. It shines like diamonds in the incandescent heat of the fire.[119]

According to Sekida, when our vision shifts to a 'meditative' mode of self-discipline,

> a blade of grass, even a stone at the roadside, begins to shine with the beauty of its essential nature. You are in closest intimacy with the 'object.'[120]

Again I want to ask: Is this mystical? What is gained by calling it that? Since the experience in question *has* been brought to words, we know that it *can* be brought to words. It is not, in fact, 'beyond words.' The stories I have collected, here, for our thoughtful reading – Rilke's, Emerson's, Jung's, Jacob Boehme's, Sekida's – have, in fact, much to say. If we read them phenomenologically, they speak the truths of lived experience. Calling them 'mystical' does not clarify anything; indeed, it is bound to obstruct understanding. This designation generally occurs in two opposite discourses: the one is reductive, since it either rejects the mystical or degrades it, defining it, for example, in strictly neurological terms; the other discourse is inflationary, since it makes the experience too great for this world. Both discourses effectively dismiss the experience as such. In the discourse, however, of phenomenological psychology – our discourse, here – we do not pass judgment on the 'reality' of the experiences human beings share with one another. We take them to heart, we work with them, we sit with them, we test them, we see for ourselves how their assimilation into our own visionary experience influences and changes the way we see things, the way things are. And the dialogue which comes from out of that experiential work continues the unfolding of the *psyche*, which is, of course, the concern of the discourse as a whole.

Sekida's words, 'closest intimacy,' say in *other* words what Merleau-Ponty was saying when he spoke of an 'intertwining.' These other words (words, moreover, of another culture) are, in some sense, talking about 'the same thing.' And yet, I would say that Sekida's words *flesh out* the felt awareness, the moodedness, which characterizes the experience of intertwining.

Sekida also gives us something else to think and live. A question: what is it for something – a blade of grass, a stone

– to shine with the beauty of its essential nature? The early Greek philosophers saw the shining of *truth* in the essential nature of things. The poet, Pindar, would surely have understood the 'shining' of beauty in Sekida's experience. Is beauty how truth shows itself, makes itself visible to the eyes? Why do we so seldom see this beauty? Why do we see beauty in (as) the shining? Why does the truth shine? Does it shine because it is beautiful, or is it beautiful because it shines? If it seems that the 'essential nature of things' comes to appearance shining with the beauty of its deep truth, is the 'essential nature' *different* from this coming-to-appearance, or are these two different ways of saying the same thing? (One is more metaphysical, the other, perhaps, more experiential.) These are questions we do not need to answer, or answer straightaway. Instead of rejecting them at once as 'too marginal' or 'too aberrant' for further consideration, instead of answering them here without further experiential work with them, we *could* simply sit with them and let them elicit, in their own good time, a response from that body of felt awareness which I have called 'subsidiary,' 'guardian,' and also 'dreambodyvision'.

If the ego-logical subject is characteristically disposed to focus on things in a way that dims them down, it cannot be expected to *see* that truth whose beauty can be seen to shine. This kind of experience requires a different channeling, a different modulation, of our visionary being. Let us read again, in its light, a passage from *Being and Time* with which we worked in Chapter One: 'In "setting down the subject," we dim entities down to focus.'[121] Heidegger's words refer to the grammatical subject; but I am suggesting that his statement may be translated, *salva veritate*, into a statement in phenomenological psychology, describing the character of the light, the luminosity of the visual field, and the visibility of things for a vision, a visionary process, in which the ego-logical subject is 'set down' as its source of identity. When an ego-logical subjectivity dominates the visual process, entities are dimmed down to focus. (In the modern epoch, enframing takes this further.) This kind of process is of course both developmentally necessary and adaptationally useful, being an inherent capacity of our visionary nature. But what happens if we practice, as a 'disciplinary practice of the Self,' a meditative *suspension* of this inclination, this *conatus*, to focus? . . . Following the logic implicit here, it should be reasonable to expect that this 'dimming down' would not happen. And what then? Perhaps this question takes us closer to the *edge* of that visionary experience for which things 'seem' to be 'shining,' 'warmly glowing': extraordinarily beautiful, somehow, just in their whatness, their isness, their simple being-there, their presence.

Arthur Deikman, a prominent experimentalist, has recently reported some noteworthy psychological experiments with 'subjects' who are doing traditional Buddhist meditation. It was discovered that, when these 'subjects' were queried about their visual experience with a blue vase they were asked to focus on, they noted, in particular, the following significant shifts:

(1) increased vividness, brightness, and richness in the colour;
(2) a sense of its 'aliveness' (not felt to be incompatible with the fact that the thing would certainly not be classified among living beings);
(3) increased sense of realness and a *visible* deepening of the thing's significance;
(4) a felt decrease in the distance, or difference, between it and them;
(5) a more explicitly affective organization of their perceptual response.[122]

And Deikman singles out three factors which he believes to contribute to this perceptual shift: first, a heightened, more appreciative attention to the sensuous, aesthetic element of the perceptual situation; second, the cessation of logically controlled, analytic, and abstract processes of thought; and third, an attitude he characterizes as 'receptivity to stimuli' (openness instead of defensiveness or suspiciousness).[123] There are, of course, other factors as well. But it is surely significant that these factors, at least, have received confirmation in controlled experimental conditions.

John Welwood, clinical psychologist, has extensively researched the phenomenon of flashes of illumination in terms of the phenomenological method. Writing on the insightful visual experience that sometimes takes place during certain meditation practices, he observes:

> Split-second flashes of this basic ground, which Buddhists have called 'primordial awareness,' 'original mind,' 'no mind,' are happening all the

> time, although one does not usually notice them. Buddha spoke about literally developing awareness in terms of fractions of a second, to awaken people to the fleeting glimpses of an open, pre-cognitive spaciousness that keeps occurring before things get interpreted [by representational, metaphysical thinking] in any particular perspective.[124]

The primordial awareness in question here flashes up like lightning, because it is an *enlightened* level of visual perception where the seer and the seen *meet* in the lighting of Being, pervasive continuum of illumination that constitutes the visual field as a whole. But this level of awareness, this level of illumination, is suppressed or forgotten:

> What is unconscious generally are patterns of organismic structuring and relating which function holistically as the background of focal attention.[125]

However, through self-disciplinary practices of the kind we could call 'practices of meditation,' it is possible for vision to come, not from the egobody, but from a much deeper (sense of) embodiment, whose subsidiary awareness structures the visionary situation in a radically different way:

> . . . awareness of the basic ground breaks through when one wears out the projects and distractions of thought and emotion. Then there is a sudden gap in the stream of thought, a flash of clarity and openness.[126]

Seeing which issues from this level of perceptivity sees things in a way which feels well-articulated by the phrase, 'as they really are,' or by a phrase like 'in their isness.' With some practice, one can form a fairly clear sense of what it might be like to see things as 'glowing blossoms in a garden of light': *We could even see ourselves that way*, and take delight in our capacity to make visible, travelling outwards, lightly, ecstatically, in all directions, travelling at the speed of light, travelling until our dreambodyvision vanishes into the egolessness of space. I recall Merleau-Ponty, quoted near the beginning of this division of the chapter: 'a being whose whole essence, like that of light, is to make visible . . . , to open itself to an other and to go outside itself.' I am not such a being; but carrying these words around with me, I find that they inhabit my vision and do influence the character of my being.

Chogyam Trungpa, Tibetan teacher of his great tradition, offers this psychogenetic account:

> Our most fundamental state of mind . . . is such that there is basic openness, basic freedom, a spacious quality; and we have now, and have always had, this openness. Take, for example, our everyday lives and thought patterns. When we see an object, in the first instant there is a sudden perception which has no logic or conceptualization to it at all; we just perceive the thing in the open ground. Then immediately we panic and begin to rush about trying to add something to it, either trying to find a name for it or trying to find pigeonholes in which we could locate and categorize it. Gradually things develop from there.[127]

What gradually develops in the individual is, of course, in our present epoch, the enframing of vision: the historically determined character of a process which, in Welwood's formulation,

> locks [the glance or gaze] into particular interpretive schemes. . . This 'locking in' is reinforced by the overlapping sequence of perceptions, thoughts, feelings which create a dense texture of mind that obscures the underlying ground of pure awareness.[128]

Because of this 'locking,' it becomes extremely difficult for our vision to inhabit an open space of luminous energy while living also through the vision of the egobody. But the wisdom of all our great civilizations celebrates that moment of vision when we experience a sudden shift, a sudden release, an opening . . . and we feel, with the special composure of deep and genuine assurance, that things really can be 'seen as they are.'

Part VI
Time of Decision

(1) 'Whither is God?' he cried: 'I will tell you. *We have killed him* – you and I. All of us are his murderers. But how did we do this? . . . What were we doing when we unchained this earth from its sun? Whither is it moving? Whither are we moving? Away from all suns? . . . Are we not wandering as though through an infinite nothing? Do we not feel the breath of empty space? Has it not become colder? Is not night continually closing in on us? Do we hear nothing as yet of the noise of the gravediggers who are burying God? . . . God is dead. God remains dead. And we have killed him.'

Nietzsche, *The Gay Science*[129]

(2) The night brought on by the death of God is a night in which every individual identity perishes. When the heavens are darkened, and God disappears, man does not stand autonomous and alone. He ceases to stand. . . The death of the transcendence of God embodies the death of all autonomous selfhood, an end of all humanity which is created in the image of the absolutely sovereign and transcendent God.

Thomas J. Altizer, *The Descent into Hell*[130]

(3) Yet the past which is negated by a revolutionary future cannot simply be negated or forgotten. It must be transcended by way of a reversal of the past, a reversal bringing to light a totally new light and meaning to everything which is manifest as the past, and therefore a reversal fully transforming the whole horizon of the present.

Mark C. Taylor, *Erring: A Postmodern A/theology*[131]

(4) The gateway of the moment is that decision in which prior history, the history of nihilism, is brought to confrontation and forthwith overcome.

Heidegger, *Nietzsche*, vol. II: *The Eternal Recurrence of the Same*[132]

In a recently published contribution to the postmodern discourse around theology, a work by Mark Taylor titled *Erring: A Postmodern A/theology*, the author offers a reflection to open his own undertaking. He writes:

> Postmodernism opens with the sense of irrevocable loss and incurable fault. This would be inflicted by the overwhelming awareness of death – a death that begins with the death of God and 'ends' with the death of our selves. We are in a time between times and a place which is no place. Here our reflection must begin.[133]

Although I have not explicitly identified the movement of our thinking in this book with the movement which calls itself 'postmodernism,' I shall now – now that we are far enough along not to be spellbound by the word – acknowledge my awareness of an important affinity, here; an 'elective affinity' based on the fact that the project of this book is a sustained reflection on the modern epoch. Insofar as our discourse takes modernity, the whole of the modern world, as its subject for thought, it is necessarily a contribution to the discourse of postmodernism. This discourse comes into being when our thinking leads us to the edge, the margin, the in-between, the *Bardo* space. To think *about* the modern world *as such*, i.e., *as* modern, or *in its modernity*, is to think in a space opened up by *the difference* this thinking makes: the difference, I mean, between that place, wherever it is, where we are standing and the perimeter of the world, the epoch, our thinking has gathered into a whole and questioned from that standpoint. To question the *whole* of modernity is, in some sense, to position oneself 'outside' its reach and range. Of course, it is difficult to say where this place *is*. It could not be otherwise.

For the most part, I agree with Taylor's narrative. But I think we must be extremely careful when we conceptualize the 'effects' of the 'death of God' in relation to 'the death of our selves' or 'the death of Man'. The raging of nihilism forces us to confront the possibility that we shall bring about our own destruction, together, perhaps, with the destruction of all but the most primitive of planetary life. But the death at issue in postmodern thought is not to be reduced to such literalism. Foucault's term, 'the death of Man,' refers to the exhaustion of the old dream of humanism, the vision of

Man which brought the modern world into being, but which this very world is now rapidly bringing to an end. Taylor's phrase, 'the death of our selves', can be understood in two different ways. In one sense, it refers to the end of the modern self and leaves open the question as to whether or not a different self is possible. In the other sense, however, it would refer to the end, or the death, of *all* historical possibilities for 'selfhood' and would deny that anything like self-development, personal growth, and individuation is possible. It should be clear that I could not agree with this second line of interpretation, and that, indeed, the project we are engaged in here is intended as a demonstration that the second, more extreme position, which I would identify as itself a form of nihilism, is not at all compelling. It is not, unfortunately, very clear where Taylor is standing in regard to this critical question. Perhaps it is indecision which generated Taylor's 'sense' of where postmodernism begins: 'the sense,' as he puts it, 'of irrevocable loss and incurable fault.' While I understand the experience behind these words, I am worried that the 'more extreme position,' as I have called it, may dominate our understanding of this 'sense'.

J. Hillis Miller has reviewed the history of metaphysics which precedes the 'death of God.' For him, as for Heidegger, the 'death of God' is related to the rise to power of ego-logical subjectivity:

> The ego has put everything in doubt, and has defined all outside itself as the object of its thinking power. *Cogito ergo sum*: the absolute certainty about the self reached by Descartes's hyperbolic doubt leads to the assumption that things exist, at least for me, only because I think them. When everything exists only as reflected in the *ego*, then man has drunk up all the sea. If man is defined as subject, everything else turns into object. This includes God, who now becomes merely the highest object of man's knowledge. God, once the creative sun, the power establishing the horizon where heaven and earth come together, becomes an object of thought like any other. . . . In this way, man is the murderer of God.[134]

If we follow Heidegger – and also the text of the *Bardo* – and think of God, or indeed of all gods, as beings who *mirror* the visionary being of mortals, then it is possible to grasp the historical significance of the 'death of God' in relation to the human potential for self-development. With the 'death of God', what happens to our cultural mirror for the Self? If we counted on the authority of God's light to reflect back to us our ideals for selfhood, the 'death of God' would seem to leave us without any higher ideal by reference to which we might steer the course of our self-development. (How is the shattering and loss of this mirror-function related to the weakening of what Freud called the 'super-ego'? I see a deep connection here.) When we look into the mirror now, we see only ourselves. Thus, the 'death of God' spells the dominance of 'narcissism' as cultural paradigm for the Self's development. This has unsettling consequences.[135]

Taylor says:

> With the death of God, a dark shadow falls over the light that for centuries illuminated the landscape of the West. Paradoxically, this eclipse begins during the period known as the 'Enlightenment' and marks the dawn of what is usually labelled 'the modern era'. . . . The humanistic atheist fails to realize that the death of God is at the same time the death of the Self.[136]

But if the Self is *not* the ego, the 'death of the self' should *only* mean: the death of the ego-logical structure with which the Self had historically identified itself. Or does Taylor want to argue that, in the wake of the 'death of God', *no* Self is possible? I agree with him that 'narcissism is finally nihilistic,' that a humanism which makes the Self into a new 'sovereign' tends to become 'inhuman,' and that, as he says,

> Carried to its conclusion, the pursuit of self-possession actually dispossesses the searching subject. When consumption becomes all-consuming, self-affirmation is transformed into self-negation.[137]

But Taylor's limitation, a limitation he shares with all philosophers and psychologists whose thinking is still under the spell of the *Freudian model* of self-development, is that he seems unable to think self-development without the myth of a regression to 'secondary narcissism.' If we want to understand the potential for self-development in a post-modern way, we must break with Freud; we must go beyond his regressive interpretation of the myth of Narcissus. We need to read the myth in a different light – as Jung does, for example, through his 'phenomenology of the Self'.

It is, in fact, precisely this regression which turns the process of self-unfolding into a pathology. There is a real experiential difference between authentic and false subjectivity: processes of insight, growth and development are *different* from processes patterned by 'narcissistic character disorders'.

Clinicians say that they can *see* deep depressions, depressions of the Self, within the heart of these narcissistic disorders. This is not surprising, because the Self which ends its journey in a regression to its 'original' Self is bound to get stuck in a 'sense of irrevocable loss and incurable fault.' But this is also, from a historical, or epidemiological point of view, the depression, ultimately, of nihilism. If it were just a question, today, of 'the death of God', we could throw ourselves into a journey beyond the ego. But today we must face into the oblivion of Being. And, in this sense, we are indeed experiencing the death of the Self – the spirit, I mean, and not, here, the ego.

In writing about Nietzsche's 'Eternal Recurrence of the Same,' Heidegger says:

> If we survey once again at a single glance our presentation of Nietzsche's thought of eternal return of the same, we cannot but be struck by the fact that our explicit discussion of the thought's *content* has receded markedly before our constant emphasis on the right *way* of approaching the thought and its conditions. The conditions may be reduced to two – and even these cohere and constitute but one.
>
> First, thinking in terms of the moment. This implies that we transpose ourselves to the temporality of independent action and decision, glancing ahead at what is assigned to us as our task and back at what is given us as our endowment.
>
> Second, thinking the thought as the overcoming of nihilism. This implies that we transpose ourselves to the condition of need that arises with nihilism. The condition requires of us that we meditate on the

> endowment and decide about the task. Our needy condition itself is nothing other than what our transposition to the moment opens up to us.[138]

In the framework of this book, 'what is given to us as our endowment' is our visionary being, or capacity for vision, our pre-ontological attunement by Being; and 'what is assigned to us as our task' is a vision of the next historical Self: a vision belonging, not to the ego, but to a Self in this 'transposition,' a Self whose vision is steered by a guardian awareness of Being. 'The condition of need that arises with nihilism' is the *pathology* of vision. This pathology constitutes our task. The most urgent task is to let this suffering speak, to help *bring* it to speech, to speak of, and not shut our eyes to, the unspeakable. To let what *needs* to be seen be *seen*.

Heidegger's text continues:

> . . . now the conditions of the thought-process as such thrust their way to the forefront. With the thought in question, *what* is to be thought recoils on the thinker because of the *way* it is to be thought, and so it compels the thinker. Yet it does so solely in order to draw the thinker into what is to be thought. To think eternity requires that we think the moment, that is, transpose ourselves to the moment of being-a-self.[139]

At the end of the passage which we read just *before* this one, Heidegger speaks of 'what our transposition to the moment opens up to us.' At the end of *this* passage, he calls this moment a 'moment of being-a-self.' We may infer, I think, that 'being-a-self' means being open to whatever the moment of thought 'opens up' to us. Should processes of self-development like this be described in terms of narcissism? I suggest that this terminology *obscures* our understanding of the 'Self' called for in the wake of the 'death of God' and the 'death of Man,' though it is undoubtedly useful in bringing the *pathology* of vision in the modern epoch to speech and thought.

The end of the paragraph we have been considering reads as follows:

> To think the recurrence of the same is to enter into

> confrontation with the 'it is all alike,' the 'it isn't worthwhile'; in short, with nihilism.

I would simply like to call attention to the fact that the speech of suffering through which nihilism appears here is a speech with which we are all, I'm sure, all too familiar. It is easily recognizable as the speech of depression, of a Self stuck in the moodedness of despair, carrying around with it a sense of some irrevocable loss, some irremediable absence: an experience of valuelessness, emptiness and nothingness at the very *centre* of its being. Such depression is an *ontological* depression: it is the 'inwardness' of nihilism, consuming the *spirit* of the modern Self like a cancer. And it is concealed, is masked, by the ego's narcissism. The pathology is both individual and collective: it is distributed throughout the population, like an epidemic, but affects different individuals diversely. As a cultural phenomenon, narcissism constitutes a cultural *defense* against the depressions of nihilism; but, as we have seen in our earlier analysis of the rise to power of the egobody in the modern world, narcissism is also involved in the *constitution* of the very conditions which have driven us into a cultural depression: conditions, namely, which deepened our forgetfulness of Being and conceal from us ways of going into the heart of this nothingness – into that encounter with nihilism from out of which a new vision, a new historical spirit, could perhaps begin to emerge.

In his 'Analysis' of Heidegger's work on Nietzsche's thought of the eternal return, David Krell writes, quoting Heidegger, that,

> With Zarathustra the tragic era begins. Tragic insight has nothing to do with either pessimism or optimism, 'inasmuch as in its willing and in its knowing it adopts a stance toward being as a whole, and inasmuch as the basic law of being as a whole consists in struggle.'[140]

Krell adds that *The Gay Science* 'stresses the essentially tragic nature of beings in general', whereas the story of 'The Convalescent' (*Zarathustra*) which we looked into earlier in this chapter, stresses 'the tragic insight gained in the glance of an eye – eternity as the Moment.'[141] But without the insight, vision is vulnerable to deep depression. Without the insight, vision is strongly tempted by the forgetfulness of Narcissus.

Like the tragic insight, the depressions of this age are amplifications of a need for wholeness-of-being. The depression may be about an individual's experience of himself – as empty or dead in the *centre* of his being, as not 'whole' in this sense – or it may come from a deeper body of vision, a collective sense of some shared historical experience: of loss, of death, of rejection and exile, absence, loss of wholeness, loss of meaning and purpose. The Self is historical: its development, its wholeness of being, is intertwined with the historical experience of the nihilism of Being, the concealment, or absence, of Being as a whole.

In the *Heraclitus Seminar*, Fink notes that,

> The beginning of our interpretation of Heraclitus by way of lightning was supposed to indicate that there is the basic experience of the outbreak of the whole. In the everyday manner of life, this experience is hidden. In everyday life we are not interested in such experience. In everyday living we do not expressly comport ourselves toward the whole, and also not when we knowingly penetrate into the distant Milky Way. But a human has the possibility of letting become explicit that implicit relationship to the whole as which relationship he always already exists. He exists essentially as a relationship to being, to the whole. For the most part, however, this relationship stagnates. In dealing with the thinker Heraclitus, one can perhaps come to such an experience in which the whole, to which we always already implicitly comport ourselves, suddenly flashes up.[142]

The stories about vision we considered in 'Vision: Our Capacity to Make Visible' (Part V, *supra*) are stories about such sudden flashes: the sudden perception of a gleam, a quiet glow, a soft shining, a reflecting. These flashes only happen, if and when they happen, for a vision deeply rooted in a sense, an awareness, an attunement, which connects it to Being as a whole. Such a vision is not the vision of the historical egobody, but the vision, rather, of the transhistorical dreambody; a vision inhabited, not by the spirit of rage, but by the spirit of peace.

'For the most part,' Fink says, our relationship to the whole 'stagnates.' 'Stagnation' is not only another word for the 'effects' of nihilism; it also figures in the epidemiological discourse which surrounds our pathologies of the Self: our narcissistic disorders, our pervasive culture of narcissism, our widespread depressions.

Krell takes up the historical need in this pathology when he writes, in his 'Analysis,' about 'the moment as a matter of and for decision'.[143] Decision: the antithesis of stagnation, depression, narcissistic indecision and inertia. The decision concerns the historical world; but it also, as Krell reports, concerns the development of the Self. When the Self feels stagnation permeating its very being, it has fallen very deeply into the depressive or narcissistic position. Krell writes:

> Such decision, Heidegger says, is a taking up of one's self into the willing act. Yet precisely *how* this is to occur Nietzsche never managed to communicate. Heidegger suggests that while such taking up is an authentic appropriation of self it is also the propriative event for historical mankind as a whole. As *Ereignis*, eternal recurrence of the same displays the covert, essential relationship of *time* to *being as a whole*. Yet it is a time, we might add, which for Heidegger *hovers somewhere between* the ecstative temporality of *individualized Dasein* and the essentially historical unfolding of *Being*.[144]

The historical unfolding of Being requires a corresponding unfolding of the Self. Thus, the danger of nihilism calls upon the Self to develop its visionary spirit, to grow beyond itself in relationship to its sense of Being as a whole.

Here, again, is Nietzsche:

> Learning to *see* . . . is almost what, unphilosophically speaking, is called a strong will: the essential feature is precisely *not* to 'will' – to be able to *suspend* decision.[145]

The 'moment of vision,' says Nietzsche, is now: this present instant, this gift, this present of the instant. The moment is instanced when, from within whatever position of time and space we may be standing in, we channel our vision into the historical task we see before us. Heidegger, too, speaks of 'decision'. It is time, here, for some thought about 'decision.' Heidegger speaks, as we have seen, of our 'endowment' and the 'task' before us. For him, 'decision,' or 'commitment,' is the third term. Our endowment projects a historical task; but the task calls for decision. The decision which is called for is a *historical* decision related to the Question of Being – the question which reminds us of the ontological difference. Earlier, I argued that the ontological difference is also 'decision': the most *primordial* decision. It is, I suggested, that which, under the sign of the thunderbolt, we saw in the constitution of the visual field as a whole: the ontological difference as primordial differentiation, primordial *Riss*, the *écart* of primordial *ecstasis*. Endowment, task, decision. The historical task depends upon an existential decision: a decision which commits our visionary being to the recollection of the ontological difference as that primordial decision which first brought forth the visual field according to the law, the *nomos*, of figure-ground differentiation. The recollection binds our visionary being to a decision which 'repeats' the more primordial decision, that 'rendering' of the field as a whole by grace of which our vision is opened up for us. Our decision to recollect the more primordial 'decision' is a historical 'repetition' of that always elusive, always invisible event, the *Ereignis*, through which the ontological difference suddenly bursts into visibility. Our visionary recollection thus becomes a *homologein*: a correspondence, a genuine taking-up of the task appropriate to our visionary endowment. The historical decision to take up this task is, as Heidegger says, 'a taking up of oneself into the willing act': it is a task through which the Self grows, unfolding and developing, on the stage of history, the ontological potential 'laid down' for it by the *legein*, the *nomos*, of the ontological difference.

The ontological difference, as primordial decision, enters into the constitution of our visionary being: it decides the conditions of our 'endowment'. With this endowment, an ontological claim against our vision is laid down in a primordial *legein*. This claim sets the Question of Being before us a historical task. The task calls for historical decision. The historical decision is up to the individual. But what is involved in such individuating decision? Nietzsche's words – the ones we just quoted – are startling and provocative: 'not to will,' he says; and by this he means 'to be able to suspend decision.' This is extremely significant when we consider it in the light of Heidegger's critique of Nietzsche, in which he argues that Nietzsche's concept of the will to power does not 'overcome' the rule of metaphysics, but rather continues it, and even deepens our submission to the spreading of nihilism. In *Gelassenheit*, Heidegger goes beyond this critique to touch on an attitude in which willfulness is suspended, neutralized.

Here, however, we see Nietzsche himself touching on this very same question. Perhaps even Nietzsche understood somehow that the historical task calls for a decision which is *no* decision, and for a will which is beyond the metaphysical dualism of willing and not willing: a willingness, we might say, in which willfulness has been suspended or neutralized.

Be this as it may, I wish at least to quote from Nietzsche's account of his inspired moods while working on the *Zarathustra* text. Here, even more so than in the preceding quotation, we see Nietzsche's flickering awareness – awareness and recognition – of an experience in which the will's position of domination is *reversed*: an experience, therefore, from out of which we might begin to think the historical task without depending on the metaphysical will. What, however, is decision without will? Nietzsche writes this:

> Has anyone at the end of the nineteenth century a clear idea of what poets of strong ages have called *inspiration*? If not, I will describe it. – If one had the slightest residue of superstition left in one's system, one could hardly reject altogether the idea that one is merely incarnation, merely mouthpiece, merely a medium of overpowering forces. The concept of revelation – in the sense that suddenly, with indescribable certainty and subtlety, something becomes *visible*, audible, something that shakes one to the last depths and throws one down – that merely describes the facts. One hears, one does not seek; one accepts, one does not ask who gives; like lightning, a thought flashes up, with necessity, without hesitation regarding its form – I never had any choice. . .[146]

It would not, I think, be very difficult for us to read Heidegger's conception of the *Ereignis* as 'historical decision' in terms of this experience. The will *is* of course involved; but only in the achievement of the *epochē*, its suspension and neutralization: the will does not destroy itself in rage; rather, it practices a spiritual discipline, a discipline of 'care' which *develops* the Self. Thus, the will could develop itself as a willing guardian of the wholeness of Being. And the ego's willfulness would be accordingly channeled into the historical task of a visionary being whose daily life gradually becomes a shining example of self-individuating commitment functioning as a medium and organ of the visibility of Being.

The ontological attunement of our visionary nature, our endowment, sets before us a difficult task. It is an individual task, requiring, as it does, the development of the deeper Self; but it is also a task in history. The task calls for decision: decision like the lightning, though; decision like the opening Nietzsche describes when he speaks of his becoming 'merely incarnation . . . merely a medium.' The more our individual commitment to the historical task becomes a decision of this kind, the more it may approximate the happening of that *other* 'decision', the ontological difference which structures the field of our vision. The *homologein*, the correspondence, can be seen wherever this approximation moves vision to make visible, through hermeneutical unconcealment, *alētheia*, the happening of the ontological difference as invisible structuring of the field.

In a startling passage, Nietzsche turns the lightning into a symbol for the *overcoming* of willfulness. His experience with 'inspiration' even moves him to tell a story in which the lightning, traditionally, at least in our culture, a symbol pointing to influences of the masculine archetype, is suddenly *turned* into its opposite: a symbol belonging to an archetypal complex which seems, in our present cultural experience, to be associated more strongly with the feminine principle. The thunderbolt not only comes *out* of the night; it also points back *into* it. It points like a sign, into the wholeness of the night, a wholeness we do see, and yet, do not.

As a final elucidation of the character of the historical decision for which our present time calls, I would like to set before our eyes some words that Heidegger repeated at the close of the *Heraclitus Seminar*.[147] These words, from Periander of Corinth: *In care, take the whole as whole.*

In caring for the whole and taking it to heart, we take

good care of ourselves. Giving up the object, giving up the subject, vision opens, and we begin to *see* with care for the whole.

CHAPTER 4

Truth

Part I

The Difference Between Alētheia and Truth as Correspondence

The English word 'philosophy' is derived from a Greek word that means 'love of wisdom'. But wisdom is nothing without a love of truth. From the most ancient times, this love of truth has provoked the philosophical mind, moving it to question what others take for granted and to think what calls for thought. The question of truth is at the center of Heidegger's philosophical reflections. Even Heidegger's earliest work expressed his concern for truth. This concern turned him around and led him back to antiquity, to the Greeks. In the texts of the Greeks, he encountered the word *alētheia*. The hermeneutical interpretation of this word, a word that for a long time has been translated as 'truth', made him question and rethink the modern experience of truth. Even before *Being and Time,* Heidegger had already been arguing that another meaning of 'truth' was concealed in the Greek word: a meaning which the Greeks themselves apparently did not experience and think. Except, perhaps, for Herakleitos, whose teachings proclaimed that 'nature loves to hide', they understood the word to mean 'correctness': truth in the sense of 'correctness', a satisfying representation of reality. On the basis of this insight, Heidegger insisted on an essential *difference* between truth as correctness and truth as opening or unconcealment – the hermeneutical meaning he saw hidden in the more common meaning of the familiar word.

In this chapter, we will take up two fundamental questions: One is this: What *is* truth as a visionary experience? In other words: how does truth manifest, how is it experienced, through vision? And the other is this: What is the *character* of our vision as, i.e., when it is, an experience of truth? In taking up these questions, we will attempt to illuminate the ontological difference between truth as correctness and truth as an opening, as unconcealment.

Although, as I said, the Greeks of antiquity seem to have used the word *alētheia* to mean 'correctness', i.e., truth in a more or less 'positivist' sense, I shall, in the following chapter, follow Heidegger's decision to reserve the Greek word for use in referring to truth in the hermeneutical sense of unconcealment, an event that is opening. (Note that, for Heidegger, *the word itself* becomes hermeneutical. It is undoubtedly correct to say that he saw the new meaning because his *reading* of the Greeks was hermeneutical. But the hermeneutical method in his reading is not the point. Rather, the point is that the new meaning is *itself* hermeneutical: 'unconcealment' refers to a hermeneutical process; it says that *truth* is hermeneutical.) Heidegger's decision was probably based on his desire to disentangle the question of truth from questions of historical fact. In any case, he lodged the Greek word in the discourse of modern philosophy, where its alien presence would immediately call attention to the culturally radical meaning by means of which he was challenging the modern experience of truth as correctness.

It seems that, for Plato and the philosophers who followed him, *alētheia* really meant only what it meant in the ordinary usage of everyday life: correctness; a *homoiosis* between *logos* and *pragma*; an *adequatio intellectus et rei*; an agreement, or correspondence, between knowledge and its object. Therefore, in 'The Origin of the Work of Art', Heidegger observes: 'Agreement with what is has long ago been taken as the essence of truth'.[1] In other words, he asserts that, according to the tradition established by Plato and maintained to this day, 'A statement is true [only] if what it means and says is in accordance with the matter about which the statement is made.'[2] (In *Der Satz vom Grund*,[3] Heidegger speaks of an 'Übereinstimmung von Vorstellen und Anwesenden' and ascribes this view of truth not only to 'the natural experience and saying' of ordinary people, but also to the experience and thought of most philosophers since Plato.)

Heidegger argues that the entire history of the discourse on truth has left unthought the experience of truth as an opening which lets unconcealment happen. Even the ancient Greeks, the very people who spoke of truth by *calling* it *alētheia*, did not experience and think the truth *as* unconcealment. Even for them, the field of Being which grants all beings the light of their presencing remained essentially unheeded:

> . . . we must acknowledge the fact that *alētheia*, unconcealment, in the sense of the opening of presence [*die Lichtung von Anwesenheit*], was originally only experienced as *orthotēs*, as the correctness of representation and statement. But then the assertion about the essential transformation of truth, that is, from unconcealment to correctness, is also untenable.[4]

It is Heidegger's contention that the primordial phenomenon of truth – the way it primordially appears, manifests, comes to light – has been 'covered up' by Dasein's 'forgetful' understanding of Being: its misunderstanding reduction of the Being of beings to the dimensionality of their being-ready-to-hand.[5] (Note that the etymology of the word 'phenomenon' relates it to an experience with the dynamism, the energy, of light: in other words, visionary experience is the source of this understanding of truth.) The enabling clearing (*gewährende Lichtung*) remains, as he says it in *Der Satz vom Grund*, 'unheeded' (*unbeachtet*).[6] Perhaps, though, considering that the 'literal' meaning of *alētheia*, the ordinary Greek word for truth, speaks of a negation of forgetfulness and concealment (*lēthē*), Heidegger is correct in suggesting that at least an unconscious experience with the primordial phenomenon of truth, and at least a rudimentary, intuitively felt understanding of it, must have been already present in the Greek world, despite the prevailing tendency to regard truth – what they called *alētheia* – as a question of correctness:

> we must not overlook the fact that, while this way of understanding Being (the way which is closest to us) is one which the Greeks were the first to develop as a branch of knowledge and to master, the primordial understanding of truth was simultaneously alive among them, even if [only] pre-ontologically – and it even held its own against the concealment implicit in their ontology – at least in Aristotle.[7]

Following Plato, the Greeks may have thought of truth in terms of correctness; but at least they continued to *call* truth by a word with etymological traces that allude to some primordial experience with unconcealment, some encounter with the phenomenon of light, an awesome event in which something comes to light.

Modernity, however, is governed by the so-called 'correspondence theory of truth': truth as correctness (*Richtigkeit*), correspondence (*Entsprechung*), agreement (*Übereinstimmung*). This theory is inherently committed to the assumption that the experience of truth is invariably – and necessarily – structured in terms of the relationship between a subject and its object: the conformity of a subject's representation to the givenness of the object represented.

By going back through the history of philosophical discourse, back to the early Greek word for truth, Heidegger is able to see and retrieve a meaning which adumbrates an historically new experience with truth: an experience different from our familiar one; different from the experience which had seemed, for so long, to be satisfactorily articulated by some historical variant of the correspondence theory.

It is not necessary, and in any case not very fruitful, to debate whether or not Heidegger has 'adequately' conceptualized what the ancient Greeks actually experienced and thought. Furthermore, such a debate would beg the question, since it would already assume at the outset that questions of truth must always be understood in terms of correspondence. This assumption is precisely what must be called, here, into question, and temporarily suspended.

What matters is not the factual truth of his claim about the Greeks and their 'truth', i.e., whether or not his claim proposes a representation which corresponds to the historical facts, but only, rather, this: that his encounter with the Greek word, his struggle to understand it, enabled him to enter into an experience with truth which the terms of the correspondence theory cannot begin to articulate – and cannot possibly do justice to. What matters is that his efforts to return to the root of the Greek word gave him, as a reward for such exertions, an experience with truth that set him to thinking: an experience, we might say, which moved him into an open space unknown to those who stay in line by thinking only of correctness. From the standpoint of correctness, Heidegger's path, the pursuit of *alētheia*, can only be errancy, subversion – at best, a futile passion. But what matters, in the final analysis, is really only this: that his interpretation of *alētheia* as unconcealment has *opened up* a dimension of truth which was not visible to the philosophers of modernity.

In the Greek word for truth, in *alētheia*, Heidegger sees truth for the first time as unconcealment, and not, as the tradition would have it, in terms of correspondence, resemblance, agreement, correctness, reliability, or certainty. Nor, as the same tradition has often insisted, as some immutable, eternally fixed, totally settled state of affairs. In 'The Question Concerning Technology', he contends that *alētheuein* should be taken to designate an event: truthing, something which happens where there is a letting be and an opening up. *Alētheuein* therefore requires an epistemological posture, or attitude, which lets unconcealment happen, lets things show themselves as they are – in a sense of 'are' which understands that facts are not isolated states and things are not unchanging, self-contained substances, but *are* in their *phainesthai*, *Alētheia*, truth as unconcealment, calls for a distinctive epistemological and ontological attitude; it takes place when there is a *practice* of truth which lets things come forth, lets things present themselves in their own way – on their own terms, and not, instead, only on ours.[8]

In *Der Satz vom Grund*, Heidegger maintains that,

> Only what *alētheia* as opening [*Lichtung*] grants [*gewährt*] is experienced and thought, not what it is as such.
>
> This remains concealed [*verborgen*]. Does this happen by chance? Does it happen only as a consequence of the carelessness [*Nachlässigkeit*] of human thinking? Or does it happen because self-concealing, concealment, *lēthē*, belongs to *alētheia*, not just as an addition, not just as a shadow to light, but rather as the heart [*Herz*] of *alētheia*?[9]

Note the reference to shadows, the presencing of which we shall be thinking in Part II of this chapter. Herakleitos, called 'The Obscure One', loved the play of shadows; but ever since Plato, shadows have been banished from metaphysics and epistemology: their status is even lower than that of images. But, for this very reason, they can, with their return to philosophical discourse, their return from exclusion, be extremely subversive, anarchic, and . . . illuminating. The status of the shadow in this passage is the status assigned to it by a metaphysical tradition Heidegger will contest – to some degree, in fact, on this very point. Note also Heidegger's reference to the heart. I am prepared to argue that this reference is not just ornamental. The experience of *alētheia*, of truth, as unconcealment demands of us an attitude of openness which can only come from the heart; it demands an exceptional capacity for caring – an attitude expressed by the German words *bewahren* and *gewähren*, which mean 'to preserve' and 'to guarantee', and which themselves preserve and guarantee the truth, the 'wahr' at their root.

Part II

Shadows and Reflections: The Play of a Negative Ontology

(1) Eyes are by nature light offered to shadow.
Roland Barthes, *Sur Racine*[10]

(2) . . . the shadow . . . is a manifest, though impenetrable testimony to the concealed emitting of light.
Heidegger, 'The Age of the World Picture.'[11]

(3) . . . an Hieroglyphical and shadowed Lesson of the whole World. . . .
Sir Thomas Browne, *The Religio Medici*[12]

(4) To the aesthetic eye, conceptual thinking obscures – or even loses – the shadow.
Patricia Berry, 'The Training of Shadow and the Shadow of Training'[13]

(5) O Lord, what is man that thou shouldst notice him?
What is mortal man that thou shouldst consider him?
Man is like a breath;
His days are like a passing shadow,

He flourishes and grows in the morning;
he fades and withers in the evening.

Psalms

(6) But the joy of mortals in a short time ripens to the full, and soon again falls to the ground, stricken by adverse fate. Creatures of a day, what is anyone? What is he not? Man is only the dream of a shadow. But when a gleam of sunshine comes as a gift from heaven, a radiant light rests on men and life is sweet.

Pindar, *Pythian Odes*[14]

(7) Nor are only green and dark colours, but shades and shadows contrived through the great Volume of nature, and trees ordained not only to protect and shadow others, but by their shades and shadowing parts, to preserve and cherish themselves. . . . Darkness and light hold interchangeable dominions, and alternatively rule the seminal state of things. . . . Light that makes things seen, makes some things invisible, and were it not for darkness and the shadow of the earth, the noblest part of Creation had remained unseen, and the Stars in heaven as invisible as on the fourth day, when they were created above the Horizon, with the Sun, for there was not an eye to behold them. The greatest mystery of Religion is expressed by adumbration. . . . Life itself is but the shadow of death, and souls departing but the shadows of the living. . . . The Sunne itself is but the dark *simulacrum*, and the light but the shadow of God.

Sir Thomas Browne, *The Garden of Cyrus*[15]

(7) Every man casts a shadow; not his body only, but his imperfectly mingled spirit. This is his grief.

Henry David Thoreau, 'A Week on the Concord and Merrimack Rivers'[16]

Herakleitos, enjoying the interplay of light and shadow, meditated on the fact that nature loves to hide. Plato, however, was not amused by this. For him, shadows are nothing but obstructions to knowledge, seducing the eyes into error. Shadows are not real and their presence is deceptive. Being unreal, they cannot be objects of knowledge; they yield no knowledge. By shading and concealing, they fall in the way of knowledge.

Plato thought of knowledge in terms of a total visibility, a total illumination, directed towards things fixed and immutable. (See *The Republic*, Book VI, 500.) The eyes of the body cannot enjoy the possession of knowledge, for their vision is entangled in the world of shadows and images. Knowledge, knowledge of essences, knowledge beyond *doxa*, is a possession of the rational mind. It requires a disciplined life, a life entirely given to pure contemplation. The gaze of pure contemplation does not see the shadows of things. It turns away from becoming and perishing (Book VI, 508), and from 'shadows and reflections' (511). It forgets that its way to knowledge passed through perception; it forgets that this way was foreshadowed by the shadows things cast. Contemplating the pure forms, not a shadow of doubt disturbs it.

Descartes repeated Plato's shadow-free vision of knowledge and set the course of modern philosophy. The history of modern philosophy is therefore a history without shadows: a history in which the ontology of the shadow has not been appreciated and understood. But, as Samuel Todes has argued in 'Shadows in Knowledge', this misunderstanding of shadows entails a misunderstanding of knowledge: a conception that regards knowledge as 'shadow-free'.[17] Taking the primacy of perception to be a fundamental tenet in his epistemology, Todes, like Merleau-Ponty, sees the presence of shadows as a significant source – and phase – of knowledge.

'My present field of consciousness', James argued, 'is a centre surrounded by a fringe that shades insensibly into a subconscious more.'[18] Before Derrida, and even before Husserl, William James introduced adumbrations into the structure of consciousness.[19] James was not as radical as Derrida, but he understood better than Husserl that this analysis of consciousness involved a fundamental revolt against Platonic and Cartesian epistemology. Modern philosophy needed his phenomenological deconstruction: of self-evidence, of certainty, of foundationalism, of the transparency of consciousness.

'Profane vision', as Merleau-Ponty terms it, has an inveterate tendency to forget its passage through penumbras and adumbrations; the circumambient regions of darkness;

the necessary concealments; its own part in the casting of shadows. Of these phenomena he writes:

> In fact they exist only at the threshold of profane vision; they are not seen by everybody.[20]

This blindness in profane vision – this impoverishing indifference – perhaps explains, in part, the persuasiveness of the traditional epistemology, which entirely abolished the realm of shadows. Profane vision 'rests', as he points out, 'upon a total visibility'.[21] This phenomenological analysis supports my contention that the enframing of modern metaphysics is implicit in the character of normal perception. The fact of the matter is that the denial of the shadow is essentially *related* to our forgetfulness of Being, for the dimensionality of beings, their way of being present, does not lend itself to total, shadowless visibility.

Instead of 'multiplying the systems of equivalences'[22] in 'a delirium which is vision itself',[23] profane vision seeks a duller light of reassuring constancy, uniformity, and stability; it resists, or avoids, the opportunities for a more poetizing, more dimensional vision: opportunities that meet the inherent promise of vision through 'the imaginary texture of the real'.[24] The shadow is cast out of awareness because it would divert the eyes into a play of light and dark. It would distract the eyes from their instrumental functions.

Heidegger's perceptions confirm those recorded by Merleau-Ponty. Heidegger writes:

> Everyday opinion sees in the shadow only the lack of light, if not light's complete denial. In [aletheic] truth, however, the shadow is a manifest, though impenetrable, testimony to the concealed emitting of light.[25]

According to Heidegger, the shadow's presence 'points', in fact, to 'something else, which it is denied to us of today to know'.[26] Some scholars may be surprised to read these passages, so convinced are they that shadows would be of no concern to the thinker of Being. But I am surprised that he said so little. The words I just quoted, for example, are like hieroglyphs, themselves taking part in the nature of shadows. What they adumbrate I cannot decipher.

It could be argued that what is 'denied' us today is an experience of concealment and unconcealment – in short, the truth of Being. In the Jewish liturgy for Rosh Hashanah, the Jewish new year, there is a passage which can be read in a way that darkly hints at this interpretation:

> Darkness is thy concealment, and thick darkness thy tabernacle.[27]

The shadow that plays in the light, moving across surfaces without leaving the slightest trace, does point, because of its relationship to the rising and setting of the sun, to a deeper and essentially impenetrable darkness: 'the concealed darkness of its origin as the unlighted'.[28] In 'The End of Philosophy and the Task of Thinking', Heidegger seems to be reminding us that shadows make visible the interplay of presence and absence:

> the clearing, the open region, is not only free for brightness and darkness, but also for resonance and echo, for sound and the diminishing of sound. The clearing is the open region for everything that becomes present and absent.[29]

Beings both are and are not their shadows; they both are and are not present, there where their shadows have fallen.

Pointing to absence as well as presence, pointing into the invisible as well as towards that which is visible, shadows display the ontological dimensionality – where, if Heidegger's words apply here, 'There remains only the play; the highest and the deepest.'[30] Shadows, like reflections, although in a different way, deepen, heighten, extend and enrich the field of visibility.[31] Themselves mere surface, without dimension, they nevertheless supplement the levels of illumination, multiply its facets, intensify its presence.

Our shadows lay claim to us, claiming our being for the lighting of Being, for the element in which we are rendered visible – visibly shadowed, visibly reflected. Shadows and reflections manifest the fact that our substance belongs to the gift of light. This claim can be seen, for example, when a shadow criss-crosses a friend's body or falls across the solid materiality of the furniture in a room: the shadow negates substantial solidity and coherence. The claim can also be seen when reflections on a lake mirror the building which stands at its edge and create an image that captures our attention and competes with solid reality. Shadows and reflections show us that we are, after all, phenomena of light.

The outcast shadow stalks us; it is always nearby. The shadow we cast. The shadow stalks us like death. The shadow we cast – or cast out – is in fact our own death, the shadow of our own death. 'I am gone like a shadow when it lengtheneth'. (*Psalm* 109:23) 'This', writes Thoreau, contemplating the peculiar cast of the mortal, 'is his grief.'[32] The shadow confirms our reality, our presence in the midst of the visible. But its confirmation is playful and ultimately ambiguous, as befits its hermeneutical, Hermes-like nature. (Hermes is a trickster.) It plays, one might say, with our ego-logical seriousness. And it alludes to our participation in the common fate of all that is visible. If we are like a shadow. . . . Crossing our field of vision, the shadow makes visible our being-in-time. It shows us ourselves – but the self it shows is insubstantial, ever-changing. It is our visibility in the light of time, confirming our mortal's finitude, our measured time, reminding us of our mutability, our continuous disintegration.

Where are we? Here? Only here? Am I not also cast out there, where my shadow falls and trembles? The shadow casts doubt on our assumption of a permanent identity as ego-logical centres; it mocks our habitual identification with the solid and fixed boundaries we take comfort in. It negates the ego's ontology. James wrote:

> What we conceptually identify ourselves with and say we are thinking of at any time is the centre; but our *full* self is the whole field, with all those indefinitely radiating subconscious possibilities of increase that we can only feel without conceiving, and can hardly begin to analyze.[33]

The shadow is one of those radiating possibilities of increase; we do not like to identify ourselves with its 'negative' nature, but it belongs, nevertheless, to our full self.

The shadow makes visible our mortal inherence, our fatal investment, in the natural elements, foreshadowing the conditional term of our visibility in unified, coherent, corporeal form. 'My days are like a lenghtening shadow'. (*Psalm* 102:12) Shadows are the signatures of time in the realm of the visible: the legibility of time, of the *passage* of time, and of the allotments cast by time, in the light of Being. They are time's way of playing with us: time's way of inscribing us, as measured, and as passing, into the textures of an ever-changing light, sacred text of life and death. Shadows cast our being into time. Silently, and with an infinitely light touch, they remind us that our being stands and falls in the wholeness of Being.

There is an experience with reflection that, like our experience with shadows, plays a more significant role in our lives than we might at first suppose. In the dynamics of interpersonal experience, the reflection, in fact, plays a *decisive* role in the formation of the ego. Sullivan, Lacan and Kohut have analyzed in detail how the ego is gradually constituted as a unified, coherent centre of experience within the dynamics of interpersonal mirroring.[34] In particular, they have shown how the mirroring that takes place between the infant and its mothering one gives the child an immediately reflexive responsiveness that enables the child to experience herself proprioceptively. This awakens in the child the functioning of a corporeal schematism, a latent sense of embodied identity and capacity, that was awaiting such interactions to bring a new phase in its development. In the process of this mirroring, the child not only enjoys an opportunity to 'see' himself 'from within' by seeing the different responses he calls forth from the mothering one; he also enjoys an opportunity to learn new dispositions, attitudes and behaviours, seeing them first in the other and receiving their implicit schemas through what Merleau-Ponty called 'postural impregnation'. In this way, an initially turbulent, disorganized body begins to develop greater unity and coherence, and its potential for a more centralized, purposive organization of movement. Gradually, a body-self begins to take shape, replicating the body image that represents the child's adaptation of an ego-logical identity.

But there is, in this shaping of the self, an inevitable alienation. Since the child's identity as an egobody is formed by a process of reflection, its constitution takes place through the other. This mediation, however, is only the beginning of a life-long dependence on the presence of reflections, confirming, approving, rejecting, correcting, taking possession of the invisible spirit. Being dependent on the mirroring of the reflection, we grow up with a tendency to be captured by the power of its peculiar ontology. But what *is* this reflection to which we are beholden? Like the shadow, it is totally insubstantial: without thickness, without weight, without sound and tangibility, its only reality is for the eyes that behold it. The reflection, like the shadow, is a symbolic paradigm of impermanence and insubstantiality.

Thus, paradoxically, the reflection that engages our narcissism also in the final analysis will deny us our dream of a body not subject to the law of impermanence and will destroy our delusions of a substantial ego-logical identity. For a vision given to thought, the presence of reflections becomes a visible reminder of our true ontological condition. Seeing ourselves reflected by the surfaces of our world – by windows, mirrors, shining metal objects, and bodies of water – we catch a glimpse of ourselves in a moment of disintegration, dispersed among the elements. We are given a chance to see ourselves as we really are: in a flash of truth, we may see ourselves in our existential *ekstasis*, a multidimensional event in the field of light. We may see ourselves in our radical openness; we may see ourselves opened up, our *Dasein* simultaneously manifesting in a multiplicity of visible forms. We may see ourselves as bodies of light, a phenomenon of the light, endlessly shaped and reshaped in the mirror-play of Being.

But this assumes a vision already to some degree *open* to the presence of the reflections: it assumes a vision willing to be responsive and thoughtful. Both shadows and reflections are occasions for learning, occasions for our self-development as visionary beings. Being nothing substantial – the mere play of light – they invite the gaze to let go, to let be. For the fact is that, when the gaze relinquishes control, the light is free to play. Shadows and reflections will really dance and play only for a poetizing gaze which is itself dancing and playful. This visible play of light can in turn, however, teach the gaze how to play. And the more playful the gaze, the more these reflections and shadows will correspondingly play. In other words, the more we learn to let go, the more the light will manifest in a play of shadows and reflections. But the more these shadows and reflections can play, the more they will elude the grasp of an objective vision – vision in the grip of the will to power. This, then, is how the presence of shadows and reflections can teach the gaze *Gelassenheit*: letting the light play. In *Delimitations: Phenomenology and the End of Metaphysics*, John Sallis reflects on the traditional hostility of metaphysics toward the play of the imagination and explores the possibility of a metaphysics constituted by such play.[35] I like his speculations. I am convinced that the play of shadows and reflections is subversive, and that an historically new ontology could emerge from a visionary attitude open to their play.

The gaze can be just as threatening to the authority of metaphysics as the play of the imagination. A gaze open to the subversions in the play of shadows and reflections implicitly inaugurates a vision of truth which deconstructs the metaphysics of presence. It is a gaze which implicitly understands the difference between truth as correctness and truth as *alētheia*. It is not from the stare, an act of direct, frontal looking fixated on its object, but rather from a playful gaze, a gaze which delights in ambiguities, uncertainties, shifting perspectives and shades of meaning, that we will learn what *alētheia* is as an experience with truth: truth not determined by the metaphysical tradition.

Heidegger introduces us to *alētheia* by spelling out the difference between truth as correctness and truth as *alēthia* in the context of language. He tries to define the difference as a difference between the assertive discourse of propositions or statements and the hermeneutical discourse of poetizing. I would like to clarify the difference phenomenologically, in terms of our visual experience. In order to do this, I must differentiate two styles or ways of seeing and make explicit how they are related to the two kinds of discourse. I will argue that truth as *alētheia* requires a playful gaze, a gaze open to the play of light, and that we can develop an understanding of *alētheia* by allowing ourselves to be more responsive to the presencing, the way of being, which is manifest in shadows and reflections.

Let us, first of all, get clear about the two kinds of discourse. In *Being and Time*, Heidegger makes the point that 'the roots of the truth of assertion reach back into the disclosedness of the understanding'.[36] And, in 'Poetically Man Dwells', he suggests that,

> The more poetic a poet is – the freer (that is, the more open and ready for the unforeseen) his saying – . . . the further what he says is from the

> mere propositional statement that is dealt with solely in regard to its correctness and incorrectness.[37]

How, and why, does the truth of poetic discourse differ from the truth of propositional discourse? How do these two kinds of discourse constitute two kinds of experience with truth?

Let us take note of what our words tell us about themselves: the *proposition* posits, puts down, settles, places, positions before, or in front, fixing securely; and the *statement* states, puts into a state, an unchanging form, stopping all movement, closing the process. Thus, when the truth that belongs to propositional discourse is allowed to regulate our poetizing, it brings the play of sounds and meanings, the interplay of words and experiencing, to a stop.

If, as Heidegger says in *Being and Time*, 'Assertion and its structure (namely, the apophantical "as") are founded upon interpretation and its structure (viz., the hermeneutical "as"). . . .',[38] could we say that this is because there is an important sense in which poetizing discourse 'founds' the discourses of propositions and assertions? Giambattista Vico seems to have thought this. And also Rousseau. But we, of course, must go deeper into the complexities of this question.

Propositions inevitably constitute, as our word for them tells us, structures of op-position, imposing a metaphysics of order; they put subjects and objects in their 'proper' place; they involve possession: possession of meaning, possession of the sensuous, the 'phonological'. They manifest our cultural logocentrism. They assume the self-possession of a Cartesian cogito, and from this position, they set up a confrontation: what they mean is always directly in front, yielding to the desire for perfect clarity and distinctness. Propositions represent a static reality, a statable state, and their truth is one which always simply says what it says and is what it is. Propositional truth is without any surrounding field of meanings; *a fortiori*, what the proposition represents is always an isolated *state* of reality. Consequently,

> When the assertion has been expressed, the uncoveredness of the entity moves into the kind of being of that which is ready-to-hand [*zuhanden*] in

> the world. But now, to the extent that, in this uncoveredness, as an uncoveredness of something, a relationship to something present-at-hand [*vorhanden*] persists, the uncoveredness (truth) becomes, for its part, a relationship between things which are present-at-hand (*intellectus* and *res*) – a relationship that is present-at-hand itself.[39]

Apophantical truth, the truth of assertions, statements and propositions, lets things appear, manifest, come forth, in a very distinctive way. The prefix attached to the word 'apophantical' tells this story itself. It is not a way which gives the thing 'free play' in its showing of itself. In *Being and Time* Heidegger remarks that, 'In "setting down the subject", we dim entities down to focus. . . .'[40] In Chapter One, we already had occasion to reflect on what this remark about our grammar implicitly, hermeneutically hints at with regard to our vision – an experience with vision controlled for the most part by a socialized ego thoroughly identified with the structure of subject and object, and with the grammatical structure that repeats it. In the following passage, too, the structure of normal discourse, and the forgetful ontology which it maintains, are characterized by reference to the character of our visual perception:

> The 'as' gets pushed back into the uniform plane of that which is merely present-at-hand. It dwindles to the structure of just letting one see what is present-at-hand, and letting one see it in a definite way. This levelling of the primordial 'as' of circumspective interpretation to the 'as' with which the present-at-hand is given a definite character is the speciality of assertion. Only so does it obtain the possibility of exhibiting something in such a way that we just look at it.[41]

Here the discursive difference between propositional assertion and circumspective interpretation is explicitly correlated with a difference between two styles or modes of vision and their respective ontologies. Assertion is associated with a vision that compresses the dimensionality of the field within which things presence: in such a vision, figure and ground are reduced to a uniform plane; and the truth of Being, its dimensionality presencing in unconcealment, is systematically occluded. Thus, if we were to depict the difference between the vision associated with the truth of assertion and the vision associated with the truth of hermeneutical discourse, we might draw the first, perhaps, as a straight line – a line of desire, going without delay, and in the most direct way, from the subject to the object of its desire. By contrast, the second might then be drawn as a web, emphasizing the circumspective character of this vision, ever alive with an encompassing sense of the wholeness of Being, and of being within a dimensionality deeper and more expansive than the totality which vision can make visible.

In vision, the most fitting analogue of the assertion is perhaps a stare. And as we noted in Chapter One, *Being and Time* does in fact give thought to the stare, together with other attitudes in looking and seeing. Both stare and assertion manifest a tendency to master and dominate: a tendency with an extremely aggressive character. The discourse of assertions is a discourse hospitable to aggressive and oppressive strategies. Heidegger uses the phrase 'pushed back into'. He also speaks of a process of 'levelling', a process making a 'uniform plane'. This often leads, however, to further discursive operations: in particular, regimentation, restriction, censorship.

For poetry, this amounts to its negation, as Plato understood: univocity, the rule of the monotone. Plato would tolerate no resonance, no play of sound, no ambiguity, no shades of meaning, no hints and forebodings, no movement of sense or sound, no polyphonic perversity, no polyphonic or symphonic subversiveness, no anarchy of sounds, no indeterminate though determinable complexity, no ungraspable richness and dimensionality. Plato, who disliked shadows and adumbrations, and wanted them excluded, along with images, from his utopian republic, also recommended very strict control over musical modes and registers. Unlike the discourse of poetizing, the discourse of assertion, propositions and statements must be a discourse of univocity, without any surprising or unregulated fluctuations in sound.[42]

In poetizing discourse, there is a sensuous, phonological field for the play of sound and sense. Truth as *alētheia* appreciates this field of play, where presences and absences are intertwined. Truth as correctness does not.

Poetizing discourse, a discourse which stays in touch with

our felt, pre-ontological pre-understanding of the world, is the locus of truth, of the experience of truth, in its primordial functioning – its 'setting itself to work' – as a bringing-forth: as disclosedness, as deeply hermeneutical. Assertions are aimed, like the look of a vision possessed by desire, at some fixed state of reality. Clarity and distinctness are their ideals; they aspire to certainty, a complete, self-contained, self-evident meaning. Since their sensuous element, the medium of their soundings, inevitably introduces echoes, and many undertones, overtones and undercurrents of meaning, along with shades and adumbrations of meanings, puns and allusions, a potential riot of ambiguities, uncertainties and even, as in Mallarmé, a vortex of confusion, the metaphysician would like to put assertions in the realm of a disembodied thinker.

I have noted that the correspondence theory situates our experience with truth in a representational relationship, a relationship determined by the structure of subject and object. As we know, however, this structure conditions the character of our involvement with things. According to Heidegger, this character is oppositional, confrontational, challenging, and marked by aggression. Thus, for example, in 'Science and Reflection',[43] he speaks of our 'challenging' and in 'The Turning',[44] of our 'injurious neglect'. In 'The Origin of the Work of Art', there is a reference to our 'assaults'. He asks:

> Can such an assault perhaps be avoided? – and how? Only, certainly, by granting the thing, as it were, *a free field* to display its thingly character directly.[45]

In propositional assertion, words state: they propose meanings; they set them down and make them stand; they are positioned in a fixed way, corresponding to a correspondingly fixed state of affairs. In poetic discourse, words work in a different way: they play, they resonate freely, they breathe. The correspondence theory holds true, and shows itself to be true, in the more monotone, more tonally disengaged discourse of assertions, statements, propositions, representations, subjects and objects. Scientific discourse is supposed to be 'rational', and that means: value-neutral, dispassionate, disinterested. In a logo-centric culture, scientific discourse is the paradigm of proper discourse, legislating the *correct* way for words to work. Do we suppose that the truth it proposes could sing like poetry?

The truth of *alētheia* requires a different kind of discourse in which to appear: a discourse in which words are allowed, made, so that they can sing, singing freely, out of control, released into the field of their own resonance. In poetic discourse, or, more generally speaking, in poetizing discourse, words can defy logical connectivity to open up hermeneutical fields for the coming forth of new meanings. Poetizing discourse is therefore not derivative from some paradigmatic propositional discourse. Rather, it is the contrary which is true, in both senses of 'true'. Propositional discourse derives from an original poetizing: it is constituted by transformation from a discourse in which words are poetizing into a discourse in which words are assigned fixed positions, clear and distinct meanings.

Assertions stop the poetizing process; they close the dimensions of sound and sense. They do not free the truth in their words, allowing them to resonate, to question, to conceal and open up. The correspondence theory cannot take into account the hermeneutics of truth: the contextuality of its happening; its field-dependence; its play with showing and hiding.

In poetizing discourse, both sound and sense require a theory of truth which understands and appreciates their 'ecstatic' play within an open field. In the *phonological* dimension, there must be a field for the play of sounds: echoes, resonances, overtones and undertones, onomatopoeia, polyphony, emotionally evocative sounds. Similarly, in its *semantic* dimension, the dimension of signifiers, there must be a field for the play of meanings: ambiguities, allusions, metaphors, shades of meaning, adumbrations of what is to come. Truth as correctness, truth represented in the discourse of statements, assertions, propositions, cannot do justice to the interactive processes essential to poetizing discourse. Truth as *alētheia* can, because it is hermeneutical: it lets sound and sense play in the interplay of presence and absence, identity and difference. To the poetizing process, the process of bringing experience as it takes shape into words that further shape it, *alētheia* gives a multidimensional field in which to unfold.

The style of vision that I would associate with the discourse of statements and propositions, and with the correspondence theory of truth, is exemplified by the direct, frontal gaze, the fixed gaze of a spectatorial 'subject' focussing on an object presumed to be independent of itself and positioned to permit it total visibility, absolute certainty, and a permanent body of knowledge. Statements require 'correct' vision. Truth as correctness requires a 'correct' gaze: a gaze correctly positioned, correctly focused, correctly pointed. The most extreme case of vision would thus be the stare.

The style of vision that I would associate with the hermeneutical discourse of poetizing, and with truth as *alētheia*, is exemplified by the playful gaze, gently relaxed, calm, centred by virtue of its openness to experience, its delight in being surprised, decentred, drawn into the invisible.

Perhaps, if we can allow our gaze to wander, to come under the spell of shadows and reflections, and can learn from *their* ways of being a more playful way to *be* with visible beings, we might at long last break the spell of metaphysics, a vision of ontology that has captured our imagination since the dialogues of Plato.

Part III
Power and Truth: Alētheia as a Gaze of Friendship and Caring

There is an extremely important *ethical difference* between correctness and *alētheia*, a difference that appears very strikingly when we analyze them as 'practices' in the context of social relationships. If there are different practices of truth, there are correlative differences in moral character: not only in that the practices are rooted in, and produced by, different types of moral character, but also in that the practices are *productive* of different character. Consider what Harold Searles, eminent American psychiatrist, has written about his experience of working with severely schizophrenic patients:

> Among the most significant steps in the maturation which occurs in successful psychotherapy are those moments when the therapist suddenly *sees* the patient *in a new light*. His image of the patient suddenly changes, because of the entry into his

> awareness of some potentiality in the patient which had not shown itself before. From now on, his response to the patient is a response to this new, enriched view, and *through such responding, he fosters the emergence, and further differentiation, of this new personality area*. [It is a process of] seeing in the other person potentialities of which even he is not aware, and helping him, *by responding to these potentialities*, to realize them.[46] (Italics added)

I value this passage because, in addition to its concrete clarification of what I mean by an ethical practice of aletheic truth, it details how the gaze can be responsive to others in a way that opens them to their ownmost, most individuating truth and is productive of character, of self-development, because of its caring, its respect for freedom. What Searles describes is a gaze decentred by its caring, a gaze which understands that processes of individuation, of self-development, are explorations of truth: truth not in the sense of 'correctness', i.e., conformity to an external norm, but truth in the sense of *alētheia*, and that helping other people be true to themselves requires a practice of truth grounded in caring. Just a glance, or a wink, may be sufficient, sometimes, to help another person into the freedom of truth.

By contrast, the gaze which corresponds to the practice of propositional (pro-positional) truth is a gaze which sets up an opposition (op-position) between the one who is looking and the one who is looked at; for the practice works by imposing a truth corresponding to the reality it sees, instead of bringing forth the truth like a good midwife – by letting others be, and setting them free to be true to themselves.[47]

Correctness *without alētheia* is embodied in a gaze that imposes on people a conformity to its predetermined representation. A gaze embodying correctness *with alētheia*, however, is a gaze that can encourage others to be true to themselves, so that they may develop their ownmost potentialities.

Here we touch on the ways in which, without the openness of *alētheia*, truth as correctness constitutes relations of power that can be extremely oppressive. Truth, in the sense of 'correctness', is always an exercise of power. So is the look. How we *see* this connection between truth and power, and how this perception affects our vision as a practice of truth are questions that direct our attention to the *character* of our visual life. Near the end of his suddenly closed life, Foucault began to study and question truth as a social practice. This led him to examine types of moral character in the telling of truth. I would like to supplement his work with some thoughts on the character of our *vision* as an experience with truth and on the clarification of truth as a visionary experience, a question of our perceptiveness. Character sets the tone for the gaze it produces; and the gaze sets the tone for the character it produces.

The 'assertoric' gaze, the 'propositional' looking that I would associate with the correspondence theory of truth, with truth as 'correctness', essentially tends to see from only one perspective, one standpoint, one and only one *position*. Such a gaze will therefore tend to be narrow, dogmatic, intolerant, rigid, fixed, inflexible, and unmoved: in sum, not very caring. By contrast, the aletheic gaze, the way of looking that I would associate with the hermeneutical theory of truth, with truth as 'unconcealment', would essentially be moved by a tendency to see from a multiplicity of standpoints and perspectives: with an awareness of contextuality, of field and horizon, of situational complexity; and with a corresponding openness to the possibility of *different* positions. The 'assertoric', 'pro-positional' gaze is essentially exclusionary, either excluding what cannot be seen from its present fixed position or else including what calls for different positions only by suppressing the differences: it allows only what can be seen from its own position. By contrast, the aletheic gaze is pluralistic, democratic: it tends to be inclusive, but in a way that does not deny or suppress the differences; and it understands the relationship between visibility and power. The gaze that is moved aletheically is a gaze that cares. This is the vision that should *steer* the 'assertoric' gaze.

The 'assertoric' gaze is ego-logical; it issues from the position of an independent and assertive egobody. By contrast, the aletheic gaze is rooted in an experience of the intertwining, a field of gazes: that communicative dimension of our bodily being where all gazes are felt to be interpenetrating, interdependent, and reciprocally responsive. This is the vision that should steer.

Faust in his study, 1652.

Part IV
Vision in the Light of Gathering

We mortals, gifted with eyes capable of turning, are free to turn our gaze where we will. In his treatise *On Free Choice of the Will*, Augustine writes:

> For turning, as it were, their backs to Thee, they are fixed down upon the works of the flesh, as it were in their own shade, and yet what even there delights them, they still have from the encompassing radiance of Thy Light. But the shade being loved, weakens the mind's eye, and makes it unequal to bear thy countenance. Wherefore a man becomes more and more darkened, while he prefers to follow what, at each stage, is more bearable to his increasing weakness. Whence he begins to be unable to see that which, in the highest degree, *is*.[48]

This story has been told many times and in many places. The Tibetan *Bardo* texts weave important spiritual teachings out of this experience of turning away. Plato's 'Myth of the Cave' likewise gives attention to our tendency to turn away from the light, the light of truth, the truth of light. The story of Moses returning to his people after communicating with God on Mt Sinai is another version of the same thought According to *Genesis*, the Jewish people, filled with fear, turned away from Moses when they beheld the extra-ordinary radiance that illumined his face. We have already had occasion to reflect on Heidegger's 'repetition' of this story in 'Alētheia', where he argued that

> [Mortals] turn [away] from the lighting, and turn only toward what is [actually] present, which is what imediately concerns them in their everyday commerce with each other. . . . [So] they have no inkling of what they have been entrusted with: presencing, which in its lighting first allows what is present to come to presence. *Logos*, in whose lighting they come and go, remains concealed from them, and forgotten.[49]

Most people, the *polloi*, appear to be as Fink noted, in a seminar on Herakleitos:

> [they are] merely lost in the many, and do not see the joining power of light. To be sure, they see the shining up [of many beings, visible] in the light. but not the unity of the lighting [as such].[50]

The stories of this turning away may be considered metaphorical; but this does not make them any less true. The metaphors, here, are not mere ornaments; they bespeak phenomenological truths. These stories are pointless unless they are understood to bear very directly on our experience as beings of vision.

Heidegger refers vision to the *Logos*, in whose lighting beings 'come and go'. According to Heidegger, in ancient times 'Logos' was a name for the gathering, the *legein*, of this lighting. He translated *legein* as 'gathering'. I suggest that, instead of turning away from the *legein* of the *Logos*, the gathering of the lighting that gathers all visible beings into its field of unity, our vision could turn towards it and itself become a gathering, a homologous gathering, inscribing in the openness cleared for events of light a gesture of gathering that repeats, and in that way makes visible, the more primordial gathering upon which it essentially depends.

Such a gesture would be, I think, a genuine *homologein*: a movement of the eyes true to the spirit of Herakleitos's words. In 'Poetically Man Dwells', Heidegger refers to 'the upward glance'. This is an example of the *homologein*: a vision whose own gathering reflects the gathering of the *Logos*. 'The upward glance', he writes, 'spans the between of the sky and earth.'[51] It is by virtue of such a gathering, a gathering of vision, that the Fourfold, *das Geviert*, is made possible. the Fourfold is a *mandala* in the field of light: a gathering mirror-play of gods and mortals, earth and sky.[52]

Heidegger's presentation of this *mandala* may be enriched by considering visionary experiences from other sources. Here, for example, is a visionary experience recorded by Carl Jung. Many years ago, traveling by train from Mombassa to Nairobi, Jung awoke one morning to glimpse through his window, on a rock high above the train, a dark-skinned man standing motionless, and leaning on a long spear. In his journal Jung wrote:

> I was enchanted by this sight – it was a picture of something utterly alien and outside my experience, but on the other hand a most intense *sentiment du déjà vu*. I had the feeling that I had already experienced this moment and had always known this world which was separated from me only by distance in time. It was as if I were this moment returning to the land of my youth, and as if I knew that dark-skinned man who had been waiting for me for five thousand years.[53]

This, I want to say, is an experience of gathering, an event taking place in the world of vision's rendering. It is not a gathering conscious of itself as a repetition of the gathering of the *Logos;* but it is a gathering which sees beyond the boundaries fixed by our standard Newtonian conception of time and space. If we must judge what Jung experienced, what he saw in terms of this paradigm, then we are compelled to say that it cannot be understood, and is therefore not possible. Later in this chapter, in the division bearing the title 'The Seer', I shall have more to say about this gathering. Suffice it to say here that Jung's experience manifests a vision capable of gathering visible beings into a non-standard light: into relationships rooted in the transpersonal and transhistorical dimensionality of the visionary matrix.

The gathering of vision may also, however, be much more 'ordinary': less of a challenge, perhaps, to our consensually validated sense of what is possible and what not. Consider, for example, what has been recorded of the ruminations of an old sage – a Lakota Indian named Lame Deer – still living, not many years ago, on tribal land in South Dakota. This old man, calling the attention of his visitor to the simple pot in which he was cooking his supper, spoke like a reincarnation of Herakleitos, whom visitors found, according to Aristotle, warming himself by his stove:

> It doesn't seem to have a message, that old pot, and I guess you don't give it a thought. . . . But I'm an old Indian. I think about ordinary, common things like this pot. The bubbling water comes from the rain cloud. It represents the sky. The fire comes from the sun which warms us all – men, animals,

> trees. The meat stands for the four-legged creatures, our animal brothers, who gave of themselves so that we should live. The steam is living breath. It was water; now it goes up to the sky, to become a cloud again. These things are sacred. We Sioux spend a lot of time thinking about everyday things. . . . We see in the world around us many symbols that teach us the meaning of life. We have a saying that the white man sees so little, he must see with only one eye. We see a lot that you no longer notice. You could notice if you wanted to, but you are usually too busy. We Indians live in a world of symbols and images where the spiritual and the commonplace are one. . . .[54]

Lame Deer's vision is nothing mystical. Being mindful of the environment, and especially of the invisible, he sees the pot at the centre of a cosmological *Gestalt*, gathering earth and sky, animals, gods and mortals. This gathering is, for him, the 'truth' of the pot – what the pot of meat *is*. He sees things in their ecological intertwining, not as the isolated, self-contained substances of our metaphysical tradition. His way of seeing things belongs to a culture different from the one we live in today; but it is not necessary for this difference to prevent us from learning from its wisdom. Our world is without those gods; but it is still dependent on earth and sky; and we, despite our science and technology, are still, in some sense, 'mortals' destined to die.

Nevertheless, if we do not shut our eyes to the world we live in, a different *Gestalt*, a different gathering, a different *Geviert*, would suggest itself. What if we attempted to see such gatherings taking place around all the things in our own lives? Guided by this question, I found myself looking at some things which happened to be near me – my television set and the tin of biscuits on top of it, attempting to see them as gathering around them, in concentric circles at greater or lesser distance, all the people involved in the political economy that made their presence in my home possible. Such a vision would, of course, gather together here many invisible people. The tin of biscuits, for example, gathered miners of bauxite, the crew of barges and freighters, makers of bakery ovens, bakers, lumberjacks and label printers. The television set gathered petroleum geologists, miners, engineers, mechanics, factory workers, factory managers, physicists, chemists, loaders, truckers, warehouse keepers, bankers, capital investors, accountants, electricians, store managers, sales-clerks, quality control testers, importers and shippers. Moreover, other people, even more invisible, must ultimately be included, because this network depends on the existence of systemic social institutions: those, for example, which govern contracts, control the flow of traffic, and protect rights in property and worker safety. Thus my vision, centred on the set and the tin, began to gather more and more people: all the people who in any way contributed to their production, transportation, financing and marketing – including all the people who planned, designed, built and continue to maintain the mines, quarries, wells, pipelines, factories, furnaces, baking ovens, offices, stores, warehouses and trucks needed by the process of production and consumption.

I soon began to see that, ultimately, because of a systemic interdependence, this vision would have to expand beyond those circles to include virtually everyone in the society. For even those people marginalized by the political economy and seemingly most unproductive – the homeless, the mad, the unskilled and unemployable – are sacrifices that contribute to the productivity of the system. But I could also see that the gathering could not be harmonious. Serious conflicts and divisions affected the people this vision gathered: inequities in wages, injustices in hiring and firing, struggles against sexual harassment, alienation and exploitation of labour, routines of coercion, dehumanizing modes of production, practices of fraud and deception. And a similar problematization appeared when I gathered all the people who contributed to the presence of the blueberries I bought for supper. Instead of seeing only the berries, I saw, gathered around them, all those who picked them. And this eventually led me to see the fact that there are many young boys and girls, the poorly nourished, poorly clothed children of migrant workers, who are not in school, forced to work from sunrise to sunset, or until they fall from exhaustion, for less than the minimum wage. Finally, I saw that, in the handling of the plants, these children are exposed to powerful insecticides damaging to their health.

Thus it happened that, in the light of this gathering, I began to develop some awareness, a capacity to see much more clearly – that is to say, more concretely – the different social positions and relationships that constitute the web of our political economy. I could see interdependencies; I could see my own complicity in the injustices; I could see my implication in the terrible logic of production and consumption, labour and capital.

Perhaps, then, in the light of this reality, so different from the gathering we may be seeing in the vision of our dreams, we will begin to see our way beyond the need, to a just and more humane restructuring of social life, rooting our visions of the future in a genuinely intersubjective sense of gathering. In the light of such gathering, utopian vision can turn critical and practical. It should be recalled that *legein*, the verb form of the Greek word for reason and rationality, meant 'gathering' and 'laying down'. In the light of its gathering, vision can make visible the unjust and dehumanizing social relations that surround the things of our world. And it can lay down an intersubjective ground, a ground of reversibilities for the achievement of our capacity to assume the viewpoint of the concrete other: a ground, therefore, for the reciprocities of rational communication – and for the possibility of progressive social change.

Rooted in this ground, vision's gathering can contribute to an immanent critique of social life. In 'Sociology and Empirical Research', published in *The Positivist Dispute in German Sociology*, Adorno proposed that such critique 'must dissolve the rigidity of the temporally and spatially fixed object into a field of tensions of the possible and the real'. Herein lies the truth, the *alētheia*, of the things we see around us. In the light of the gathering, the 'correct' way of seeing things, maintaining the invisibility of this field of tensions, can be problematized and resisted.

Part V

Lumen Naturale: Tracing the Repression of the Body of Light

Gershom Scholem, author of *On the Kabbalah and Its Symbolism*, tells us that 'Adam was by nature a purely spiritual figure, a "great soul", whose very body was a spiritual substance, an ethereal body, a body of light.'[55] I want to argue that the concept of the *lumen naturale* is a residue, is in fact the last discursive trace, of what I take to be an acknowledged experience of the body of light. The body of light – the fact that we are essentially bodies of light – is frequently evoked in the teachings of many 'primitive' spiritual traditions. It is also spoken of in traditions still in touch with their experiential origins.[56] In the course of Western civilization, however, the body of light has been subject to a process of distortion and repression. Nevertheless, Heidegger and Merleau-Ponty seem to have recollected this primordial experience of our visionary being, for there are places in their written work where it returns from its long repression, recovered for the discourse of philosophical thought.

In 'Alētheia', an essay concerned, as the title already informs us, with the question of truth, Heidegger affirms his conviction that we human beings 'are not only illuminated in the lighting, but are also enlightened from it and toward it'.[57] This point he clarifies further, claiming that, because of this, human beings

> can, in their way, accomplish the lighting (bring it to the fullness of its essence) and thereby protect it. Gods and men are not only lighted by a light – even if a supersensible one – so that they can never hide themselves from it in darkness; they are luminous in their essence. They are alight [*erlichtet*]. . . .[58]

What does he mean by this 'erlichtet'? If he is not speaking in words meant to be merely ornamental, then to what experiences is he referring? What does he mean by 'accomplish the lighting'? How – in what way – could we 'accomplish the lighting'? As always, I presume that these questions are phenomenological, intended to provoke in us some awareness of actual or possible experiences.

We are luminous, we are alight, in that we have been given the lighting of Being through a gift of nature. This lighting is, as it were, sealed within the protection of our flesh: inscribed, like a secret text of nature, into the scroll of our flesh. Because of this gift, we have the capacity to make visible – and to shed on our world the light of human meaning. This capacity to make visible, to illuminate, is of course a cause of some delight, and our eyes are alive, alight, with the joy we take in their own functioning.

But Heidegger declares that

> men belong in the lighting not only as lighted and

> viewed, but also as invisible, bringing the lighting with them in their own way, preserving it and handing it down in its endurance.[59]

What do we make of this relation to the lighting? Perhaps what Heidegger means by 'accomplishing the lighting' is spelled out further in the passage just cited: bringing the lighting with us, preserving it and handing it down. How we understand this, and whether and how we comport ourselves with such understanding is not only a *test* of our character; it is essentially constitutive, determinative, of our basic nature, our character.

How, then, might we understand this task? Recalling our beholdenness to the lighting, I suggest that we could be said to bring the lighting with us and to preserve or protect it insofar as we are mindful of its self-concealing presence and aware of our primordial dependency, aware that it is only by grace of this lighting that we are enabled to see. To recollect the elemental lighting *as* the lighting of Being is to gather it into ourselves and bring it with us, to preserve and protect it. And to make this lighting visible, manifest, *as* that by grace of which our eyes can see, is to light up this invisible lighting, to let it shine forth as such: this, I think, is to 'hand it down', to pass it on to others, so that they too may give it the protection of their thought.

Since the lighting cannot accomplish this manifestation, this becoming visible, by itself, but only insofar as our visionary organs function as its medium or channel, it may be said, as Heidegger does indeed say in 'Recollection in Metaphysics', that 'Being needs the reflection of a radiance of its essence in truth.'[60] We are richly rewarded for serving in this way, because there is a relationship of reflective co-responsiveness, of reciprocity, between our gaze and the lighting it discloses. Since the lighting of Being is lit up and becomes radiant when we make it visible as such, our responsiveness to the ontological task that claims us and gathers us into its mirror-play becomes an experience full of enchantment and joy; an experience through which, in communion with the lighting, we may feel ourselves somehow gathered up into the 'fulness', the 'truth', of our natural capacity for vision.

In *Being and Time*, Heidegger observes that,

> When we talk in an ontically figurative way of the

> *lumen naturale* in man, we have in mind nothing other than the existential-ontological structure of this entity, that it *is* in such a way as to be its 'there'. To say that it is 'illuminated' [*erleuchtet*] means that, as Being being-in-the-world, it is cleared [*gelichtet*] in itself, not through any other entity, but in such a way that it is itself the clearing.[61]

But is the 'lumen naturale' nothing but an ontical figure of speech? Since it is a literal truth that we are visionary beings, how could this language avoid making contact with our visionary experience? How could it have a merely ornamental meaning, when it says something that actually addresses our visionary experience? Even if the language of *Being and Time* is to be understood as merely figurative, the language of 'Alētheia', and of Heidegger's other writings on the pre-Socratics, clearly is not. To read the later essays in a figurative way would reduce them at once to nonsense. The fact is that they make sense, and only make sense, when read as referring to familiar, or at least possible, experiences with vision.

I do not believe that, in his later thinking, the concept of 'lumen naturale' was merely an ornament of rhetoric. I submit, moreover, that it is part of the ontic and everyday *misunderstanding* of our visionary nature to believe that the concept of 'natural light' is nothing but a figure of speech, rather than an accurate phenomenological description of a deeply realized experience of being human. If we suppose that 'lumen naturale' is only a metaphor, then we are left totally in the dark concerning what Heidegger really means, in a more literal sense, to be saying. If the language is merely figurative, why does he persist in using it? Wouldn't it be much better for him to say what needs to be said in a language governed by more stringent rules of clarity and directness?

The fact of the matter is that what the Latin phrase, 'lumen naturale', and its subsequent translations into the languages of later philosophical discourse, really 'preserved' and 'handed down' is but a paltry light, an extremely reduced and diminished *experience* of light, when compared to the ecstatic experience which anthropological knowledge tells us must originally have informed the term. Indeed, it now seems clear that, by the Eighteenth Century, this luminosity of being, awesome in its mystery, this strange light which made the flesh of mortals glow, and which, once upon a time, seems to have been *really visible* as a luminosity taken into the flesh, borrowed from the field of light, was no longer regarded as anything more than an ornament of speech. But the term was *not*, originally, an ornament. And the testimony of 'primitives' who continue to speak of their experience of luminosity just as did the ancient Greeks and Hebrews, should serve to remind us – although we always have at our disposal so many ways of defending ourselves against unsettling reminders. Would the Israelites have recoiled in fear if they had thought the radiance in Moses's face to be a metaphor, an optical illusion? The fact that visionary locutions have become, for us, mere figures of speech simply reflects the historical fact that we are no longer familiar with the *real experience* in question. – It is much easier to accuse our language than to blame ourselves; much easier to take the meaning out of words than to take the criticism in that meaning to heart.

I am arguing that the concept of the 'lumen naturale' originally referred to actual experience – an experience of the body as a phenomenon of light, an experience of the radiance sometimes visible in the human face, and especially visible in the eyes, when there is rapt attention or rejoicing – and that, in the discourses of Western history, this reference has been increasingly, systematically denied, although, even today, it continues to be experientially significant in numerous systems of 'excluded knowledges' flourishing on the margins of our culturally dominant discourse.

Because this argument could easily be misunderstood as a narrative in the tradition of Romanticism, as a metaphysical drive to recover some postulated origin, a metaphysically 'original' experience, it is necessary to observe that, in point of fact, there are many systems of discourse in which we can find references to a 'natural light', a 'body of light', that explicitly bear experiential meaning. In Book VII (532) of *The Republic*, Plato writes of 'that faculty which is the very light of the body'.[62] But evidence points back to much older experiences: in Orphism, for example, and in the Eleusinian Mysteries. An *experience* of the 'body of light' is also at the heart of the Tibetan Dzogs-chen teachings, and is no less significant in the shamanic wisdom of the Indians living in North and South America, the Dogon of Africa, the Eskimos of the Arctic, and the aboriginal peoples of the South Pacific. In sum, then, the 'original reference' is not a metaphysical projection, a desire for the origin, but an authentic visual experience.

The systematic repression of all experiences relating to the luminosity of the flesh, and in particular, the radiance of the face and the eyes, indicates a metaphysical closure to dimensions of experience that do not readily permit 'normal' conceptual control and fixation. The repression of the experiences to which 'lumen naturale' originally referred, culminating in the displacement of the concept and its relocation in a discourse of metaphors, of myths and fictions, is essentially consistent with the sequence of repressions that Heidegger documents in his critical interpretation of the history of metaphysics. According to this interpretation, there has been a progressive closure to the ontological dimensionality of experience, and this closure, traced with great scholarly care, is reflected in the history of the basic concepts of metaphysical discourse: concepts such as *energeia, physis, logos, alētheia, aitia, dikē, nomos, telos, archē* and *ousia*. (See Heidegger's essays on 'Metaphysics as History of Being' and 'Sketches for a History of Being as Metaphysics', assembled under the title, *The End of Philosophy*.)

Why this repression? As a beginning, I suggest that we consider the story, in *Genesis*, narrating what happened when Moses returned to his people after communion with God on the heights of Mt. Sinai. According to this story, the people saw his face shining with an awesome, preternatural radiance – and they took flight, overpowered by their fear. But such fear, as Freud has taught us, is the cause of repression. Is the history of our 'natural light' in the discourse of philosophy a *reflection* of this fear? I think it is. I think that this history has been a long history of repression.

The experience of our natural luminosity, our radiance as bodies of light, is an *ontological* experience, an ek-static experience of human being that takes in our ontological dimensionality, our primordial inherence in a field of light. Like all other ontological experiences, it eventually got reduced, until it was visible only as the 'natural light of reason'. And even this was reduced, by reason itself, to a mere figure of speech, the embarrassing trace of ancient myths and superstitions. Having successfully occluded the *actual* experience which originally gave rise to the concept, the metaphysics of reason could then argue very persuasively that the concept *never was* more than a metaphor. But reason's presumed 'clarification' is really just a conceit. We need to see that 'rationality' is a diminished, a partial light: a light which has been denied, or suffered to lose, its heartfelt, bodily reality, its visible truth.

The reflective luminosity of human being is much more than what is visible as the light of reason. In fact, the natural light of our being, being multidimensional, radiates not only as the light of reason, but as bodily warmth; the glow of health; warmth of personality; the visibility of joy; the warmth of benevolence and compassion; and also enlightenment, the glow of deepest mortal wisdom, the visible 'presence' of the sage.

In his essay on 'Spiritual Laws', Emerson observes that,

> When a man speaks the truth in the spirit of the truth, his eye is as clear as the heavens. When he has base ends, and speaks falsely, the eye is muddy and sometimes asquint.[63]

What Emerson is describing is an experience of truth: how truth appears, how it makes itself visible, in the eyes of an other. This is not fiction, not metaphor, but rather the phenomenology of vision, of truth as an event in the visible world.

The 'natural light' of reason, understood, now, as nothing but a metaphor, or as a trace of primitive folklore, is all that remains of that wondrous aureole of light which the early Christians saw and painted, true to their experience, in the ikons depicting their saints. Before the aureole of spiritual accomplishment became a convention of religious representation, it was a living visual experience, a phenomenological truth. The history of our conceptualization of this experience is a history of spiritual decline; and it represents our declining of the gift of light.

It is unfortunate that, as Heidegger says in a seminar on Herakleitos, 'The truth of the immediate experience of the world disappears by reason of the scientific interpretation of the world.'[64] Nevertheless, it is encouraging to realize, as Heidegger points out in 'Moiraͅ', an essay on Parmenides, that 'Knowledge and the evidence of knowledge cannot renounce their essential derivation from luminous disclosure, even where truth has been transformed into the certainty of self-consciousness.'[65]

Part VI
The Seer

(1) The eye accomplishes the prodigious work of opening the soul to what is not soul – the joyous realm of things and their god, the sun.

Merleau-Ponty, 'Eye and Mind'[66]

(2) Here the vision is free, the spirit exalted.

Goethe, *Faust* II

(3) The visionary capacity is basic to man, but it is intertwined with the capacity to cultivate the vision and to turn it into a reality by acting in its light.

Herbert V. Guenther, 'On Spiritual Discipline'[67]

(4) . . . vision which is a mirror or concentration of the universe. . . .

Merleau-Ponty, 'Eye and Mind'[68]

The 'seer' is one who can 'accomplish the lighting' and 'bring it to the fullness of its essence'. The 'seer' is one who exemplifies in some way an individual – that is to say, self-individuating – realization of the human potential, the human capacity to see. It is essential, however, that we demystify, making the accomplishments of the 'seer' accessible to experience. Mystification can assume two forms. One is an *inflation* which so glorifies and exalts the seer's accomplishments that they are put out of reach, totally beyond us. Such inflation has the effect of mythologizing the accomplishments. But, if they are not real, then we do not have to measure ourselves against them: we can twist ourselves free of their initial claim upon our character. The other form is a deflation. Though apparently the opposite, it has the same effect. Deflation works on the assumption that the accomplishments in question cannot be understood experientially, and concludes from this that the accomplishments of the seer must be fictitious. Demystification is the task of phenomenology. Our hermeneutical phenomenology will attempt to understand the seer's accomplishments in a way that 'retrieves' them for our experiential work, our continuing self-development as visionary beings.

In 'The Origin of the Work of Art', Heidegger articulates his view that, 'In the midst of beings as a whole, an open place occurs: there is a clearing, a lighting [*Lichtung*].'[69] And he holds that it is within this clearing, this openness, that human vision is born and takes place, being dependent upon it, but also necessary for its occurrence.[70] Merleau-Ponty puts it this way: the human being is 'a being whose whole essence, like that of light, is to make visible . . . , to open itself to another and to go outside itself'.[71] ('To make visible' does not mean a metaphysical drive to make everything fit a 'ready-to-hand' or 'present-at-hand' ontology. It does not mean 'conquering' the invisible, but rather letting it be seen, and protected, *as* invisible.) Merleau-Ponty's phenomenological interpretation of our luminous nature fleshes out the characterization Heidegger suggests in *Der Satz vom Grund*, where he writes of 'the ekstatic sojourn of human beings in the openness of presencing'.[72]

In reading Homer's *Iliad*, which he turns to in his essay on Anaximander, Heidegger sees an opportunity to spell out his own thoughts concerning the development of our visionary capacity. The seer, he writes, is 'one who makes present and belongs in an exceptional sense to the totality [*in das Ganze*] of what is present'.[73] This sentence calls for some clarification before we can think with it. The translator's word, 'totality', here, can be very misleading. What Heidegger has in mind, I think, is an organic 'whole', because wholes, unlike totalities, cannot be summed up, part by part, and apprehended, or grasped, in objective closure. Likewise, 'one who makes present' (*ein Anwesender*) does not mean that the seer's vision is a Cartesian gaze, a gaze moved by the metaphysical drive to totalize, to possess, to secure with absolute certainty; on the contrary, it is a gaze which can enjoy the whole precisely because it lets what is absent be absent, and lets the invisible be (seen as) invisible.

This is why Heidegger says that the seer 'belongs in an exceptional sense' to the whole of the visual world. But we need to go into this 'exceptional sense' with more of an eye for its experiential bearing. There is a descriptive passage in the *Phenomenology of Perception* that sheds light on the distinctive *spatiality* of the seer's gaze, making explicit the character of this gaze as a gesture of 'recollection', gathering all visible and invisible beings into an embrace of co-existence:

> . . . to look at an object is to inhabit it, and from this habitation to grasp all things in terms of the aspect which they present to it. But insofar as I see those things too, they remain abodes open to my gaze, and, being potentially lodged in them, I already perceive from various angles the central object of my present vision. Thus, every object is the mirror of all others.[74]

Merleau-Ponty speaks in a way which seems to suggest that such panoramic awareness is typical, is in fact a universal law of everyday perception; but this obviously is not the case. The gaze he is describing is an experience that most people would not call 'familiar'. Can we resolve the conflict between the vision in his account and the vision of common experience? I think we can, if we remember that ego-logically centred vision, the normal, socialized vision of everyday experience, developed out of a more global vision, a vision with an implicit pre-understanding (or a pre-ontological understanding) of a field of simultaneous co-existences, and that an implicit sense of the mirror play remains with us, carried by the visionary body, the body of visionary feeling, even after the ego-logical structure of vision is firmly in place. Because we normally 'sublimate' this mode of vision, abandoning it for a more ego-centred structuring of our experience, the gaze Merleau-Ponty describes does not seem familiar. But, because it continues to function below the threshold of consciousness and is at any time retrievable, Merleau-Ponty speaks of it in the grammar of essential universality. I suggest that the seer is someone who has retrieved and developed the panoramic vision of pre-ontological understanding – and kept it alive in the gaze he actually experiences.

'Presence', of course, is not only an experience of our spatiality; it is also an experience that is deeply rooted in the temporality of our vision. According to Heidegger, 'The seer sees [only] inasmuch as he *has seen* everything as present.'[75] The seer, then, is 'the one who has *already seen* the whole of what is present in its presencing'.[76] This 'whole' is 'all that becomes present, whether at the present time or not'.[77] How is this possible? Since it seems, at first, utterly impossible, there is a strong temptation to suppose that it must be a phantasy of the metaphysical drive to achieve, or bring into view, a totalized presence. But if we succumb to this temptation, we will miss an opportunity to develop our visionary capacity: we will not see that the seer's experience exemplifies a new existential possibility for every human gaze.

But how could anyone, howsoever developed, get to see the *whole* of the past and the *whole* of the future? This suggestion seems like madness, until we understand that 'seeing the whole' does not mean seeing every detail in a summed totality, but rather a certain capacity to feel, to sense, the whole significance of our historical existence, our being-in-time. (See a letter Rilke wrote on August 11, 1924 at Muzot.)[78]

Among the Cheyenne, the most spiritually developed people are called 'Keepers of the Shields of Light'.[79] The sense in which the seer is 'one who has already seen the whole of what is present in its presencing' is to be defined in terms of an attitude, a developed discipline, of caring: the seer's vision is, literally speaking, a *Wahrnehmung*, a guardian awareness of temporality as the meaning of Being. 'The seer', writes Heidegger, 'speaks from the preserve [*Wahr*] of what is present.'[80] The seer's gaze is a gesture of gathering, an embrace of the passage of time, rooted in the wholeness of heartfelt caring.

The seer's achievement is the consummation of a relationship to the passage of time in which he has succeeded in freeing himself from the normal pathologies. Whereas *our* vision tends to be motivated, all too often, by what Freud called 'the compulsion to repeat', the vision of the seer is free: it does not cling to the past. The seer moves on; she lets the past be past, neither repeating its pain nor craving its pleasures. Paradoxically, however, it is precisely because she 'forgets' the past that she gets it back again as a present. As Adorno points out, in writing about the process of recollection: 'Only because it brings back the past as irretrievable does it grant it to the present.'[81] The seer likewise experiences a different relationship to the future, neither spoiling his enjoyment of the present with anxieties about an unknowable future nor missing what pleasures the present has to offer because he lives in a state of *perpetual* deferment, always hoping for something much better. (In his *Essays*, Montaigne reflected on our tendency to 'lose the present through fear of the future'. And, in his *Pensées*, Pascal wrote: 'we never live, but we hope to live; and as we are always preparing to be happy, it is inevitable we should never be so'.)[82] For the seer, whose vision is *released* from the standard re-presentational patterning of time,

> even what is absent is something present, for, as absent from the expanse, it [still] presents itself in concealment. What is past and what is to come are also *eonta* [i.e., in being].[83]

Thus there is a moment, a 'moment of vision', when Being as a whole becomes visible *as* a whole; and this happens because the seer's vision is rooted in a wholehearted discipline of caring: a 'thoughtful maintenance of Being's preserve'.[84]

This view of Being as a whole is therefore not at all a gaze driven by the metaphysical desire for presence; it is not at all a gaze rooted in what Michael Heim has described as the 'temporalizing mood peculiar to Leibniz's analytical formalism', namely, the 'all-at-once simultaneity of totalizing presentness' modelled after the *visio Dei*, the deity's 'intuitive cognition'.[85]

Indeed, the seer's gaze achieves a vision of the whole *as* whole precisely because it *refuses* the temptation to totalize and instead embraces the temporal *ekstasis* which spans our past, present, and future. Unlike the multitude, which does not know the *ekstasis* and does not live in the openness of its present, the seer is one who understands and enjoys the intertwining of protentional and retentional experiences, living to the fullest its opening up of the present:

> The present still holds on to the immediate past without positing it as an object, and since the immediate past similarly holds its immediate predecessor, *past time is wholly collected up and grasped in the present*. The same is true of the imminent future, which will also have its horizon of imminence. But with my immediate past, I have also the horizon of futurity which surrounded it, and thus I have my actual present seen as the future of the past. With the imminent future, I have the horizon of a past which will surround it, and therefore my actual past as the past of that future.[86]

The past is not simply past, the future not simply future; and the present, therefore, 'is not shut up within itself but transcends itself towards a future and a past.'[87] Most people, not understanding this, live in a present which is more or less closed. Their 'present' is *only* a present: it is a graspable totality, but not an experience of time as a whole. The seer's experience is different from this, because it is grounded in a thoughtful maintenance of the clearing. The seer's experience rests in a presence that vision keeps open. Thus it is an experience of time as a whole precisely because it is *not* a drive to grasp or master the serial totality. Understanding the sense in which the past and future are gathered into their presence, the seer, holding herself open, may accordingly say that she has 'already seen' what is to come.

The seer's vision is a keeping of time and a keeping of beings in the light of time as a whole. It is a preserving and protecting, a *Wahrnis*, a *Gewähren*; and it is in *this* sense a vision of truth. But in which sense of 'truth'? If it is supposed that the sense of 'truth' in question here is truth as correctness, truth as a correspondence or *adequatio*, then we would indeed have reason to say (i) that such vision is not possible and (ii) that it unquestionably exhibits the narcissistic delusion of the metaphysical drive to possess a totalized presence. But this supposition would be a tragic misunderstanding. For the 'truth' of the seer's vision is aletheic; it is not a matter of correctness. This is the significance of Heidegger's distinction between *das Gegenwärtige* and *das Anwesende*. It is a question of visionary openness, grounded in a felt sense of the whole of time, time gathered into the wholeness of the seer's care. It is not a question of the correspondence between the seer's vision and time-points in a fixed temporal reality, a permanent presence of temporal moments, an unchanging state of Being. And this is why the seer's *sense* of presence is not entangled in the 'metaphysics of presence' that has dominated the history of Western thinking – and the life of Western society.

In an essay called 'Heidegger/Derrida: Presence', John Sallis notes that 'What is common to metaphysics is the positing of Being (*Sein*) as presence, its self-positing as metaphysics *of presence*.'[88] And he clarifies this notion of 'presence' in terms of *Being and Time*:

> In *Being and Time*, presence means predominantly, though not exclusively, *Vorhandenheit* (presence-at-hand, in the usual translation). This is to be understood in its correlation with pure seeing, with *noein*, with intuition (*Anschauung*): When something gives itself to one's sheer gaze, when it is simply there for one's looking, displaying itself before and for apprehension, then it has the character of being-present-at-hand. Such a character is to be contrasted with that of things with which one deals concernfully, when one manipulates things and puts them to use. The contrast between *Vorhandenheit* and *Zuhandenheit* (readiness-to-hand, in the usual translation) is well-known. . . .[89]

But the seer is, as Heidegger says, exceptional; and what makes the seer's gaze 'exceptional' is precisely the fact that it is neither instrumental nor detached, neither a gaze which sees things in terms of their *Zuhandensein*, nor a gaze of 'pure seeing', a seeing of things as *Vorhandensein*, which posits things in terms of what Sallis describes, at another point in his study, as a 'sheer perceptual presence', a 'simple sensory presence in the living present'.[90]

In 'Hegel und die Griechen', Heidegger argued that

> *Alētheia* is that which, most worthy of thinking, is [still] unthought. . . . For this reason, *alētheia* remains for us that which above all is to be thought – to be thought as released [*gelöst*] from any metaphysically bequeathed reference to a notion of 'truth' in the sense of correctness and [released] from 'Being' in the sense of actuality [*Wirklichkeit*].[91]

This means that we must also attempt to think *alētheia* free of the metaphysics of ego-subjects and their objects – a metaphysics which installs them in the objectivity of a linear, serial time-order, composed of a succession of self-contained, externally related 'nows'. The seer's vision is 'exceptional' in this regard, for the seer is one who has achieved a certain freedom from this way of thinking and seeing. Here, then, we should bear in mind what Heidegger observed in 'The End of Philosophy and the Task of Thinking':

> In the Greek language, one is not only speaking about the action of seeing, about *videre*, but about that which gleams and radiates. But it can radiate only if openness has already been granted. The beam of light does not first create the opening, openness; it only traverses it. It is only such openness that grants to giving and receiving . . . what is free, in which they can remain and move.[92]

I agree with Heidegger: the 'essence' of our being is to exist as 'an ecstatic inherence in the truth of Being'.[93] But what does this mean in terms of our vision? The sense of 'truth', here, is of course *alētheia*. *Alētheia* is an experience with truth that is radically open to the presencing of the absent, the invisible: it is, in this sense, ecstatic. By contrast, correctness involves an experience with truth that sees it as a posited state; it is a vision of truth that denies shadows, adumbrations, the presencing of the invisible, and unconcealment. It recognizes only two modes of being: *Zuhandensein* and *Vorhandensein* – in other words, it can see only totalized presence.

The seer's way of seeing is hermeneutical: hermeneutically circumspective. Instead of seeing things one-dimensionally, as most of us do today, in this age of reductionism, his gaze sees things in terms of a hermeneutical 'as'.[94] Just as there are two modes of discourse – the apophantic and the hermeneutical, the assertive and the disclosive, so there are two modes of vision, describable in similar terms. Here is Heidegger's discussion of the different discursive modes in *Being and Time*:

> The 'as' gets pushed back into the uniform plane of that which is merely present-at-hand. It dwindles to the structure of just letting one see what is present-at-hand, and letting one see it in a definite way. This levelling of the primordial 'as' of circumspective interpretation to the 'as' with which the present-at-hand is given a definite character is the specialty of assertion. Only so does it obtain the possibility of exhibiting something in such a way that we just look at it.[95]

We should note that Heidegger himself implies the homology between vision and discourse by giving an account of discourse in the language of vision. Heidegger's analysis continues:

> Thus, assertion cannot disown its ontological origin in an interpretation which understands. The primordial 'as' of an interpretation (hermeneia) which understands circumspectively we call the 'existential-hermeneutical "as" ' in distinction from the 'apophantical "as" ' of the assertion.[96]

Likewise, the seer's way of seeing things is more primordial than our everyday way: its ecstatic openness, and its corresponding *sense* of things in the dimensionality of their wholeness, though not understood, and not consciously practiced, by more 'ordinary' mortals, in fact underlies *all* human perception, and not only that of the seer. This is what I think Merleau-Ponty's phenomenological explorations of perception and its temporality enable us to appreciate.

Since the seer's capacity for openness is crucial, here, for our understanding of *alētheia* as an experience with vision, we must briefly return to the fact that the seer's gaze is not ego-logical. That is to say, it is neither egocentric nor logocentric. As a spiritually developed being, the seer is a self, not an ego. The difference is important, so I will briefly define it. The ego is the self limited to its social identifications: its roles, practices, and socially adaptive routines. The ego is the active pole in a structure of subject and object. The self, however, is not identified with any one structure; structurally speaking, it is a process always open to further structuring. Even when the self functions like an ego, it is not totally identified with it. The self is a sense of living in which all identifications are subject to deconstruction. Thus we may say that *there is no self* in the sense of a substance, a fixed identity, and a rigid closure to processes of change. Instead, there are only different styles, types, and dimensions of experiencing – and different styles, types, and dimensions of integration, unity and coherence.

With the emergence of ego, there is an inevitable agitation of mind: anxieties and tensions determine the shaping of our visual intentionalities; inveterate tendencies prevail, structuring the field of our vision in very rigid, narrow, and restricted ways. Ego-logical vision, an assertive mode of vision, always tends to follow the straight line of desire, the shortest, most direct distance between subject and object. For such a vision, a 'circumspective' experience with *alētheia* is not possible. Ego-logical vision is adaptively necessary, of course. Without its conformity to 'objective' truth, its relationship to correctness, we mortals could not survive. The ego-logical gaze constitutes the ground of our experience with truth – truth, that is, as correspondence. But the seer has achieved a different vision, and he enjoys a different experience with truth. Without rejecting the ego-logical experience of vision and its corresponding truth, he has chosen to *develop* his visionary capacities beyond their ego-logical stage.

To see aletheically, i.e., to experience *alētheia* in vision, the seer must first of all learn to relax, to lessen the grip of normal anxieties and tensions. This relaxation will in turn alter the character of her visual intentionality, allowing new and very different tendencies to come into play, and restructuring the visual process, the formation of the visual *Gestalt* – the figure/ground, centre/periphery, focus/diffusion relationships.[97] Without the control, the constant, obsessive monitoring of the ego, the seer's gaze is radically decentred, centred in a calm, more restful, more receptive relationship to the openness of the visual field as a whole. This openness, this visual clearing, is what makes the seer's gaze 'ecstatic'.

If *alētheia* is the *ekstasis* of truth, then the vision of this truth, a vision capable of experiencing it, must also be ekstatic. It is significant that those who have experienced *alētheia* all want to describe it in terms of its radiance, its shimmering, its gleaming, glowing, and reflecting. They speak of *alētheia* as the 'splendour of radiant appearing'. These words are words for ekstatic phenomena in the field of vision, and only an ekstatic gaze would therefore be capable of seeing them.

In his work on *Gelassenheit*, Heidegger invites us to consider what happens when we 'place the appearance of objects, which the view within a field of vision offers us, into this [gathering light of] openness'.[98] I would suggest this answer: that this kind of clearing is made possible by virtue of the openness a gaze has achieved, and that this openness is one which lets beings be, making it possible for them to shimmer and glow, and appear in all their radiance.

If we acknowledge the fact that things normally do *not* presence with such radiance, we need to see that this is because the character of our vision is such that there is no clearing for the ekstatic appearing, the *phainesthai* of things: that we have granted to visible beings, the luminous beings which appear, and which we are given to behold, no field of openness. But without this openness, how can their radiance take place? Since the gaze and what it sees are simultaneously and interdependently co-emergent, the radiance of things will be visible only to a gaze which has developed its visionary capacity beyond the requirements of 'correct' vision. The seer's gaze, thus developed, experiences the truth of the world ekstatically – experiences it, therefore, aletheically.

Augustine said that, 'With its light, truth gives joy to the men who turn toward it, and punishes with blindness those who turn away.'[99] And Marcuse, writing many centuries later, opined that 'There are modes of existence which can never be "true" because they can never *rest* in the realization of their potentialities, in the *joy* of being.'[100] There is joy in the seer's aletheic experience of radiance. This joy is essential. Since joy is opening and opening is necessary for an experience with truth as *alētheia*, rejoicing, an ecstasy of light, is the *ground* in which the seer's gaze is rooted. Rejoicing opens us; but the experience of this opening, and the vision of radiance it makes possible, themselves bring joy, increase the joy. Medard Boss, a psychiatrist influenced by Heidegger, describes this process very well:

> Another major emotional mode is composed, joyous serenity. It can give human existence the kind of receptivity that allows it to see in the brightest light the meaningfulness and connections of every phenomenon that reveals itself. Such a serenity is a clearness and openness in which a human being is emotionally connected to everything he meets, wanting not to have things in his own power but content to let them be and develop on their own. Because this composed, joyous serenity opens a human being to the broadest possible responsiveness, it constitutes happiness as well. For every human existence is by itself attuned to happiness whenever all its innate potential ways of being stand open to it.[101]

As Heinz Kohut, American psychoanalyst, once observed,

'Joy relates to experiences of the total self': that is to say, it is both 'cause' and 'effect' of a process of self-development, and is related, in particular, to the self's journey towards an openness that would make it whole.[102] The seer's vision, a vision of 'the essential richness of Being',[103] is rooted in a joyful experience of living in a 'forgetting' release of the past and an openness to the future. The seer is one who *embraces* whatever time has to offer. It is not that he *knows* the future, in a predictive or prophetic sense, although mystification often understands the visionary capacity in this way, but rather that he has developed a deep understanding of our protentional-retentional structuring of time – the structural intertwining of the temporal ecstasies – and overcome the psychopathology in egologically centred time-experience. Living thus, the seer's vision is centred in a felt sense of the whole – a felt gathering of time as a whole. The seer has 'already seen' what is yet to happen because she *understands* the ecstatic intertwining, is *free* of pathological relationships to time, and is *open* to whatever may come to pass in the dimensions of the visible.

Part VII

Beauty: The Evidence of Truth

Opening Conversation

(1) The perception of beauty is a moral test.

Thoreau, *Journal*[104]

(2) Color is the evidence of truth.

Charles Olson, *Archaeologist of Morning*[105]

(3) . . . a delirium which is vision itself. . . .

Merleau-Ponty, 'Eye and Mind'[106]

(4) It was when I said,
'There is no such thing as the truth,'
That the grapes seemed fatter.

Wallace Stevens, 'On the Road Home'[107]

(5) . . . the splendour of radiant appearing. . . .

Heidegger, *The Question Concerning Technology*[108]

As an experience with vision, 'correctness', an 'adequate' apprehension, calls for, and belongs in, a normal lighting. But in comparison with the lighting of *alētheia*, this lighting is dull, not very alive. *Alētheia* is an experience with truth in which the gaze is *drawn out of itself* into an upsurge, an ecstasy of radiance. This experience with truth is only possible when the gaze is relaxed, restful, free of agitation, open, and genuinely 'present', i.e., undistracted and steady, gently maintaining good felt contact. Its openness, receptivity, and responsiveness clear a space for radiant appearing.

Alētheia is 'beautiful' because of this radiance. We see 'beauty' in such radiance, and regard such luminosity as a manifestation of truth: it is the way truth appears, when it appears within the clearing of clear eyes. 'Beauty', Heidegger says, 'is the presencing of Being. Being is the true preserve of beings [*das Wahre des Seienden*]'.[109] The 'beauty' of radiance therefore 'belongs to the advent [*Ereignis*] of truth, truth's taking place'.[110]

It is not difficult to understand why this event of luminosity is seen as 'beautiful'. But what accounts for the fact that it is seen as manifesting the visibility of truth? In his *Lectures on Russian Literature*,[111] Vladimir Nabokov points out that *istina*, one of the Russian words for 'truth', still evokes, for him, its etymological relationship to luminous appearing. Why? Many valuable answers are possible here, if genuinely brought forth from the visionary experiences of different people.

Consider once again the glow, the radiance, that has been seen in the eyes and faces of people: in Moses, for example, and in children opening presents, and in the schizophrenic patients whom Searles was able to help with his friendly gaze. And consider the glow or gleam that has been seen coming from things: 'ordinary' things like some cups and saucers, a room, a vase of flowers, a tree, a hinge on the gate, a bunch of grapes in a bowl; the things about which William Blake wrote, and William Wordsworth, in *The Prelude*. I would like to suggest that such luminosity is as much an 'effect', a 'confirmation' of our attitude and our approach to them as it is a 'cause' of our sense that we are seeing them in their truth – or seeing their truth. We learn from them, we grow in our knowledge of them, we see something we had not seen before, something that had been invisible, something striking, something that seems especially meaningful: when we approach them with patience and care; when we let them be, but also stay *with* them in a hermeneutical, openly enquiring attitude; when our gaze is quiet, and not imposing; and when we are responsive to the dimensions of their presence, their ways of being visible, and in touch with our own responsiveness. The luminosity is inseparable from our quiet joy in being with them, open to their presence. It is a visible reflection of our sense of fulfillment in getting to 'know' them with our eyes. We call this luminosity 'truth' because the knowledge and understanding that the character of our gaze made possible evoke in us the feeling that an especially meaningful encounter took place.

Merleau-Ponty writes: 'Sight is achieved, and fulfills itself in the thing seen.'[112] This is true; but we need to see that there are qualitatively different achievements – and correspondingly different fulfillments. Sight is achieved and fulfills itself when I walk into my bedroom, turn on the light, and finally see the notebook I couldn't find elsewhere and believed to be lost. But this is a question of truth as correspondence. What about achievement and fulfillment when vision is an experience of *alētheia*? The luminosity of *alētheia* takes place when vision has achieved an unusually deep understanding and is drawing its capacity to make visible from the rejoicing it feels in this fulfillment of the visible situation.

In 'The Anaximander Fragment', Heidegger tells us that 'Presence within the lighting articulates all the human senses.'[113] This also is true, in that our eyes are always responsive to what Merleau-Ponty calls the 'solicitations' of light.[114] However, the initial articulation of the lighting inaugurates a process the further development of which is largely a question of our own response-ability. Thus I would like to conclude our meditation on beauty – and this chapter on truth – with some words from a philosopher whom I greatly admire. In *Art as Experience*, John Dewey gives us his thoughts on an ugliness in the modern world that his eyes would not avoid:

> The ugliness, for example, of most factory buildings, and the hideousness of ordinary bank buildings, while it depends upon structural defects on the technically physical side, reflects as well a distortion of human values. . . . No mere technical

> skill can render such buildings beautiful as temples once were. First there must be occur a humane transformation, so that these structures will spontaneously express a harmony of desires and needs that does not now exist.[115]

Dewey was one who did not shut his eyes to the ugliness; nor was he blind to the sufferings – the needs and delusions of needlessness – made visible to him through this ugliness. In his work on Nietzsche, in the volume concerned with nihilism, Heidegger brooded over the fact that, in an age such as ours, an 'age of confusion, of violence and despair in human culture, of disruption and impotence of willing', an age, therefore, of immeasurable need, what prevails is an ideology of needlessness, a strongly rooted delusion.[116]

What Dewey's diagnosis suggests is that, if somehow we could begin, as a collective body, to see the ugliness, the spreading of the industrial wasteland, perhaps we could also begin to make contact with our historical needs and see through the ideology, the delusion, of needlessness. I say: 'if somehow we could begin to see the ugliness. . . .' The 'if' is a small word; but the question it introduces is an individual and collective challenge of great proportions. Like Dewey, though, I believe that thoughtful people should persist in calling attention to the ugliness, for I am not entirely without faith in our senses, and therefore write with the hope that, if the presence of this ugliness can be evoked and the danger it portends be sensed, a different lighting would articulate within us, within the body politic, the next historical steps in our response-ability. In view of the nihilism determinative of our postmodern situation, I see the awakening of this response-ability as an obligatory collective task.

Since the advent of nihilism portends an historical closure of Being, it is a response to the danger for us to develop the openness of our capacity for vision. To be radically open to Being – an extremely difficult practice – is to let beings be seen in their open dimensionality. It means, therefore, that we protect their invisibility, letting it be seen in unconcealment: that is to say, *as* invisibility. 'The visible', as Dewey observes in *Experience and Nature*, 'is set in the invisible; and in the end what is unseen decides what happens in the seen', just as 'the tangible rests precariously upon the untouched and ungrasped'.[117]

We are beings of light, not only because we belong to the light and are, as visionary beings, essentially dependent on it, but also because our 'substance' is light – luminous energy. Being ourselves made of light, we are capable of making visible. We are made capable of seeing by grace of the lighting which surrounds us. But it is we who make Being visible; without us, the dimensionality of beings would not be visible. Being thus beholden, we are to be held responsible for developing our capacity to let the Being of beings be visible in the presencing of light and to make our relationship to this presencing luminously visible. We also *need* to understand Being as such in terms of our experience with vision, i.e., in terms of the clearing made for the light we see by. This calls for our self-development: a new body of understanding, a body which understands itself as a body of light.

I see this new body of understanding, that for which our post-modern situation calls, emerging from the play of shadows in an intertwining of texts. This book is a 'readerly' book, and, by design, a texture quilted with materials from many different visions, different points of view, all stitched together with convictions derived from my own experience with vision. It was written, I believe, with uncommon respect for the position of the reader – and for the absence of the authors with whose texts I myself have been engaged as reader. As befits my commitment to *alētheia*, to truth as a clearing, I have attempted to open our vision to a multiplicity of visions – a field of difference. Many texts have been woven together here: illuminations from many sources. Some of them, unfamiliar and surprising, belong to bodies of knowledge marginalized or excluded by our prevailing social discourse. The vision we need is a collective project, and even 'primitive' systems can contribute to our understanding of what is humanly possible.

This book follows the logic of an experiential enquiry. The four chapters take the reader through four reflective movements in a process of self-examination. First, we attempted an immanent critique of modernity, a diagnosis of the epidemic pathology visible in the hegemony of the optical paradigm in the modern empire of everyday vision. Following Nietzsche and Heidegger, we looked at this pathology and interpreted what we made visible in terms of the metaphysics of subjectivity and the advent of nihilism. Then, in Chapter Two, we looked into the roots of this vision, making explicit a historically unrealized potential for development already schematized in the inherently communicative, inherently protosocial intertwinings and reversibilities of the flesh. Then, in Chapter Three, we passed through the night of the soul, confronting a moment of decision and change: a moment symbolized by a flash of lightning breaking into the night. Finally, in Chapter Four, in the process of reflecting on truth as an experience with vision and on the character of our vision as an experience of truth, we looked into the formation of *das Geviert*, a new *Gestalt*, a vision of truth released from the enframing *Gestell*, gathering us, with care for our differences, into the time of our response-ability.

I have attempted to show why I think our present historical situation, an age of nihilism, calls for the opening of vision. Only, I believe, with a more open vision, a more ontologically developed body of understanding, can we make the openness of Being, the truth of Being, luminously visible, genuinely taking the Being of all beings into our thought and care. It is only by *virtue* of our vision, our capacity to see, that Being is made visible: visible as the lighting in the care of which we and all the other beings on this planet are destined to live and die.

APPENDIX

Dzogchen Dark Retreat An Abbreviated Phenomenological Diary

(1) For this queen of colours, the light, bathing all which we behold, wherever I am through the day, gliding by me in varied forms, soothes me when I am engaged by other things, and not noticing it. And so strongly doth it entwine itself, that if it be suddenly withdrawn, it is with longing sought for, and if absent long, saddeneth the mind.

Augustine, *Confessions*[1]

(2) Demeter's most important shrine was at Eleusis. There the great Mysteries, based upon Korē's seasonal death and resurrection, were performed. . . . The climax of the Mysteries was the day-long procession which danced and sang its way from Athens, wound through the pass of Daphni, debouched into the Eleusinian plain, and circled the shore of the gulf of Eleusis, to arrive at Eleusis by night. Every step of this route is marked by the appearance, disappearance, and reappearance of the sacred landscape symbols. . . . Directly ahead of the outer propylaia is the grotto of Hades, a natural cave in the rock toward which the procession first directly leads. . . . Having come first to the cave of death, the [Sacred] Way passes beyond it and curves snakelike upward to the left around the side of the hill. One can imagine the path of the torches, a sinuous trail of fire, as they approached the great hollow bulk of the Telesterion, . . . a columned hall surrounded by windowless walls and pierced by narrow entrances. The Mysteries were to take place inside, with the initiates crowded together . . . in a shadowy grove of columns.

Vincent Scully, *The Earth, The Temple and The Gods: Greek Sacred Architecture*[2]

> (3) Thus the hierophant began the drinking; the initiates then followed his example, waiting, as they listened to his chanting in the darkened telesterion, for the moment of revelation. . . . [S]eated on the tiers of steps that lined the walls of the cavernous hall, in darkness they waited. From the potion they gradually entered into ecstasy. . . . Ancient writers unanimously indicate that something was seen in the great telesterion, or initiation hall, within the sanctuary. . . . The experience was a vision whereby the pilgrim became someone who saw, an *epoptēs*. . . . What was witnessed there was no play by actors, but *phantasmata*, ghostly apparitions, in particular, the spirit of Persephone herself, returned from the dead. . . . Then there came the vision, a sight amidst an aura of brilliant light that suddenly flickered through the darkened chamber. Eyes had never before seen the like. . . . Even a poet [Pindar, for example] could only say that he had seen the beginning and the end of life and known that they were one, something given by god. The division between earth and sky melted into a pillar of light.
> R. Gordon Wasson, Carl A. P. Ruck, Albert Hofmann, *The Road to Eleusis: Unveiling the Secret of the Mysteries*[3]

On August 20, 1986, I went into Dark Retreat at Tsegyelgar, the Dzogchen community center in Conway, Massachusetts. After dark, I entered the isolation of a hut in the woods. This hut was designed, built and equipped for the special conditions of the Dark Retreat, during which time the practitioner lives continuously in the dark, totally cut off from contact with light. I remained for seven nights and seven days, isolated in the darkness of the hut. The practices of the Dark Retreat are at the heart of the Dzogchen teachings brought to this country by Dr. Professor Namkhai Norbu. (Norbu, Rinpoche is a Tibetan Buddhist who teaches Tibetan language and culture at the Oriental Institute of The University of Naples, Italy. He is a meditation master, a physician trained in Tibetan, Chinese and Indian medicine, a scholar, and the author of books on Tibetan culture, medicine, and the psychology of meditation.) The *Yang-thig*, the teachings of the Dark Retreat that he has recently begun to communicate – for the first time in Western history – were given to him by his principal teacher, Chang-chub dorje, and by a very old woman, Ayo Khandro, who at the time he studied with her had lived continuously in the darkness of a hut for more than fifty years.[4]

The teachings in question, together with their practices, are intended to develop clarity of vision and clarity about the essential nature of vision.[5] The ultimate realization of these teachings and practices is the body of light, a non-dual existence perfectly integrated into the presencing of the elemental lighting.

The *Yang-thig* teachings are concerned with the 'external' manifestations of 'internal' processes, 'internal' energy. They call for specific experiential practices: in particular, an extremely difficult set of visualizations. Although these practices are not at all 'secret', it would not be appropriate for me to discuss them here. In any case, however, it is not necessary for our present purposes that I do this. Suffice it to say that the visualizations are extraordinarily demanding, and that they presuppose some degree of accomplishment in other meditative practices: practices which are themselves difficult and exacting. These prior practices, belonging to the *'Kregschod* system, are called *zhi-gnas* (to develop *gnas-pa*, a calm and relaxed state of being in which vision is steadied in awareness by virtue of its non-duality, its neutralization of emotive energy) and *lhaktong* (to develop an experiential insight, working with the movement of mental activity, into *stong-pa nyid*, the ultimate nothingness of what we call 'reality', and into *lhundrub*, the 'self-perfectedness' of our primordial state).

Before I begin the phenomenological account, I would like to give a very brief summary of Abhidharma psychology and Madhyamika philosophy. Although the Dzogchen teachings and practices existed in Tibet long before the Abhidharma and Madhyamika systems were brought there from India, one can see, today, some very deep affinities and similarities. Since the texts of the Abhidharma and Madhyamika systems have been available for some time in the West and are therefore more accessible, if not also more familiar, a brief introduction to the Dzogchen practice of the

Dark Retreat by way of the Abhidharma and Madhyamika systems might therefore be useful.

Abhidharma is concerned with the analysis of our psychophysical development.[6] In particular, it analyzes the emergence of perception as a process governed by the ego. According to its analysis, ego-logical perception manifests six stages of 'normal' development:

(1) primordial openness, space, formlessness
(2) the advent of bodily felt forms, or *Gestalten*
(3) global feelings, moods of ontological attunement
(4) motivating intentionalities of desire: attraction, aversion, indifference
(5) object-oriented ego-logical perception
(6) conceptual articulations that further shape and structure the 'perceptual situation' (*yul-can* in Tibetan).

The experiential realization of this analysis sets the stage, however, for an experiential 'deconstruction' of the inveterate reification tendencies inherent in the perceptual process. In other words, Abhidharma psychology shows us the possibility of 'undoing' our self-limiting fixation in the ego-logical condition and experiencing a primordial openness.

Abhidharma psychology is inseparable from its philosophical interpretation.[7] In Madhyamika philosophy, two concepts are considered to be fundamental for our understanding of human psychology: one is called, in Sanskrit, *pratityasamutpada* (*rten'brel* in Tibetan), the other *shunyata* (*stong-pa nyid*, in Tibetan). The first concept refers to the fact of functional interdependence: the fact that subject and object, figure and ground, form and field, self and other are always interdependently co-emergent and co-functional. It is essentially equivalent to Merleau-Ponty's concept of 'intertwining'. The second concept refers to the fact that nothing is inherently permanent or essentially substantial: all things that come to appearance in our world (*snang-ba*) are ultimately 'empty' or 'void', in the sense that there is no eternal and immutable substratum or *hypokeimenon* underlying them. Nor is there any thing-in-itself organizing the various appearances. In the final analysis, all forms are utterly transitory, phenomenal, insubstantial, empty. Moreover, since even 'emptiness' itself, however, is a form of interpretation, Madhyamika logic calls for the deconstruction of 'emptiness' as well.

Without further introduction, let me now report in phenomenological terms my experience with the Dzogchen practice. The first night and first day were extremely exciting. I suddenly realized, by direct experience, that light is a stimulant, exciting the activity of vision and drawing it out. But I also began to undersand that the absence of light – deprivation of light – is an equally powerful stimulant, revealing and provoking the movement of our eyes. I had expected to find the darkness restful, but instead it aroused me. I was tense, overexcited. An incessantly changing display of forms kept me enthralled, entertained, and on the look-out: forms, like clouds, making their appearance, lingering a while, and then vanishing without any enduring trace. By the second night, I understood that this ceaseless play of light, this constantly changing display of shapes and patterns, sometimes suggesting familiar objects and fantastic landscapes, was a *reflection* of my state of mind. The display was functioning like a mirror, showing me the inner nature of my mind. Because of a dynamic, functional interdependence (*pratityasamutpada*), the ever-changing forms corresponded to the nervous, agitated movements of my gaze. Instead of resting, my eyes were constantly moving about, rapidly darting and jumping about. These movements were extremely fine vibrations or oscillations – quite different from the slower, grosser REM's.

Was all this movement caused by curiosity? Perhaps at first. But the room was totally dark and objectively uneventful: nothing really was happening in the surrounding space – nothing other than the darkness itself. There was, after all, nothing (objective) to see. I did experience some waves of anxiety from time to time, but I do not believe that this anxiety, nor even occasional projections of paranoia, can explain the incessant movement. (Experimental psychology has established that, even during sleep, there are rapid eye movements, REM's, which seem to be correlated with the processes of dreaming.) By the end of that second night, I reached the conclusion that the movement was basically habitual, manifesting an inveterate tendency (in Tibetan, *bag-chags*) of embodied consciousness.

I was reminded of a remark Heidegger makes in 'Moira', his essay on Parmenides. He observes that 'ordinary perception certainly moves within the lightedness of what is present and sees what is shining out . . . in colour', and then comments that it is 'dazzled by changes in colour' and 'pays no attention [at all] to the still light of the lighting. . . .'[8] Most scholars pay no attention to this brief analysis: their eyes glide right over the words, unchallenged by their significance. I myself missed much of it; but at least I took his words to heart, i.e., I gave them an experiential reference. Remembering the text did not immediately help me. By the end of the second day, my eyes were strained, tired, and occasionally hurt. I rubbed them gently and allowed tears to come. This brought some temporary relief.

My visionary experiences during the third night and third day were not much different. But, by the end of the third day, it was clear to me that the visualization practices I was attempting to perform were only increasing the eyes' strain and mental agitation. And since this condition of strain and agitation was reflected back by the restless changing of forms, the more intense my exertions, the more these displays of light agitated and pained me. On the fourth night, I finally realized that I was caught in a vicious cycle, a wheel of suffering, unable to break out of the dualistic polarizations characteristic of my normal, habitual, routinized patterns of ego-logical vision. I was, in fact, shifting back and fourth, interminably caught in one of four possible visionary attitudes in relation to the display of forms presencing in the dark:

(a) seduction, i.e., attraction, involvement, grasping and clinging
(b) resistance, i.e., attempts to fixate and control the wrathful movements of light by rigidly staring into the space before me
(c) disengagements that involved withdrawing into inner monologue, i.e., continuous conceptualization
(d) disengagements that resulted in drowsiness, i.e., a withdrawing into the 'unconsciousness' of sleep.

The first two attitudes only intensified the movements of light; consequently they increased my inner agitation – which, in turn, increased the play of light. Furthermore, both styles of interaction inflicted on my eyes a strain which always at some point became unbearable. But the second two attitudes were equally unsatisfactory as ultimate solutions: the monologue became repetitive, compulsive and boring; nor could I withdraw into continuous sleep for the duration of the retreat.

The third night and the following third day were extremely difficult. They tried me to the limit. As it turned out, these were in fact the most difficult hours of the week-long retreat. I could not accomplish the principal visualization. I felt discouraged and depressed. The displays of light no longer frightened, enthralled, amused, or entertained me. They no longer had the power to divert me from an extremely negative process of self-examination. I was tired, bored, impatient, skeptical. My body ached. I tried to sleep but couldn't. I began to feel like a mouse or a mole, and wanted to escape the cold, the damp, the oppressive darkness. But I was determined to remain in the retreat for at least one week: seven nights, seven days.

The fourth night and the following day, I began to feel somewhat different. I was in the process of developing a very different attitude: toward the practices I had been struggling with and myself in relationship to them, toward the darkness, and toward the interminable displays of light. And these changes in me were immediately reflected in corresponding changes in the environment. Briefly described, this environment was gradually beginning to feel less wrathful and more friendly – more like a nurturing, gently encompassing presence. And, as I found myself able to put into practice the meditative disciplines I had been learning for many years prior to the retreat (primarily the practice of *zhi-gnas*, calming and quieting the mind, and the practice of *lhaktong*, developing the deconstructive clarity of my insight into the ultimate emptiness of all passing forms), I began to see a decisive change in the phenomenal displays. The transformations of the lighting became slower, less violent; and in between the displays of forms, I saw more of a clear space. There were more frequent times when I was surrounded by large curtains, or regions, of relatively constant and uniform illumination, sometimes brownish red, sometimes pale green, sometimes a dull white. Sometimes, I found myself looking out into an infinite expanse of clear, dark blue space, punctuated here and there by tiny stars of intense white light.

During the fourth day and fifth night, I gradually experienced the fact that there *is* a fifth attitude: a way out of the vicious cycle of suffering. The way out was to be found in the teachings and practices I had brought with me into the retreat. And finally, I knew this through direct experience, my own experience – and not by a leap of faith. The calmness and relaxation I was beginning to achieve was reflected back to me by corresponding qualities in the luminous presencing of the darkness. This different lighting in turn helped me to deepen my state of calm and relaxation and continue developing a non-dual (*dbyer-med*) visionary presence.

Beginning with the fifth day, then, it became progressively easier for me to experience what the Tibetans call *rig-pa*: the simple presence of awareness. Staying in this non-duality, I could begin to experience my integration into the element of light. I *felt* the truth of the Dzogchen teaching that I am by nature a body of light: that I *am* light; that I and the phenomenal displays of light are really one. Correspondingly, the darkness became a warm, softly glowing sphere of light, an intimate space opening out into the unlimited. I felt bathed in its encompassing luminosity, an interplay of softly shimmering grey-white and blackish-red lights. I experienced a kind of erotic communion with the light, as if the light and I were entwined in a lover's embrace.

With the development of more neutralized, non-dualistic awareness, my vision was less caught up in the antithesis of movement (*gyu-ba*) and non-movement (*gnas-pa*). With the development of my capacity for letting go and letting be, my gaze was less troubled by forms in movement. There was less need to withdraw into sleep, because *rig-pa* is a *restful* aliveness. There was less need for painful staring, less need to stare the forms into fixity, because the greater tranquillity of my gaze effortlessly stabilized the inevitable display of moving, changing forms. There was less visual jumping and darting about, because the gaze was not so readily seduced by the play of light into forming attachments to its transformations that would disturb my becalmed presence. And there was less compulsion to withdraw into conceptual interpretation, because the gaze, more inwardly quiet, could let me begin to enjoy simply being in and with the lighting of the dark.

On the seventh night, just as I was drifting into sleep, but still in a state which is half way *between* waking and dreaming, and which the Tibetans call *Bardo*, I was suddenly jolted back into full consciousness, eyes wide open. I had been lying down, of course: in the normal position I assumed for sleeping. But there was suddenly a peremptory 'call' to me, and, simultaneously, I lifted my head up – so fast, in fact, that I almost jumped out of bed. Confronting my raised eyes was a visionary phenomenon for which my comfortable categorial scheme was completely inadequate.

Until this unnerving event, I had experienced only *three* essentially distinctive categories of visionary event. (a) I could 'see' my own body, especially when I moved: it had a ghostly presence, luminous, yet also dark, like a shadow; but I had no doubt whatsoever that I could 'see' it – clearly and distinctly. Although this contests our normal constructs, Merleau-Ponty's phenomenology of the body calls attention to a corporeal schematism that makes it entirely understandable.[9] (b) Pure luminosities: dots, spots, jigzag and straight lines, sudden explosions, tiny cones and pyramids, irregularly shaped regions and curtains of light and, near the end of the retreat, an embracing atmosphere of softly glowing, relatively constant illumination. And (c) Rorschach phantoms: because of all the involuntary eye movements, the luminosity of the dark manifested in a continuously changing display of shapes and forms; and because of the participation of consciousness in the process of the spectacle, these apparitions were subject to continuous, and more or less effortful interpretation. (I was reminded of Wittgenstein's observations, in his *Philosophical Investigations*, in regard to the 'dawning' of an aspect.)

But the visionary apparition which compelled me to rethink my understanding of vision was fundamentally different from these fugacious Rorschach phantoms. Unlike the phantoms, it was totally spontaneous, i.e., involuntary, without any antecedent, and more or less effortful, participation by consciousness. It was sudden, instantaneous, without any gradual 'dawning' or 'unfolding'. It was totally unrelated to earlier conscious thought. And, finally, it was clear and distinct, intensely vivid, luminously present. Indeed, what made it obviously 'apparitional', rather than 'real', was precisely its *extraordinary* luminosity: colours of incredible, 'supernatural' purity, intensity, aliveness, and clarity. Otherwise, I might have been taken in by it, since it had the sharpness of outline, the distinctness, the steady duration, and also the three-dimensionality, the compelling appearing of volume and solidity, which are characteristic of the 'real' things in our normal, consensually validated world. Yet I was not at all, except for an instant, perhaps, *deceived* by what I saw. Were it not for the peculiar luminosity, it might perhaps have been, or seemed to be, quite 'real'; but I looked directly at it without any doubt that it was in truth 'only' an apparition – or a manifestation of some *other* dimension of our reality. It *looked* real – or rather, it looked, in fact, *more* than real, and I saw it *as* a vision, a vision of something which 'objective reality testing' would not confirm. (It was therefore different from the visions of Eleusis, which were induced, as we now know, by the ingestion of a drink containing pulverized ergot, a hallucinogenic substance derived from barley.) By contrast, my experience was not induced by any psychotropic substance, nor was I in some radically altered state of mind, e.g., deprived of sleep. Like the Eleusinian visions, however, it was determined by the traditional symbolic associations of the Dark Retreat. What I saw was the ornamental pelt worn by Sengē Dong-ma, one of the female dakkinis and a supernatural being of light associated with the Dark Retreat teachings.

There are, then, four epistemologically distinct visionary processes, and it is essential that we not confuse them:

(a) *hallucination*: a spontaneous, i.e., unwilled projection of consciousness taken for real
(b) *phantom*: a Gestalt in whose process of formation consciousness participates, but in a relatively passive or receptive attitude, in the sense that it lets whatever configurations begin to appear (perhaps in response to its own unconscious projections) suggest the interpretation that completes and stabilizes the Gestalt
(c) *visualization*: different from the phantom in that the participation of consciousness in the process of its formation is less passive and receptive; in other words, a deliberately produced image
(d) *an authentic vision*: different from hallucination in that the projection is not deceptive, but, on the contrary very deeply understood (this understanding of the projective process is in fact a necessary condition of its possibility); different from a phantom in that its formation is instantaneous and spontaneous, and does not involve the participation of consciousness in an unfolding process of formation; and different from visualization in that it does

> not appear while, or so long as, one's attention is absorbed in a process of willfully trying to produce it.

The 'authentic vision' is *like* the hallucination, however, in that its appearance is *not* immediately connected to conscious attention, willful exertions of a greater or lesser degree (as in the formation of phantoms and visualizations). And it is *like* the visualization in that a necessary condition of its possibility *is* the kind of exertion, the kind of work, that goes into the production of the image in the process of visualization. *A fortiori*, in this respect authentic vision is *unlike* the hallucination, despite the spontaneity of its actual appearance.

Let me add, as part of a final note, that the darkness profoundly altered my sense of spacial distance and my sense of the passage of time. The first of these I expected; but I was surprised to find that time passed very rapidly. The nine-hour stretch from breakfast to supper, for example, often seemed to be no more than a few hours. At no time, however, was I confused or disoriented. I maintained a 'normal' sense of reality, of being grounded in the 'reality' of the world outside.

When I emerged from the retreat at the beginning of my eighth night, I found even the tree-filtered moonlight overwhelming. My eyes had developed a tremendous sensitivity to light, and even the faintest flickering concentration of illumination seemed at first almost unbearably intense. This I expected. I was surprised, however, by the nausea and dizziness which overcame me during the first few minutes in the relatively dark night outside the hut. (The moon was waning, and I was, after all, in the woods). For one week, the eyes had been attuned by the peculiar conditions of the darkness; they needed some time – about 48 hours – to readjust and conform to the conditions of the world into which I had re-emerged.

Roger Levin, who was himself in the Dark Retreat for the same length of time, has pointed out that the sensations of dizziness and nausea seem to be related to the luminous pulsations and vibrations that make the edges and contours of things indistinct, blurred, uncertain. Arguing that the Dark Retreat makes us aware of eye movements which are *also* taking place in the world of light, he suggested that, after one week in the dark, the cortical processes of vision were no longer making the stimulus-constancy corrections that normally function to compensate for – and successfully conceal – the movements of our eyes. In other words, our eyes are in constant motion in the light as well as in the dark; but normally, we are not made aware of this motion, since, in the presence of stimuli that make us see differences in illumination, there is a compensatory process at work in our vision, and this process reifies the light, imposing a certain level of constancy and uniformity at the edges and boundaries of things. But in the dark, we 'see nothing': the differentiating stimuli that activate the compensatory process are not present. Coming out of a prolonged retreat into darkness, the eyes cannot immediately resume their normally automatic and unconscious work of correction. Thus, what is normally concealed from awareness – the motion of the eyes and the constancy adjustments we make – is rendered visible.

The retreat was a rich and deeply therapeutic experience for me. I emerged from the archetypal womb of darkness feeling nourished in spirit and more deeply integrated, more whole and complete, than when I entered it. Conceivably, this sense of inner growth is nothing but an emotional rationalization. I am familiar with the psychological theory of cognitive dissonance. But, after much critical thought and self-examination, I have reason to believe that the benefits I have noted are real and that their significance for my life – and in particular, for my visionary propensities and habits – will be enduring.

Visionary habits are not easily broken – especially not when the prevailing social consensus continuously reinforces them. The Dark Retreat is an extension of the Dzogchen practice of the Chöd. In Tibetan, 'chöd' refers to a process of cutting off. The Dark Retreat helped me to cut myself off from the inveterate tendencies that bind human vision to the karmic wheel of endlessly reproduced suffering.

Notes

Introduction

1 Martin Heidegger, *Nietzsche*, vol. IV: *Nihilism* (New York: Harper & Row, 1982), p. 28.
2 See Carl G. Jung, 'Mind and Earth', in Read, Ford and Adler (eds), *The Collected Works of Carl G. Jung*, vol. 10: *Civilization in Transition* (Princeton: Princeton University Press, 1970), p. 49, and 'The Spiritual Problem of Modern Man', in Joseph Campbell (ed.), *The Portable Jung* (New York: The Viking Press, Penguin Books, 1971), p. 479.
3 Heidegger, *Being and Time* (New York: Harper & Row, 1962), p. 49.
4 *Ibid.*, p. 187.
5 Heidegger, 'Letter on Humanism', in David Farrell Krell (ed.), *Basic Writings* (New York: Harper & Row, 1977), p. 230. For the German, see *Holzwege* (Frankfurt am Main: Vittorio Klostermann, 1963, p. 272. Also see Heidegger's 'What Are Poets For?', in Albert Hofstadter (ed. and trans.), *Poetry, Language, Thought* (New York: Harper & Row, 1971), p. 117.
6 Heidegger, 'Letter on Humanism', *op. cit.*, p. 237.
7 Maurice Merleau-Ponty, 'Working Notes', in Claude Lefort (ed.), *The Visible and the Invisible* (Evanston: Northwestern University Press, 1968), p. 227.
8 Heidegger, *Being and Time*, p. 37.
9 *Ibid.*
10 Heidegger, *Nietzsche*, vol. II: *The Eternal Recurrence of the Same* (New York: Harper & Row, 1984), pp. 182-3.
11 *Ibid.*
12 Merleau-Ponty, 'Eye and Mind', in James M. Edie (ed.), *The Primacy of Perception* (Evanston: Northwestern University Press, 1964), p. 188.
13 Heidegger, *Nietzsche*, vol. IV, p. 218.
14 John Welwood, 'Principles of Inner Work: Psychological and Spiritual', *Journal of Transpersonal Psychology*, vol. 16, no. 1 (1984), p. 63.
15 Joel Kovel, *The Age of Desire: Reflections of a Radical Psychoanalyst* (New York: Pantheon, 1982), p. 249.
16 See the interview published in the second edition of Hubert Dreyfus and Paul Rabinow, *Michel Foucault: Beyond Structuralism and Hermeneutics* (Chicago: University of Chicago Press, 1983), pp. 229-52.
17 Raymond Geuss, *The Idea of Critical Theory: Habermas and the French*

School (New York: Cambridge University Press, 1981), p. 76.

18 Theodor Adorno, *Minima Moralia: Reflections from Damaged Life* (London: New Left Books, 1974; Verso edition, 1978), pp. 17-18.

19 Merleau-Ponty, 'Cézanne's Doubt', in Hubert Dreyfus and Patricia Dreyfus (eds), *Sense and Non-Sense* (Evanston: Northwestern University Press, 1964), p. 19.

20 See Mark Warren, *Nietzsche and Political Thought*, presently available only in manuscript.

21 See David Michael Levin, 'Psychopathology in the Epoch of Nihilism' and 'Clinical Portraits: The Self in the Fury of Being', in David M. Levin (ed.), *Pathologies of the Modern Self: Postmodern Studies on Narcissism, Schizophrenia, and Depression* (New York: New York University Press, 1986). A common complaint from people fitting the so-called 'narcissistic character disorder' is helplessness, powerlessness, and an incapacitating lack of self-esteem. Nietzsche already saw this very clearly, and understood it, moreover, as a symptom of the nihilism epidemic in our society.

22 See Michel Foucault, 'On the Genealogy of Ethics: An Overview of Work in Progress', in Dreyfus and Rabinow, *Michael Foucault: Beyond Structuralism and Hermeneutics*, pp. 229-52. Also see Michel Foucault, *Le Souci de Soi* (Paris: Gallimard, 1984).

23 See Eugene Gendlin, 'A Philosophical Critique of the Concept of Narcissism', in David M. Levin (ed.), *Pathologies of the Modern Self*.

24 See Foucault, 'Why Study Power? The Question of the Subject', in Dreyfus and Rabinow, *op. cit.*, p. 216.

25 See Foucault, *Language, Counter-Memory, Practice: Selected Essays and Interviews* (Ithaca: Cornell University Press, 1977), p. 228.

26 See Heidegger's 'Letter on Humanism', *op. cit.*, pp. 201-14.

27 *Ibid.*, p. 213.

28 *Ibid.*, p. 205.

29 *Ibid.*, p. 201.

30 *Ibid.*, pp. 201-2. Italics added.

31 *Ibid.*, p. 202. Italics added. Also see George E. Marcus and Michael M. J. Fischer, *Anthropology as Cultural Critique: An Experimental Moment in the Human Sciences* (Chicago: University of Chicago Press, 1986).

32 William Blake, *The Marriage of Heaven and Hell* (New York: Oxford University Press, 1975), p. 5.

33 Merleau-Ponty, 'The Intertwining – The Chiasm', in *The Visible and the Invisible*, p. 151.

34 Heidegger, *What is Called Thinking?* (New York: Harper & Row, 1968), p. 110.

35 Heidegger, 'The Question Concerning Technology', *Basic Writings*, p. 310.

36 *Ibid.*

37 Friedrich Nietzsche, 'On the Uses and Disadvantages of History for Life', in *Untimely Meditations* (Cambridge: Cambridge University Press, 1983), p. 85.

38 Heidegger, *Being and Time*, pp. 274-5. Also see David C. Hoy, *The*

Critical Circle: Literature and History in Contemporary Hermeneutics (Berkeley: University of California Press, 1978).

39 Heidegger, *Being and Time*, p. 49.
40 *Ibid.*, p. 276.
41 See Nietzsche, *The Will to Power* (New York: Random House, 1967), p. 262. But the translation I am following is David Farrell Krell's. His translation appears in Heidegger's *Nietzsche*, vol. II, p. 38.
42 Heidegger, *Nietzsche*, vol. II, p. 167.
43 *Ibid.*, pp. 117-18.
44 *Ibid.*, p. 118.
45 *Ibid.*, p. 119.
46 *Ibid.*
47 *Ibid.* p. 102.
48 *Ibid.*
49 *Ibid.*
50 Merleau-Ponty, *Humanism and Terror* (Boston: Beacon Press, 1969), xlii.
51 *Ibid.*, pp. 94-5.
52 *Ibid.*, p. 13.
53 *Ibid.*, p. 167.
54 Heidegger, *Being and Time*, pp. 188-9.
55 *Ibid.*, p. 44.
56 *Ibid.*
57 *Ibid.*, p. 37.
58 *Ibid.*, pp. 42-3.
59 *Ibid.*, p. 25.
60 *Ibid.*, p. 37.
61 *Ibid.*, p. 35.
62 *Ibid.*, p. 25.
63 *Ibid.*
64 *Ibid.*, p. 33.
65 *Ibid.*
66 *Ibid.*
67 *Ibid.*, p. 32.
68 *Ibid.*
69 *Ibid.*, p. 175.
70 *Ibid.*, p. 276.
71 See Eugene Gendlin's 'Analysis', following the text of Heidegger's *What Is A Thing?* (Chicago: Henry Regnery Co., 1967), p. 283.
72 *Ibid.*, p. 285.
73 *Ibid.*, p. 284.
74 See the recent work of Eugene Gendlin, especially: 'Experiential Phenomenology', in Maurice Natanson (ed.), *Phenomenology and the Social Sciences* (Evanston: Northwestern University Press, 1973), pp. 281-319; 'Experiential Psychotherapy', in Raymond Corsini (ed.), *Current Psychotherapies* (Itasca, Illinois: F. E. Peacock Publishers, 1973); '*Befindlichkeit*: Heidegger and the Philosophy of Psychology', *Review of Existential Psychology and Psychiatry*, vol. xvi, nos 1-3 (1978-9), pp. 43-71; *Focusing* (New York: Bantam Books, revised edition, 1981); 'A

Philosophical Critique of the Concept of Narcissism', in David M. Levin (ed.), *Pathologies of the Modern Self* (New York: New York University Press, 1986); and 'Non-Logical Moves and Nature Metaphors', in *Analecta Husserliana* (Dordrecht: D. Reidel Publishing Co., 1985), vol. 19, pp. 383-400.

Chapter 1 Das Ge-stell: The Empire of Everyday Seeing

1 William Wordsworth, *The Prelude*, J. C. Maxwell, ed. (New York: Penguin, 1972), Book XI, lines 170-6, p. 474.
2 Ralph Waldo Emerson, 'Nature', in *Essays* (New York: A. L. Burt, 1936), vol. II, p. 63. Also see Lame Deer and Richard Erdoes, *Lame Deer: Seeker of Visions* (New York: Simon & Schuster, 1972), pp. 39-44, where Lame Deer, who tries to see with what the Lakotas, his people, call 'the eye of the heart' (*cante ista*), speaks with the most audible sadness of the blindness and cruelty which are so visible, to the Indian, in the white man's way of seeing things. The contrast between these two modes of vision, and the profound differences in their worlds, should be given more philosophical thought.
3 Heidegger, 'What Are Poets For?', *Poetry, Language, Thought* (New York: Harper & Row, 1975), p. 138. Also see Heidegger's *Holzwege* (Frankfurt am Main: Vittorio Klostermann, 1963), p. 297. The German speaks of a 'begehrendes Sehen der Dinge'.
4 See Northrop Frye, *Fearful Symmetry: A Study of William Blake* (Princeton: Princeton University Press, 1947).
5 Heidegger, *Being and Time* (New York: Harper & Row, 1962), p. 175. Italics added.
6 Michel Foucault, *The Birth of the Clinic: An Archaeology of Medical Perception* (New York: Random House, Vintage Books, 1975), p. 84.
7 Erich Neumann, *The Origins and History of Consciousness* (Princeton: Princeton University Press, Bollingen Series, 1970), p. 341. Italics added.
8 *Ibid.*, p. 340.
9 Ludwig Wittgenstein, *Philosophical Investigations*, trans. G. E. M. Anscombe (New York: Macmillan Co., 1953), p. 31, para. 66.
10 Ronald Blythe, *Akenfield: Portrait of an English Village* (New York: Dell Publishing Co., 1969), p. 72.
11 *Ibid.*, p. 80. Vision has a history in everyday life: it has gone through changes; it will pass through many more. What is our responsibility in this historical process?
12 Heidegger, 'Moira', *Early Greek Thinking* (New York: Harper & Row, 1975), p. 100.
13 Emerson, 'Character', *Essays*, vol. II, p. 97.
14 See William James, *A Pluralistic Universe* (New York: Longmans, Green & Co., 1958), p. 165.
15 Heidegger, *Being and Time*, p. 213.
16 Heidegger, 'Der Anfang des abendländischen Denkens', *Heraklit*,

Gesamtausgabe, Abteilung II, Vorlesungen 1923-1924 (Frankfurt am Main: Vittorio Klostermann, 1979), Band 55, p. 23.
17 Herbert V. Guenther and Chogyam Trungpa, *The Dawn of Tantra* (Berkeley: Shambala Publishing, 1975), p. 26.
18 See Heidegger, *Being and Time*, pp. 197 and 213-17.
19 *Ibid.*, p. 214.
20 Heidegger, 'The Anaximander Fragment', *Early Greek Thinking*, p. 36.
21 Maurice Merleau-Ponty, *Phenomenology of Perception* (London: Routledge & Kegan Paul, 1964), p. 450.
22 *Ibid.*, p. 55.
23 Heidegger, 'Andenken', *Erläuterungen zu Hölderlins Dichtung* (Frankfurt am Main: Vittorio Klostermann, 1971), p. 96.
24 Mikel Dufrenne, *The Phenomenology of Aesthetic Experience* (Evanston: Northwestern University Press, 1973), pp. 11, 86.
25 This moving utterance is told in a legend about Indrabhuti, king of Uddiyana (Odiyan), who wanted, out of gratitude, to give his teacher the wish-fulfilling gem he had gotten from the deities. See *Crystal Mirror*, Annual Journal of the Tibetan Nyingma Meditation Center (Berkeley: Dharma Publishing, 1975), vol. IV, p. 7.
26 Quoted by Tarthang Tulku in 'The Life and Liberation of Padmasambhava', *Crystal Mirror*, vol. IV, p. 26.
27 Heidegger, 'The Age of the World Picture', in *The Question Concerning Technology and Other Essays* (New York: Harper & Row, 1977), p. 154.
28 Amedeo Giorgi, *Psychology as a Human Science* (New York: Harper & Row, 1970), p. 121.
29 *Ibid.*
30 *Ibid.*, p. 122.
31 *Ibid.*
32 Viktor Shklovsky, 'Art as Technique', in *Russian Formalist Criticism: Four Essays* (Lincoln: University of Nebraska, 1965), p. 13. Italics added.
33 Merleau-Ponty, *Phenomenology of Perception*, p. 316.
34 Heidegger, 'What Are Poets For?', *Poetry, Language, Thought*, p. 111. Also see *Ibid.*, 116-17 and 120-1.
35 Heidegger, 'The Thing', *Poetry, Language, Thought*, p. 181.
36 Heidegger, 'The End of Philosophy', in *Basic Writings* (New York: Harper & Row, 1977), p. 374.
37 Heidegger, 'The Word of Nietzsche: "God is dead" ', *The Question Concerning Technology and Other Essays*, pp. 82-4.
38 Heidegger, 'The Age of the World Picture', *op. cit.*, p. 150.
39 Perhaps the prohibition of religious imagery in Judaism is an implicit acknowledgement of the potential for pride and violence inherent in the visual situation.
40 Medard Boss, *The Existential Foundations of Medicine and Psychology* (New York: Jason Aronson, 1979), p. 222.
41 See Heidegger's discussion of stereotyped vision in *Discourse on Thinking* (New York: Harper & Row, 1966), p. 62.
42 Concerning shadows and reflections, see Robert Romanyshyn, 'Science and Reality: Metaphors of Experience and Experience as Metaphorical',

in Ronald Valle and Rolf von Eckartsberg (eds), *Metaphors of Consciousness* (New York: Plenum Publishers, 1980); 'Psychological Language and the Voice of Things', *Dragonflies: Studies in Imaginal Psychology*, Part I, no. 1 (Fall, 1978), pp. 74-90, and Part II, no. 2 (Spring, 1979), pp. 72-9.

43 Heidegger, *Being and Time*, p. 88. For John Dewey, the stare occurs in 'the abnormality of a situation in which bodily activity is divorced from the perception of meaning. . . .' See his *Democracy and Education* (New York: Macmillan, The Free Press, 1966), p. 141. I agree.

44 See George Berkeley, *An Essay Towards a New Theory of Vision* (London: Dent, 1934), p. 13.

45 See Katsuki Sekida, *Zen Training: Methods and Philosophy* (New York: John Weatherhill, 1975), pp. 141-4. Also see Lloyd Kaufman, *Perception: The World Transformed* (New York: Oxford University Press, 1979), p. 65: 'Today we know', he says, 'that if the eye does not make small movements, after several seconds the objects viewed may disappear.'

46 Plato, *The Republic*, in *The Dialogues of Plato*, B. Jowett, trans. (New York: Random House, 1937), vol. I, Book VIII, 532, p. 791.

47 Heidegger, *The Basic Problems of Phenomenology* (Bloomington: Indiana University Press, 1982), p. 284.

48 Plato, *Timaeus*, 47a-b, in Jowett, *The Dialogues of Plato*, pp. 27-8.

49 Immanuel Kant, *The Critique of Practical Reason* (New York: Library of Liberal Arts, 1956), p. 166.

50 Nietzsche, *The Birth of Tragedy* (New York: Doubleday, 1956), p. 136.

51 Emerson, 'Nature', op. cit., p. 169.

52 *Ibid.*, p. 172. Italics added. Also see Emerson's essay on 'Education', in which he reflects on 'the disturbing effect of passion and sense, which by a multitude of trifles impede the mind's eye from the quiet search for that fine horizon-line which truth keeps. . . .' See Mark Van Doren (ed.), *The Portable Emerson* (New York: Viking, 1946), p. 259.

53 Emerson, 'Circles', in *Essays*, vol. I (New York: A. L. Burt, 1936), p. 299.

54 William James, 'The Continuity of Experience', *A Pluralistic Universe*, pp. 288-9.

55 Nietzsche, *Daybreak: Thoughts on the Prejudices of Morality* (Cambridge: Cambridge University Press, 1982), Book II, Note 117, p. 73.

56 Nietzsche, *The Will to Power* (New York: Random House, 1968), Book III, Note 616, p. 330.

57 Heidegger, *Being and Time*, p. 49.

58 Heidegger, *Discourse on Thinking*, p. 64.

59 Samuel Beckett, *Endgame* (New York: Grove Press, 1958), p. 31.

60 Heidegger, 'Letter on Humanism', *Basic Writings*, p. 208.

61 Descartes, 'Rules for the Direction of the Mind', in *The Philosophical Works of Descartes*, trans. E. S. Haldane and G. R. T. Ross (New York: Dover Publishing Co., 1955), p. 3.

62 Louis Althusser, *Lire le capital* (Paris: Maspero, 1968), vol. I, pp. 26-8. In Proverbs (25:2) we read: 'The glory of God is to conceal a thing'.

63 Heidegger, 'The Question Concerning Technology', *Basic Writings*, p. 309. Italics added.

64 St. Augustine, *On Free Choice of the Will* (New York: Bobbs-Merrill, Library of the Liberal Arts, 1964), p. 75.
65 Heidegger, 'What Are Poets For?', *Poetry, Language, Thought*, p. 110.
66 Merleau-Ponty, *Phenomenology of Perception*, p. 68.
67 Merleau-Ponty, 'Interrogation and Intuition', *The Visible and the Invisible* (Evanston: Northwestern University Press, 1969), p. 123.
68 *Ibid.*
69 Merleau-Ponty, 'The Intertwining – The Chiasm', *The Visible and the Invisible*, p. 148.
70 Heidegger, *Discourse on Thinking*, p. 64. Italics added.
71 *Ibid.* Italics added.
72 *Ibid.* Italics added.
73 Heidegger, 'The Anaximander Fragment', *op. cit.*, p. 36.
74 *Ibid.*, p. 56.
75 Merleau-Ponty, 'Working Notes', *The Visible and the Invisible*, p. 227.
76 *Ibid.*, p. 192.
77 *Ibid.*, p. 197.
78 Merleau-Ponty, *Phenomenology of Perception*, p. 67.
79 Frederick Perls, Ralph Hefferline and Paul Goodman, *Gestalt Therapy: Excitement and Growth in the Human Personality* (New York: Delta Books, 1951), p. 56.
80 Heidegger, 'The Age of the World Picture', *op. cit.*, p. 145. Italics added.
81 *Ibid.*, p. 146. Italics added.
82 *Ibid.*, p. 147. Italics added.
83 Heidegger, 'The Word of Nietzsche: "God is dead" ', *op. cit.*, pp. 106-7.
84 *Ibid.*, p. 107. Italics added.
85 Merleau-Ponty, *Phenomenology of Perception*, p. 310.
86 Gershom Scholem, *The Kabbalah and Its Symbolism* (New York: Schocken, 1969), p. 114.
87 Heidegger, 'The Question Concerning Technology', *The Question Concerning Technology and Other Essays*, p. 28.
88 Heidegger, 'The Anaximander Fragment,' *Early Greek Thinking*, p. 26.
89 *Ibid.*, p. 36.
90 Merleau-Ponty, *Phenomenology of Perception*, p. 311.
91 *Ibid.*, p. 310. Also see Alphonso Lingis, 'The Elemental Background', in James M. Edie (ed.), *New Essays in Phenomenology* (Chicago: Quadrangle Books, 1969), pp. 24-38.
92 See Heidegger's discussion of Parmenides in 'Moira', *Early Greek Thinking*, p. 99.
93 Heidegger, 'Alētheia', *Early Greek Thinking*, p. 122. Italics added.
94 Heidegger, *Was ist Metaphysik?* (Frankfurt am Main: Vittorio Klostermann, 1965), pp. 7-8. Italics added.
95 Heidegger, *Being and Time*, pp. 196-7.
96 Heidegger, *Nietzsche*, vol. I: *The Will to Power as Art* (New York: Harper & Row, 1979), p. 80.
97 Heidegger, 'Moira', *Early Greek Thinking*, p. 97. Italics added.
98 Mircea Eliade, *Patterns in Comparative Religions* (New York: Sheed & Ward, 1958), pp. 150-1.

99 *Ibid.*
100 Heidegger, 'The End of Philosophy', *Basic Writings*, p. 286. I have substituted the word *lighting* for the translator's word *opening*.
101 James Hillman, *The Myth of Analysis* (New York: Harper & Row, 1978), p. 45.
102 Heidegger, *Being and Time*, p. 422.
103 Emerson, 'Nature', *op. cit.*, pp. 179-80.
104 Foucault, *The Birth of a Clinic*, p. 89.
105 Nietzsche, *The Will to Power*, Note 470, p. 262.
106 Heidegger, *Nietzsche*, Vol. II: *The Eternal Recurrence of the Same* (New York: Harper & Row, 1984), p. 62.
107 Descartes, *Meditations on First Philosophy*, p. 155.
108 Heidegger, 'The Turning', *The Question Concerning Technology and Other Essays*, pp. 36-49.
109 *Ibid.*
110 Heidegger, *Being and Time*, p. 177.
111 Concerning the new physics of light and the eye of physiology, see Robert Romanyshyn, 'Copernicus and the Beginnings of Modern Science', *Journal of Phenomenological Psychology*, vol. 3, no. 2 (Spring, 1973, pp. 187-200, and 'Science and Reality: Metaphors of Experience and Experience as Metaphorical', in Ron Valle and Rolf von Eckhartsberg (eds), *Metaphors of Consciousness* (New York: Plenum Publishing Co., 1981), pp. 3-19. Also see Drew Leder, 'Medicine and Paradigms of Embodiment', *The Journal of Medicine and Philosophy*, vol. 9 (1984), p. 30 and Richard Selzer, *Mortal Lessons: Notes on the Art of Surgery* (New York: Simon & Schuster, 1978), pp. 24-36.
112 See Hans Jonas, 'The Nobility of Sight: A Study in the Phenomenology of the Senses', in Stuart Spicker (ed.), *Philosophy of the Body* (New York: Quadrangle Press, 1970), pp. 47-83 and Erwin Straus, 'Born to See, Bound to Behold: Reflections on the Function of the Upright Posture in the Esthetic Attitude', also reprinted in Spicker, *op. cit.*, pp. 334-61; Wolfgang Metzger, *Gesetze des Sehens* (Frankfurt, 1953); Ralph M. Evans, *Perception of Color* (New York: John Wiley, 1974); Erwin Straus, *The Primary World of the Senses* (London: The Free Press, 1963); and J. M. Heaton, *The Eye: Phenomenology and Psychology of Function and Disorder* (London: Tavistock, 1968).
113 Heidegger, 'What Are Poets For?', *op. cit.*, p. 110.
114 Heidegger, 'Science and Reflection', in *The Question Concerning Technology and Other Essays*, pp. 155-82. Also see David C. Lindberg, *Theories of Vision from Al-kindi to Kepler* (Chicago: University of Chicago Press, 1976), and David C. Lindberg and Nicholas H. Steneck, 'The Sense of Vision and the Origins of Modern Science', in Allen G. Debus (ed.), *Science, Medicine and Society in the Renaissance: Essays to Honor Walter Pagel* (New York: Science History Publications, 1981), Vol. I, 29-45.
115 See Edmund Husserl, *The Crisis of European Sciences and Transcendental Phenomenology* (Evanston: Northwestern University Press, 1970), esp. pp. 3-74.

116 Heidegger, 'Science and Reflection', *op. cit.*, p. 179. Also see 'The Age of the World Picture', *op. cit.*, p. 149 and 'The Word of Nietzsche: "God is dead" ', *op. cit.*, p. 100.
117 Heidegger, 'Science and Reflection', *op. cit.*, p. 163.
118 *Ibid.*
119 *Ibid.*
120 *Ibid.*, p. 164.
121 *Ibid.*
122 *Ibid.*, p. 166.
123 Heidegger, *Introduction to Metaphysics* (New York: Doubleday, 1961), p. 52. I have modified the translation after consulting the original (German) text.
124 Heidegger, 'Science and Reflection', *op. cit.*, p. 167.
125 *Ibid.*
126 Samuel Y. Edgerton, Jr., *The Renaissance Rediscovery of Linear Perspective* (New York: Harper & Row, 1976), p. 27.
127 Heidegger, 'The Age of the World Picture', *op. cit.*, p. 119.
128 Heidegger, 'Moira', *Early Greek Thinking*, p. 99. Also see his 'Alētheia' study, *op. cit.*, p. 116. In *Democracy and Education*, Dewey has this to say about 'home': 'The earth, as the home of man, is humanizing and unified; the earth *viewed as* a miscellany of facts is scattering and imaginatively inert.' (p. 212. My italics.) Heidegger's most important discussion of 'home' is to be found in 'Building Dwelling Thinking', in *Poetry, Language, Thought*, pp. 145-61. Also see Vincent Scully, *The Earth, the Temple and the Gods: Greek Sacred Architecture* (New York, Praeger, 1969) for an account of building and dwelling in the ancient Greek Lebenswelt. Scully says (private conversation) that he is not familiar with Heidegger's work. Nevertheless, what he shows us of the ancient Lebenswelt strongly supports Heidegger's interpretation of the history of Being.
129 See David Michael Levin, 'Psychopathology in the Epoch of Nihilism', in Levin (ed.), *Pathologies of the Modern Self: Postmodern Studies on Narcissism, Schizophrenia, and Depression* (New York: New York University Press, 1986).
130 Heidegger, *Introduction to Metaphysics*, pp. 53-5.
131 Heidegger, 'Science and Reflection', *op. cit.*, p. 172.
132 *Ibid.*, p. 173. Erwin Schroedinger asserts that 'Subject and object are only one. The barrier between them cannot be said to have broken down as a result of recent experience in the physical sciences, for this barrier does not exist.' See his thoughts on *Mind and Matter* (London: Cambridge University Press, 1969), p. 137.
133 Heidegger, 'The Age of the World Picture', *op. cit.*, p. 174.
134 John Dewey, *Democracy and Education*, p. 213.
135 Tarthang Tulku, *Knowledge and Freedom* (Berkeley: Dharma Publishing, 1984), p. 70.
136 Heidegger, 'The Word of Nietzsche: "God is dead" ', *op. cit.*, pp. 80-1.
137 Heidegger's Zollikon Seminars are reported in Medard Boss, *Erinnerung an Martin Heidegger* (Pfullingen: Guenther Neske, 1977). Some of this

material has been translated by Brian Kenny and published in the *Review of Existential Psychology and Psychiatry*, vol. XVI, nos 1-3 (1978-79), p. 11.

138 Samuel Y. Edgerton, *The Renaissance Rediscovery of Linear Perspective* (New York: Harper & Row, 1976), p. 104.

139 Heidegger, 'The Turning, *op. cit.*, p. 48.

140 Max Horkheimer, *Dawn and Decline: Notes 1926-1931 and 1950-1969* (New York: The Seabury Press, Continuum Books, 1978), p. 162.

141 Paul Breines, 'From Guru to Spectre: Marcuse and the Implosion of the Movement', in *Critical Interruptions: New Left Perspectives on Herbert Marcuse* (New York: Herder & Herder, 1970), p. 3.

142 Henri Lefebvre, *La vie quotidienne dans le monde moderne* (Paris: Gallimard, 1968), p. 96.

143 Heidegger, *Being and Time*, p. 88.

144 *Ibid.*, p. 129.

145 Horkheimer, *op. cit.*, p. 161.

146 John O'Neill, *Sociology as a Skin Trade: Essays Towards a Reflexive Sociology* (London: Heinemann Educational Books, 1972), pp. 58-9.

147 *Ibid.*, p. 64.

148 *Ibid.*, p. 75.

149 Norman Bryson, *Vision and Painting: The Logic of the Gaze* (New Haven: Yale University Press, 1983), p. 10.

150 Foucault, *The Birth of the Clinic*, p. 166.

151 Raymond Geuss, *The Idea of Critical Theory: Habermas and the Frankfurt School* (New York: Cambridge University Press, 1981), p. 63.

152 Foucault, 'Body/Power', in *Power/Knowledge: Selected Interviews and Other Writings 1972-1977*, edited by Colin Gordon (New York: Pantheon, 1980), p. 62.

153 Foucault, 'The Politics of Health in the Eighteenth Century', *op. cit.*, pp. 170-1.

154 Jean Paris, *Painting and Linguistics* (Pittsburgh: Carnegie-Mellon University, 1975), pp. 31-72.

155 *Ibid.*, p. 39.

156 *Ibid.*

157 *Ibid.*

158 *Ibid.*, p. 43.

159 *Ibid.*

160 *Ibid.*

161 *Ibid.*, p. 45.

162 *Ibid.*, p. 49.

163 *Ibid.*, p. 50.

164 *Ibid.*, p. 52.

165 *Ibid.*, p. 61.

166 *Ibid.*, p. 62.

167 *Ibid.*, p. 70. Italics added.

168 *Ibid.*, p. 69.

169 *Ibid.*

170 *Ibid.*, p. 72.
171 Francis M. Cornford, *From Religion to Philosophy* (New York: Harper & Row, 1957), p. 198. Italics added.
172 *Ibid.*, p. 200. Italics added.
173 Foucault, *The Birth of the Clinic*, p. 84.
174 *Ibid.*, p. 83.
175 *Ibid.*, p. 9.
176 *Ibid.*, p. 166.
177 O'Neill, *op. cit.*, p. 69.
178 Heidegger, 'The Age of the World Picture', *op. cit.*, p. 130.
179 *Ibid.*
180 *Ibid.* Italics added.
181 *Ibid.*, p. 131.
182 *Ibid.*
183 *Ibid.*
184 *Ibid.*, p. 129. Italics added.
185 *Ibid.*, p. 131. I have substituted the words 'to remain in' for the translator's words 'back into'.
186 See, for example, *ibid.*, p. 149.
187 See my essay on 'Psychopathology in the Epoch of Nihilism', *op. cit.*
188 Susan Sontag, *On Photography* (New York: Dell Publishing, 1977), p. 111.
189 *Ibid.*, p. 106.
190 *Ibid.*, p. 80.
191 *Ibid.*, pp. 22-3.
192 *Ibid.*, p. 9.
193 *Ibid.*, p. 64.
194 On the difference between vision and hearing, see my forthcoming book, *The Listening Self*. Also see Johannes Fabian, *Time and the Other: How Anthropology Makes Its Object* (New York: Columbia University Press, 1983) and Erwin Straus, *The Primary World of the Senses*, pp. 376-8.
195 See *The Chicago Tribune*, March 10, 1983, section 1, p. 3. A very different matter is the visionary situation discussed recently by David Kehr in 'The End of the Line', his review in *Chicago Magazine* (January, 1986, p. 87) of Claude Lanzmann's documentary on the Holocaust. The title of the film is 'Shoah', which means 'Holocaust' in Hebrew. Writing about the filming of a barber who, in the midst of his talking about the work he did in the death camp, suddenly breaks down, Kehr says: 'Lanzmann keeps the camera rolling as the barber struggles to collect himself, but this isn't the emotional exploitation of the six-o'clock news. Lanzmann's refusal to cut away is itself an act of courage, as much as the barber's is in daring to remember. The decent thing to do would be to cut away, but decency counts for very little against the need to put these things on record'. I think he is right. But this is very different from the refusal to look away in the filming of the Duke of Windsor, which I discuss later in the chapter, and different, too, from

the filming of the families of the hostages held in Iran at the end of the Carter administration. On Sunday, November 9, 1980, the following article appeared in *The New York Times* (p. 25): 'On Christmas day last year, the mother of an American hostage being held in Teheran agreed to let a television network spend two hours filming the hostage's family at home. Later, when members of the family gathered to see themselves on the air, they wondered why other families appeared on the program but nothing was used about them. Then suddenly it became clear: Unlike the other families, they had not cried on camera. "They wanted our tears and we didn't give them any", the mother explained later. The families of the hostages have learned a lot about the news media in the past year.'

196 See Homer, *The Odyssey* (New York: Penguin Books, 1950), Book VIII, lines 83-93.

197 Foucault, 'Discourse and Truth: The Problematization of Parrhēsia', a seminar which took place Fall, 1983 at the University of California, Berkeley. In 1985, tapes of this seminar were transcribed and published by Joseph Pearson, graduate student in Philosophy at Northwestern University.

198 Nathan Schwartz-Salant, *Narcissism and Character Transformation: The Psychology of Narcissistic Character Disorders* (Toronto: Inner City Books, 1982), p. 106.

199 *Ibid.*, p. 72.

200 Christopher Lasch, *The Culture of Narcissism: American Life in an Age of Diminishing Expectations* (New York: W. W. Norton, 1979), p. 50.

201 Jurgen Habermas, 'Moral Development and Ego Identity', in *Communication and the Evolution of Society* (Boston: Beacon Press, 1979), pp. 70-1.

202 Nietzsche, *The Will to Power*, Note 18, p. 16. Also see Heidegger, 'Letter on Humanism', *Basic Writings*, p. 210.

203 Schwartz-Salant, *op. cit.*, p. 36.

204 Heidegger, 'The Word of Nietzsche: "God is dead" ', *op. cit.*, p. 68.

205 Sigmund Freud, 'On Narcissism: An Introduction', *The Collected Works of Sigmund Freud*, Standard Edition, edited by James Strachey (London: Hogarth Press, 1953), vol. 14, pp. 67-102.

206 Heinz Kohut, *The Restoration of the Self* (New York: International Universities Press, 1977), pp. 267-91.

207 *Ibid.*, p. 193.

208 *Ibid.*, pp. 192-3.

209 *Ibid.*, p. 130. Italics added.

210 *Ibid.*, p. 137.

211 *Ibid.*, p. 152.

212 Heidegger, 'The Age of the World Picture', *op. cit.*, p. 140.

213 Schwartz-Salant, *op. cit.*, p. 40.

214 *Ibid.*, p. 67.

215 *Ibid.*, p. 99. Also see pp. 91-107.

216 Merleau-Ponty, 'The Intertwining – The Chiasm', *The Visible and the Invisible*, p. 159. Going more deeply than the tradition into the phenomenology of the body, Merleau-Ponty brings to light, here, a narcissism whose intercorporeal reversibility *contests* the Cartesian version. (See pp. 139, 141, 249, 255-6). Merleau-Ponty shows that Cartesianism (especially, for example, its metaphysical isolation and privileging of the subject) is false precisely *because* things mirror our gaze.
217 Schwartz-Salant, *op. cit.*, p. 37.
218 Nietzsche, *The Will to Power*, Note 20, p. 16.
219 Theodor Adorno, *Minima Moralia: Reflections from Damaged Life* (London: New Left Books, 1974; Verso Editions, 1978), p. 50.
220 Jules Henry, *Pathways to Madness* (New York: Random House, 1973), p. 109.
221 Heidegger, 'Metaphysics as History of Being', *The End of Philosophy* (New York: Harper & Row, 1973), p. 18.
222 *Ibid.*, pp. 21-2.
223 Heidegger, 'Sketches for a History of Being as Metaphysics', *The End of Philosophy*, p. 69.
224 *Ibid.*
225 Henry, *op. cit.*, p, 109.
226 *Ibid.*, p. 110.
227 Schwartz-Salant, *op. cit.*, p. 49.
228 Heidegger, 'The Question Concerning Technology', *op. cit.*, p. 27.
229 Heidegger, 'The Word of Nietzsche: "God is dead" ', *op. cit.*, pp. 83-4.
230 *Ibid.*
231 Heidegger, 'Overcoming Metaphysics', *The End of Philosophy*, p. 107.
232 *Ibid.*
233 *Ibid.*
234 Joel Kovel, *The Age of Desire: Reflections of a Radical Psychoanalyst* (New York: Pantheon, 1982), p. 106.
235 *Ibid.*, p. 125.
236 See Horkheimer, *Dawn and Decline*, p. 172 and Adorno, *Minima Moralia*, p. 237.
237 Henry, *op. cit.*, p. 110.
238 Heidegger, 'The Age of the World Picture', *op. cit.*, p. 142.
239 Ronald D. Laing, *The Divided Self: An Existential Study in Society and Madness* (London: Tavistock Publications, 1960), p. 57.
240 Schwartz-Salant, *op. cit.*, p. 98.
241 See Bryan S. Turner, *The Body and Society: Explorations in Social Theory* (Oxford and New York: Basil Blackwell, 1984), p. 111.
242 See Charles H. Cooley, *Human Nature and the Social Order* (New Brunswick, New Jersey: Transaction Books, 1983).

Chapter 2 Crying for a Vision

1 Merleau-Ponty, *Phenomenology of Perception*, p. 127.
2 Heidegger, *Nietzsche*, vol. II: *The Eternal Recurrence of the Same*, p. 55.
3 *Service for the Day of Atonement: Yom Kippur* (New York: Hebrew Publishing Co., 1935), p. 297 and p. 377.
4 Joseph Epes Brown (ed.), *The Sacred Pipe: Black Elk's Account of the Seven Rites of the Oglala Sioux* (New York: Penguin, 1971), p. 46. 'Humility' is considered essential for an authentic rite of lamentation.
5 *Ibid.*, p. 85.
6 Heidegger, *Nietzsche*, vol. II, p. 91.
7 Gershom Scholem, *On the Kabbalah and Its Symbolism* (New York: Schocken, 1969), p. 112. According to Erich Neumann, the vessel is 'the central symbol of the feminine'. See *The Great Mother: An Analysis of the Archetype* (Princeton: Princeton University Press, 1972), pp. 39-63.
8 Scholem, *op. cit.*, p. 114.
9 *Service for the Day of Atonement*, p. 310.
10 Meister Eckhart, *An Introduction to the Study of his Works, with an Anthology of His Sermons* (London: Nelson & Sons, 1957), p. 237.
11 Arnold Mindell, *Dreambody: The Body's Role in Revealing the Self* (Santa Monica: Sigo Press, 1982), pp. 94-6.
12 Michel Foucault, 'Nietzsche, Genealogy, History', in Donald F. Bouchard (ed.), *Language, Counter-Memory, Practice* (Ithaca: Cornell University Press, 1977), p. 50.
13 See Hubert V. Guenther's commentary on the *Rig-pa rang-shar chen-po'i rgyud* and other texts, in his translation of Long-chen-pa, *Kindly Bent to Ease Us* (Berkeley: Dharma Publishing Co., 1976), Part II, p. 16.
14 Heidegger, 'The Essence of Truth', in *Basic Writings*, p. 141.
15 John Rawls, *A Theory of Justice* (Cambridge: Harvard University Press, 1971), p. 415.
16 Heidegger, *Vom Wesen der Wahrheit* (Frankfurt am Main: Vittorio Klostermann, 1943), p. 18. Italics added.
17 Heidegger, *Discourse on Thinking*, p. 64.
18 Merleau-Ponty, *Phenomenology of Perception*, p. 216.
19 *Ibid.*, pp. 278-9.
20 See the translation of Xenophanes, fragment 24, in Phillip Wheelwright, *The Presocratics* (New York: Odyssey, 1966), p. 32, and in Kathleen Freeman, *Ancilla to the Pre-Socratic Philosophers* (Cambridge: Harvard University Press, 1978), p. 23. She translates it as: 'He sees as a whole, thinks as a whole, and hears as a whole.' The resonances between this thought and thoughts recorded in the Hindu *Upanishads* are really very striking and should be given more philosophical attention.
21 Heidegger, 'What Is Metaphysics?', in *Basic Writings*, p. 102. Also see 'On the Essence of Truth', *op. cit.*, p. 132: 'as a whole' does not describe a mere feeling or a merely subjective experience, although it certainly can be defined in other ways. To his credit, Heidegger appreciates that

it is only 'from the point of view of everyday calculations and preoccupations, [that] this "as a whole" appears to be incalculable and incomprehensible'.

22 Heidegger, 'What Are Poets For?', *Poetry, Language, Thought*, p. 140.

23 Heidegger. 'The Anaximander Fragment', *Early Greek Thinking*, p. 39. I have changed the word 'totality' to 'wholeness'. This is a crucial point, where failure to make the distinction causes philosophers, in effect, to abandon all concern for 'wholeness' as a way of avoiding metaphysical totalism.

24 *Ibid*.

25 Heidegger, 'On the Essence of Truth', *Basic Writings*, p. 128.

26 *Ibid*., p. 134.

27 Herbert V. Guenther, *The Tantric View of Life* (Boulder: Shambhala, 1976), p. 93.

28 Heidegger, 'Language', *Poetry, Language, Thought*, p. 204.

29 Primo Levi, *Survival in Auschwitz: The Nazi Assault on Humanity* (New York: Collier Macmillan, 1961), p. 49.

30 Heidegger, 'The Anaximander Fragment', *Early Greek Thinking*, p. 17. For the German, see *Holzwege* (Frankfurt am Main: Vittorio Klostermann, 1950), pp. 300-1.

31 See Elie Wiesel, *Night* (New York: Hill & Wang, 1960).

32 Nietzsche, *Beyond Good and Evil*, p. 89.

33 Heidegger, 'The Anaximander Fragment', *op. cit*., p. 57.

34 Nietzsche, *The Genealogy of Morals*, Third Essay, note 1, p. 231; also see note 28, p. 299.

35 Nietzsche, *The Will to Power*, note 55, p. 35.

36 *Ibid*., note 18, p. 16.

37 *Ibid*., note 5, p. 10.

38 See Eugene Gendlin, 'A Philosophical Critique of the Concept of Narcissism: The Structural-Political Significance of the Awareness Movement', in David M. Levin (ed.), *Pathologies of the Modern Self* (New York: New York University Press, 1987).

39 Heidegger, 'The Age of the World Picture', in *The Question Concerning Technology and Other Essays*, p. 142. Also see Neumann's discussion of the emergence of the ego in *The Origins and History of Consciousness* (Princeton: Princeton University Press, 1980), pp. 109-27.

40 See Nathan Schwarz-Salant, 'The Dead Self in Borderline Personality Disorders', in Levin (ed.), *Pathologies of the Modern Self*.

41 On depressions, see Roger Levin, 'Cancer and the Self: How Illness Constellates Meaning'; Robert Romanyshyn, 'Depression and the American Dream: The Struggle with Home'; and Cisco Lassiter, 'Relocation and Illness: The Plight of the Navajo', in Levin (ed.), *Pathologies of the Modern Self*.

42 See Nietzsche, *Genealogy of Morals*, Third Essay, pp. 256-8, 171-2, 277-8.

43 See David M. Levin, 'Psychopathology in the Epoch of Nihilism', in Levin (ed.), *Pathologies of the Modern Self*.

44 Heinz Werner, *Comparative Psychology of Mental Development* (New York: International Universities Press, 1957), especially pp. 67-86.

45 Norman O. Brown, *Life Against Death: The Psychoanalytic Meaning of History* (Middletown: Wesleyan University Press, 1959), p. 52. Also see Heinz Kohut, *The Restoration of the Self* (New York: International Universities Press, 1977), p. 188. See also *op. cit.*, pp. 156-60 on 'primary object-loss'.
46 Sigmund Freud, *The Ego and the Id* (New York: W. W. Norton, 1960), p. 19.
47 See Freud, *Inhibition, Symptom, and Anxiety* (New York: W. W. Norton, 1963) and Heidegger, 'Overcoming Metaphysics', *The End of Philosophy*, pp. 102-3.
48 Nietzsche, *Beyond Good and Evil*, note 278, pp. 223-4.
49 *Ibid.*, p. 83.
50 See Kathleen Freeman, *Ancilla to the Pre-Socratic Philosophers*, p. 28. The saying in question is fragment 52.
51 Heidegger, 'Der Anfang des abendländischen Denkens', in *Heraklit*, 'Gesamtausgabe', Bd. 55 (Frankfurt am Main: Vittorio Klostermann, 1979), p. 23.
52 Nietzsche, *Beyond Good and Evil*, note 57, pp. 68-9.
53 Jean-Paul Sartre, *The Transcendence of the Ego* (New York: Noonday, 1957), p. 100.
54 Heidegger, 'The Age of the World Picture', *op. cit.*, p. 154.
55 Nietzsche speaks of 'numbness' and 'anaesthesia' in *The Will to Power* (Note 23, p. 18 and Note 47, p. 30). Heidegger speaks of the same phenomenon in 'Overcoming Metaphysics', *op. cit.*, p. 110, where he refers to our being 'cut off from pain'. Also see Robert Jay Lifton, *The Life of the Self: Towards a New Psychology* (New York: Basic Books, 1983) and Edith Wyschogrod, *Spirit in Ashes: Hegel, Heidegger and Man-Made Mass Death* (New Haven: Yale University Press, 1985).
56 Jeremy Rifkin, *Declaration of a Heretic* (London and Boston: Routledge & Kegan Paul, 1985), p. 100.
57 Heidegger, 'What Are Poets For?', in Hofstadter (ed.), *Poetry, Language, Thought*, p. 120.
58 Heidegger, *Discourse on Thinking*, p. 78. I characterize his point as 'Nietzschean' because it is reminiscent of the passage in *The Will to Power* (Note 55, p. 35) that I quoted earlier, i.e., in the discussion immediately following passages from Elie Wiesel.
59 Nietzsche, *The Genealogy of Morals*, Second Essay, Note 6, p. 160.
60 Merleau-Ponty, *Phenomenology of Perception*, p. 430.
61 *Ibid.*, p. 241. Italics added. Also see Wolfgng Köhler, *Gestalt Psychology* (New York: Mentor Books, 1947), p. 80ff.
62 Mindell, *Dreambody*, p. 143.
63 See Northrop Frye, *The Anatomy of Criticism* (Princeton: Princeton University Press, 1957).
64 Heidegger, *Discourse on Thinking*, p. 61.
65 See *The Royal Song of Saraha*, translated with commentary by Herbert V. Guenther (Berkeley: Shambhala, 1973), p. 100. Also see David M. Levin, 'Freud's Divided Heart and Saraha's Cure', *Inquiry*, vol. 20, nos 2-3 (Summer, 1977), pp. 165-88.

66 Sartre, *The Transcendence of the Ego*, p. 93.
67 *Ibid.*, p. 99.
68 Iris Murdoch, *The Sovereignty of Good* (New York: Penguin, 1970), p. 84.
69 Heidegger, 'The Turning', *The Question Concerning Technology and Other Essays*, p. 37.
70 Merleau-Ponty, *Phenomenology of Perception*, p. 377.
71 Murdoch, *op. cit.*, p. 341. Also see *op. cit.*, p. 91. And see Stanley Cavell, *The Claim of Reason* (New York: Oxford University Press, 1979), p. 372: 'For justice to be done, a change of perception, a modification of seeing, may be called for.' That conviction is what motivated me to write this book.
72 See Phillip Wheelwright, *op. cit.*, p. 32.
73 Merleau-Ponty, *Phenomenology of Perception*, p. 250.
74 See Jane B. Greene and M. D. Herter Norton (eds), *Letters of Rainer Maria Rilke 1892-1910* (New York: W. W. Norton, 1945), p. 336. In this letter to his wife, Clara Rilke, the poet speaks of a vision whose sense of the whole (whose sense of wholeness) leaves all things free in every way. This is a vision that I would characterize as *Gelassenheit*.
75 Heidegger, 'Overcoming Metaphysics', *op. cit.*, p. 87.
76 Merleau-Ponty, *The Visible and the Invisible* (Evanston: Northwestern University Press, 1979), p. 452.
77 *Ibid.*, p. 240.
78 *Ibid.*, pp. 187-8.
79 Hubert Benoit, *The Supreme Doctrine* (New York: Viking Press, 1955). p. 180.
80 William James, *The Principles of Psychology*, vol. I (New York: Dover Publications, 1950), p. 284.
81 See John J. McDermott (ed.), *The Writings of William James* (New York: Random House, 1968), p. 157. Also see Merleau-Ponty, 'The Intertwining – The Chiasm', in *The Visible and the Invisible*, p. 139: '. . . to be seduced, captivated, . . . so that the seer and the visible reciprocate one another, and we no longer know which sees and which is seen'.
82 Wei Wu Wei, quoted by Kenneth Wilber, *The Spectrum of Consciousness* (Wheaton, Illinois: Theosophical Publishing House, 1979), p. 332.
83 *Brihadaranyaka Upanishad*, III, 4, ii, in *The Upanishads*, trans. F. Max Muller (New York: Dover Publications, 1962), vol. II, p. 129.
84 G. Spencer Brown, *Laws of Form* (New York: Julian Press, 1972), v.
85 Merleau-Ponty, *The Visible and the Invisible*, p. 216.
86 Neumann, 'Creative Man and Transformation', in *Art and the Creative Unconscious* (Princeton: Princeton University Press, 1971), p. 164.
87 Heidegger, *The Basic Problems of Phenomenology* (Bloomington: Indiana University Press, 1982), p. 314.
88 Neumann, 'Creative Man and Transformation', *op. cit.*, p. 177. Italics and bracketed words added.
89 Heidegger, 'Logik: Heraklits Lehre vom Logik', in *Heraklit*, 'Gesamtausgabe', Bd. 55 (Frankfurt am Main: Vittorio Klostermann, 1979), p. 337. He also speaks (p. 336) of 'Absonderung' and

'Scheidung' as interpretations of *chōrizein*.

90 Daisetz T. Suzuki, *The Zen Doctrine of No Mind* (New York: Samuel Weiser, 1973), p. 133. Italics added.

91 John P. Muller and William J. Richardson, *Lacan and Language* (New York: International Universities Press, 1982), p. 63.

92 Jacques Lacan, *Le Séminaire*, Livre I: *Les Écrits techniques de Freud* (Paris: Éditions du Seuil, 1975), p. 22.

93 See G. E. Bentley (ed.), *The Works of William Blake*, 2 vols. (Oxford: Oxford University Press, 1979).

94 Merleau-Ponty, 'Eye and Mind', in *The Primacy of Perception* (Evanston: Northwestern University Press, 1964), p. 188.

95 Merleau-Ponty, *Phenomenology of Perception*, p. 219. The duality of the subject-object structure can be overcome in bodily movement, but only if the movement is experienced together with its entire situation, and through the channels of feeling, i.e., with felt awareness. See Henry David Thoreau, 'The Bean Field', in Carl Bode (ed.), *The Portable Thoreau* (New York: Viking Press, 1964), p. 408. Also see Nagarjuna's *Mulamadhyamiḳakarika*, trans. Kenneth K. Inada (Tokyo: Hojuseido Press, 1970), and Śantideva, *Entering the Path of Enlightenment* (*Bodhicaryavatara*), trans. Marion L. Matics (New York: Macmillan, 1970). The Buddhists deconstructed this dualism long ago.

96 Merleau-Ponty, *Phenomenology of Perception*, p. 430.

97 *Ibid.*, p. 320. Italics added.

98 *Ibid.*, p. 329.

99 Heidegger, 'Alētheia', in Krell and Capuzzi (eds), *Early Greek Thinking*, p. 103.

100 Heidegger, *What Is Called Thinking?*, p. 110.

101 Heidegger and Fink, *Heracleitus Seminar 1966–1967* (University of Alabama Press, 1979), p. 126.

102 See Carl G. Jung's 'Psychological Commentary' to *The Tibetan Book of The Dead*, trans. W. Y. Evans-Wentz (London: 1954), pp. xxix-lxiv. Also published in Joseph Campbell (ed.), *The Portable Jung* (New York: Penguin Books, 1976), p. 491.

103 See Herbert V. Guenther, *Buddhist Philosophy in Theory and Practice* (Baltimore: Penguin, 1971), and *Philosophy and Psychology in the Abhidharma* (Berkeley: Shambhala, 1974). The Buddhist scholar Nagarjuna brilliantly expounded a method for deconstructing all metaphysical positions. In his *Mulamadhyamikakarika*, a work written many centuries ago, he applied this method to such topics as the nature of motility.

104 Jung, 'Consciousness, Unconscious, and Individuation', *The Archetypes and the Collective Unconscious*, in *The Collected Works of Carl G. Jung* (Princeton: Princeton University Press, 1968), vol. 9, Part I, p. 280. Italics added.

106 *Ibid.*, p. 283.

106 Merleau-Ponty, *Phenomenology of Perception*, p. 351.

107 Jung, 'Consciousness, Unconscious, and Individuation', *op. cit.*, p. 283.

108 Merleau-Ponty, *Phenomenology of Perception*, p. 216. Italics added. Also see *op. cit.*, pp. 12, 267, 327, 352.
109 On the relation between experiencing and interpreting, see David Michael Levin, 'The Poetic Function in Phenomenological Discourse', in William McBride and Calvin O. Schrag (eds), *Phenomenology in a Pluralistic Context: Selected Studies in Phenomenology and Existential Philosophy* (Albany: State University of New York Press, 1983), vol. 9, pp. 216-34.
110 Merleau-Ponty, *The Visible and the Invisible*, p. 191. On the subject of regions and fields, also see Lloyd Kaufman, *Perception: The World Transformed* (New York: Oxford University Press, 1979), p. 302.
111 Merleau-Ponty, *The Visible and the Invisible*, p. 197.
112 Heidegger, 'Logik: Heraklits Lehre vom Logik', *op. cit.*, p. 337.
113 See Frederick Perls, Ralph Hefferline and Paul Goodman, *Gestalt Therapy: Excitement and Growth in the Human Personality* (New York: Delta Books, 1951), p. 56. Also see pp. 320-32.
114 William James, *The Principles of Psychology* (New York: Henry Holt, 1890), p. 244.
115 John Welwood, 'Exploring Mind: Form, Emptiness, and Beyond', in *The Journal of Transpersonal Psychology*, vol. 8, no. 2 (1976), p. 91.
116 Heidegger and Fink, *Heraclitus Seminar 1966-1967*, p. 85.
117 See the *Kitab al-Manazir* (*Perspectiva*), a work by Al-hazen, a eleventh century optical scientist living in Cairo. His work was eventually passed on to Ghiberti in Florence. See Samuel Edgerton, *op. cit.*, p. 74. The 'axis' of vision is, I suggest, the line of greatest desire: a line imposed by the will to power.
118 See Guenther, *The Royal Song of Saraha*, p. 45.
119 Merleau-Ponty, *Phenomenology of Perception*, p. 219. Also see *The Visible and the Invisible*, pp. 122-3. Bracketed words added.
120 Heidegger and Fink, *op. cit.*, p. 140.
121 Merleau-Ponty, *The Visible and the Invisible*, p. 123.
122 Heidegger and Fink, *op. cit.*, p. 140.
123 Merleau-Ponty, 'Interrogation and Dialectic', in *The Visible and the Invisible*, p. 76.
124 See Tarthang Tulku, 'The Life and Liberation of Padmasambhava', in *Crystal Mirror* (Berkeley: Dharma Publishing, 1977), vol. IV, p. 27.
125 Heidegger, *Der Satz vom Grund* (Pfullingen: Gunther Neske, 1957), p. 193. Also see Guenther, *Buddhist Philosophy in Theory and Practice*, pp. 196-8.
126 Heidegger and Fink, *op. cit.*, p. 126.
127 Merleau-Ponty, *The Visible and the Invisible*, p. 192.
128 *Ibid.*, p. 221. Also see p. 207, regarding the experience of 'polymorphism'.
129 See Gerardo Reichel-Dolmatoff, *The Shaman and the Jaguar: A Study of Narcotic Drugs Among the Indians of Colombia* (Philadelphia: Temple University Press, 1975), p. 181. The Desana Indians ritually drink a psychotropic potion called *yahé*, made from a vine which grows in the

rainforest, and commonly experience visions of intertwining, spiritual communions, connecting them to their tribal ancestors.

130 Freud, *Civilization and Its Discontents* (New York: Doubleday, 1930), pp. 1-11.
131 *Ibid.*, p. 1.
132 *Ibid.*, p. 2.
133 *Ibid.*, pp. 11 and 27.
134 Merleau-Ponty, *Phenomenology of Perception*, p. 156.
135 Freud, *Civilization and Its Discontents*, p. 105.
136 Rilke, *The Notebooks of Malte Laurids Brigge* (New York: W. W. Norton, 1949), pp. 14-15.
137 D. W. Winnicott, *Playing and Reality* (London: Tavistock, 1971), p. 2.
138 Merleau-Ponty, *Phenomenology of Perception*, p. 154.
139 Jung, 'The Concept of Libido', *Symbols of Transformation*, in *The Collected Works of Carl G. Jung* (New York: Pantheon, 1956), p. 137.
140 See Neumann, *The Origins and History of Consciousness*, p. 113.
141 See Merleau-Ponty, *Phenomenology of Perception*, p. 288.
142 *Ibid.*, p. 214. Italics added.
143 *Ibid.*, p. 320. Italics added. Also see p. 212.
144 *Ibid.* Italics added.
145 See Freud, 'The Psycho-Analytic View of Psychogenic Disturbances of Vision', *The Complete Works of Sigmund Freud*, Standard Edition (London: Hogarth Press, 1953), vol. 11, pp. 211-17.
146 Merleau-Ponty, *Phenomenology of Perception*, p. 214. Italics added.
147 Mark Taylor, *Erring: A Postmodern A/Theology* (Chicago: University of Chicago, 1984), p. 162.
148 Heidegger, 'Building Dwelling Thinking', in Hofstadter (ed.), *Poetry, Language, Thought*, p. 157.
149 Sartre, *Being and Nothingness* (New York: Philosophical Library, 1956; London: Methuen, 1957), p. 325.
150 Henri Bergson, *The Two Sources of Morality and Religion* (New York: Greenwood, 1935), p. 246.
151 Primo Levi, *Survival in Auschwitz: The Nazi Assault on Humanity* (New York: Collier-Macmillan, 1961), pp. 22-3.
152 Giambattista Vico, *The New Science* (Ithaca: Cornell University Press, 1970), p. 88.
153 Robert Romanyshyn, 'Science and Reality: Metaphors of Experience and Experience as Metaphorical', in Ronald Valle and Rolf von Eckartsburg (eds), *The Metaphors of Consciousness* (New York: Plenum Press, 1981), p. 10.
154 Mindell, *Working with the Dreaming Body* (Boston and London: Routledge & Kegan Paul, 1985), pp. 43-5.
155 Merleau-Ponty, 'The Intertwining – The Chiasm', in *The Visible and the Invisible*, pp. 137-38. The flesh, he says, is 'Visibility sometimes wandering and sometimes reassembled'.
156 Sir Thomas Browne, *The Religio Medici and Other Writings* (London and

Toronto: J. M. Dent & Sons, Ltd; New York: E. P. Dutton & Co., 1928), Part I, p. 42. What Merleau-Ponty sees as the 'intertwining' character of the flesh was surprisingly anticipated by Sir Thomas Browne's observation that 'All flesh is grass'.

157 See David M. Levin, 'The Poetic Function of Phenomenological Discourse', cited earlier in note 109.

158 Concerning the symbolic meaning of weaving, spinning, mother goddesses and the spider, see Helen Diner, *Mothers and Amazons: The First Feminine History of Culture* (New York: Doubleday Anchor Books, 1973), p. 16: 'All mother goddesses spin and weave. . . . Everything that is comes out of them: They weave the world tapestry out of genesis and demise, "threads appearing and disappearing rhythmically".' Also see Mary Daly, 'Spinning Cosmic Tapestries', in *Gyn/Ecology: The Metaethics of Radical Feminism* (Boston: Beacon Press, 1978), pp. 385-424: 'The mindbinders and those who remain mindbound do not see the patterns of the cosmic tapestries, nor do they hear the labyrinthine symphony, for their thinking has been crippled and tied to linear tracks. Spiraling/Spinning is visible/audible to them only where it crosses the straight lines of what they call thinking. Hence the integrity of Spinning thought eludes them, and what they perceive is merely a series of fragmented breaks/crosses, which might appear like an irregular series of dots and dashes.' The word 'chiasm', which appears in Merleau-Ponty's late work, comes from the old Greek word for 'crossing' and 'crossroads'. With his formulation of the concept of the 'intertwining', Merleau-Ponty's thinking comes under the spell of the Feminine Archetype and challenges logocentric, patriarchal philosophy to acknowledge a neglected wisdom.

159 Neumann, *The Great Mother*, p. 226.

160 See David M. Levin, 'The Living Body of Tradition', in *The Body's Recollection of Being*, pp. 167-223.

161 Merleau-Ponty, 'Eye and Mind', *op. cit.*, p. 183.

162 Rilke, *Letters 1910-1926* (New York: W. W. Norton, 1947), p. 116.

163 Nietzsche, *Daybreak: Thoughts on the Prejudices of Morality* (New York: Cambridge University Press, 1982), note 506, p. 203.

164 Stephen Beven, 'Bases of Cognition', in *Gesar* (Berkeley: Tibetan Buddhist Nyingma Institute, Summer-Fall, 1979), vol. 6, no. 2, p. 40.

165 Welwood, 'Meditation and the Unconscious: A New Perspective', in *The Journal of Transpersonal Psychology*, vol. 9, no. 1 (1977), p. 20.

166 Donald L. Philippi (ed.), *Songs of Gods, Songs of Humans: The Epic Tradition of the Ainu* (San Francisco: North Point Press, 1982), p. 79. The Ainu are an aboriginal Caucasian people inhabiting the northernmost islands of 'Japan'.

167 Heidegger *Being and Time*, p. 176.

168 Heidegger, 'The Origin of the Work of Art', in Hofstadter (ed.), *Poetry, Language, Thought*, p. 25.

169 Nietzsche, *Twilight of the Idols* (New York: Penguin Books, 1968), p. 5. Also see the discussion of practices of the self for the development of awareness in Herbert V. Guenther and Chogyam Trungpa, *The Dawn of Tantra* (Berkeley: Shambhala, 1975). These practices *teach* seeing. Also see Namkhai Norbu, *The Cycle of Day and Night* (Berkeley: Zhang Zhung Editions, 1985).
170 See Merleau-Ponty, *The Visible and the Invisible*, p. 197.
171 *Ibid.*, p. 207. In a note to himself, Merleau-Ponty writes: 'Show that, since the *Gestalt* arises from polymorphism, this situates us entirely outside the philosophy of subject and object.'
172 Heidegger, *Discourse on Thinking*, p. 55.
173 Heidegger, 'The Anaximander Fragment', *Early Greek Thinking*, p. 36.
174 *Ibid.*, p. 38.
175 Heidegger *Being and Time*, p. 187.
176 Heidegger, 'The Anaximander Fragment', *op. cit.*, pp. 33-6.
177 See Yeshé Tsogyal, *The Life and Liberation of Padmasambhava*, a translation of the *Padma bKa'i Thang*, edited by Tarthang Tulku (Berkeley: Dharma Publishing, 1978), vol. I, p. 68. Also see Tarthang Tulku, *Gesture of Balance* (Berkeley: Dharma Publishing, 1977) and *Kum Nye Relaxation*, vols. I and II (Berkeley: Dharma Publishing, 1978).
178 On the stillness of the eyes, see Heidegger, 'Der Anfang des abendländischen Denkens', in *Heraklit*, p. 123.
179 Foucault, 'Discourse and Truth: The Problematization of Parrhēsia', lectures delivered at the University of California, Berkeley, Fall, 1983, and published privately in 1985, from a transcript of the tapes, by Joseph Pearson.
180 See Jean Paris, *Painting and Linguistics* (Pittsburgh: Carnegie-Mellon University, 1975), p. 12. Paris is summarizing what J. F. Lyotard has to say about vision in his book, *Discours Figure* (Paris: Klincksieck, 1972).
181 Jean Paris, *op. cit.*, p. 12.
182 See O. Poetzl *et al.*, 'Preconscious Stimulation in Dreams, Associations, and Images', *Psychological Issues*, vol. 2 (1960), pp. 1-18. The authors argue that the dream processes tend to take over peripheral vision and that more rational processes control the focal centre of the visual field. On this same question, also see Aron Gurwitsch, 'Aspects of Gestalt Psychology', in *Studies in Phenomenology and Psychology* (Evanston: Northwestern University Press, 1966), p. 14, and Wolfgang Köhler, *Dynamics in Psychology* (New York: Washington Square Press, 1965), pp. 72-109 and 155-5, and *Gestalt Psychology* (New York: Mentor Press, 1947), pp. 100-1.
183 See Richard Wallen, 'Gestalt Therapy and Gestalt Psychology', and Frederick Perls, 'Four Lectures', in Joen Fagan and Irma Shepherd (eds), *Gestalt Therapy Now* (New York: Harper & Row, 1970). Also see Namkhai Norbu, *The Mirror: Advice on Presence and Awareness* (Arcidosso, Italy: Shang Shung Edizioni, 1983) and Tarthang Tulku, *Time, Space and Knowledge* (Berkeley: Dharma Publishing, 1977).
184 Tarthang Tulku, 'The Ocean of Knowledge', in Stephen Randall and Ralph Moon (eds), *Dimensions of Thought* (Berkeley: Dharma, 1980), p. xliv.

185 See John Sallis, 'Heidegger/Derrida – Presence', *Journal of Philosophy*, vol. 81, no. 10 (October, 1984), pp. 594-601. The quotation is to be found on p. 597.
186 *Ibid.*
187 See the interview with Jacques Derrida in Richard Kearney (ed.), *Dialogues with Contemporary Continental Thinkers: The Phenomenological Heritage* (Dover, New Hampshire: Manchester University Press, 1984), p. 115.
188 *Ibid.*, p. 128. Kearney's phrase.
189 Heidegger, 'Andenken', in *Erläuterungen zu Hölderlins Dichtung* (Frankfurt am Main: Vittorio Klostermann, 1971), p. 96. The concept of an 'authentic greeting', does not have to remain shrouded in mystification. There are behavioural criteria which enable us to see the difference between a situation where two people are 'present' in their being with one another and a situation where they are not. See, for example, Nancy Henley, *Body Politics: Power, Sex and Nonverbal Communication* (Englewood Cliffs, New Jersey: Prentice-Hall, 1977), pp. 10-12.
190 Merleau-Ponty, 'The Child's Relations with Others', in *The Primacy of Perception*, pp. 131-2.
191 Nietzsche, *Daybreak: Thoughts on the Prejudices of Morality*, Note 506, p. 205.
192 See 'Now That I Come to Die', the last will and testament of the great fourteenth century Tibetan scholar, Long-chen-pa, translated by Herbert V. Guenther, published in *Crystal Mirror*, vol. V (Berkeley: Dharma Publishing, 1977), pp. 334-5.
193 See the Preface to Claude Lévi-Strauss, *Tristes Tropiques* (New York: Athaneum, 1964).
194 David Kehr, 'Shoah: The End of the Line', in *Chicago Magazine* (January, 1986), p. 86. 'Shoah', which means 'Holocaust' in Hebrew, is a film about the Nazi death camps made by Claude Lanzmann.
195 Edward Bellamy, *Looking Backward* (New York: New American Library, Signet Classics, 1960), p. 214.
196 Harold Searles, *Collected Papers on Schizophrenia and Related Subjects* (New York: International Universities Press, 1965), p. 428.
197 Foucault, *The Birth of the Clinic*, p. 40.
198 Freud, *Three Essays on the Theory of Sexuality* (New York: Basic Books, 1975), p. 22.
199 Heidegger, 'The Origin of the Work of Art', *op. cit.*, p. 25.
200 Merleau-Ponty, *Phenomenology of Perception*, p. 353. Here we have a phenomenological explanation for the experience of the seer Heidegger discusses in his study on Anaximander. The seer, he says, has 'always and already' seen the 'totality' (i.e., the wholeness) of the visible.
201 Merleau-Ponty, 'The Child's Relations with Others', *op. cit.*, p. 146.
202 Merleau-Ponty, *Signs* (Evanston: Northwestern University Press, 1964), p. 239.
203 Merleau-Ponty, 'Eye and Mind', *op. cit.*, p. 186.
204 Plato, *The Republic*, Book VI: 508, in Benjamin Jowett, *The Dialogues of*

Plato (New York: Random House, 1937), vol. I, p. 769.
205 Mindell, *Working with the Dreaming Body* (London: Routledge), p. 77.
206 Merleau-Ponty, *Phenomenology of Perception*, p. 223. Italics added.
207 *Ibid.*, p.229.
208 *Ibid.*, p. 317. Italics added.
209 Merleau-Ponty, 'The Intertwining – The Chiasm', *op. cit.*, p. 133.
210 *Ibid.*
211 Merleau-Ponty, *Phenomenology of Perception*, p. 223.
212 Merleau-Ponty, 'The Intertwining – The Chiasm', *op. cit.*, p. 134.
213 Stanley Cavell, *The Claim of Reason*, p. 414.
214 Merleau-Ponty, *Le visible et l'invisible* (Paris: Gallimard, 1964), p. 153.
215 Merleau-Ponty, 'The Intertwining – The Chiasm', *op. cit.*, p. 134.
216 Merleau-Ponty, *Phenomenology of Perception*, p. 229.
217 G. W. F. Leibniz, *The Monadology*, Statement 61, p. 545, in Philip P. Wiener (ed.), *Leibniz: Selections* (New York: Charles Scribner's Sons, 1951). Also in G. Parkinson (ed.), *Leibniz: Philosophical Writings* (London: J. M. Dent & Sons, 1973), p. 188.
218 Meister Eckhart, *An Introduction to the Study of His Works, With an Anthology of His Sermons*, pp. 226-7.
219 Daisetz Suzuki, *Studies in Zen* (New York: Dell Publishing Co., 1955), p. 85.
220 Cavell, *The Claim of Reason*, p. 202.
221 Heidegger, 'The Age of the World Picture', *op. cit.*, p. 131.
222 *Ibid.* Italics added.
223 *Ibid.* Also see Jean Paris, *Painting and Linguistics* (Pittsburgh: Carnegie-Mellon University Press, 1975) and *L'espace et le regard* (Paris: Editions du Seuil, 1965).
224 Merleau-Ponty, *Phenomenology of Perception*, p. 146.
225 Ralph Waldo Emerson, 'Education', in Mark van Doren (ed.), *The Portable Emerson* (New York: Viking, 1946), p. 270.
226 See Carl Rogers, *On Becoming A Person* (Boston: Houghton Mifflin, 1961). Also see Harry Stack Sullivan, *The Interpersonal Theory of Psychiatry* (New York: W. W. Norton, 1953), p. 147. He makes some important observations on the face as perceived.
227 See Marilyn B. Rosanes-Berrett, 'Gestalt Therapy as an Adjunct Treatment for Some Visual Problems', in Joen Fagan and Irma Shepherd (eds), *Gestalt Therapy Now* (New York: Harper & Row, 1970), p. 257.
228 *Ibid.*
229 *Ibid.*, p. 258.
230 *Ibid.*, p. 259.
231 *Ibid.*
232 *Ibid.*
233 *Ibid.*, p. 260.
234 *Ibid.*, p. 261.
235 *Ibid.*
236 *Ibid.*, p. 262.
237 F. W. J. Schelling, *Of Human Freedom* (Chicago: Open Court, 1936),

p. 42. James Gutman's translation has been slightly modified: 'so that' has been replaced by 'insofar as'.

238 Harold F. Searles, 'Transference Psychosis in the Psychotherapy of Chronic Schizophrenia', in *Collected Papers on Schizophrenia and Related Subjects*, p. 682. Also see 'The Contributions of Family Treatment to the Psychotherapy of Schizophrenia', *op. cit.*, p. 735.

239 Adorno, *Minima Moralia: Reflections from Damaged Life*, p. 68.

240 Stanley Cavell, *The Claim of Reason*, p. 430.

241 Merleau-Ponty, 'The Intertwining – The Chiasm', *op. cit.*, p. 137.

242 Yeshé Tsogyal, *The Life and Liberation of Padmasambhava*, vol. I, p. 69.

243 Heidegger, 'The Origin of the Work of Art', *Poetry, Language, Thought*, p. 83.

244 See Stanley Cavell, 'The Avoidance of Love: A Reading of *King Lear*' and 'Knowing and Acknowledging', both published in his *Must We Mean What We Say?* (New York: Charles Scribner's Sons, 1969).

245 See Lucien Lévy-Bruhl, *The Soul of the Primitive* (London: George Allen & Unwin, 1965).

246 Merleau-Ponty, 'The Child's Relations with Others', *op. cit.*, pp. 115-18.

247 Merleau-Ponty, *The Visible and the Invisible*, pp. 187-8. Italics added. In 'The Philosopher and His Shadow' (*Signs*, p. 174), he argued that 'The constitution of others does not come after that of the body; others and my body are born together from an original ecstasy.' This 'original ecstasy' plays a crucial role in our present study.

248 Merleau-Ponty, *Phenomenology of Perception*, p. 353. Italics added.

249 *Ibid.*, p. 215.

250 *Ibid.*

251 Merleau-Ponty, 'Nature and Logos: The Human Body', in *Themes from the Lectures at the Collège de France* (Evanston: Northwestern University Press, 1970), p. 131.

252 Yeshé Tsogyal, *op. cit*, p. 69.

253 Lewis Carroll, *Through the Looking Glass* (New York: Random House, 1946), p. 6.

254 Nietzsche, 'On the Uses and Disadvantages of History for Life', *Untimely Meditations* (Cambridge: Cambridge University Press, 1983), pp. 83-4.

255 Stéphane Mallarmé, 'Prose', in *Oeuvres Complètes*, ed. Henri Mondor and G. Jean-Aubry (Paris: Éditions Gallimard, 1945), p. 57.

256 Nietzsche, 'On the Uses and Disadvantages of History for Life', *op. cit.*, p. 85.

257 Foucault, *Discipline and Punish: The Birth of the Prison* (London: Allen Lane, 1977; New York: Pantheon, 1978), p. 193.

258 Jung, 'Anima and Animus', in *Two Essays on Analytical Psychology* (Cleveland and New York: The World Publishing Co., Meridian Books, 1956), p. 200; reprinted in *Aspects of the Feminine* (Princeton: Princeton University Press, 12982), p. 79. Also see *The Collected Works of Carl G. Jung*, vol. 7: *Relations Between the Ego and the Unconscious* (Princeton: Princeton University Press, 1953), para. 300.

259 Heidegger *Being and Time*, p. 44.

260 See Freud, 'Draft M' (May 25, 1897), in Marie Bonaparte, Anna Freud and Ernst Kris (eds), *The Origins of Psycho-Analysis: Letters to Wilhelm Fliess* (New York: Basic Books, 1977), p. 203. Also see Marie Balmary, *Psychoanalyzing Psychoanalysis: Freud and the Hidden Fault of the Father* (Baltimore: Johns Hopkins University Press, 1982).

261 Jung, 'Women in Europe', in *Aspects of the Feminine*, p. 55. This essay is also published in *The Collected Works of Carl G. Jung*, vol. 10: *Civilization in Transition* (Princeton: Princeton University Press, 1964).

262 Rita Mae Brown, 'The New Lost Feminist', in Mary Daly, *Gyn/Ecology*, p. 135.

263 Seyla Benhabib, 'The Generalized and the Concrete Other: Toward a Feminist Critique of Substitutionality Universalism', in Eva F. Kittay and D. Meyers (eds), *Proceedings of the Women and Moral Theory Conference* (New Jersey: Rowman & Allenheld, 1986), p. 3.

264 Mary Daly, *Gyn/Ecology*, pp. 2-3. Bracketed words added. Daly is not saying that women are in the foreground in a sense which would contradict Jung's observation that women have traditionally stood in the shadows of men. Indeed, I am sure that she would agree with Jung that women have always been kept in the background, unseen, unheard, except in relation to male needs and male desires. Women certainly have been kept in the background of a male-structured society, excluded, denied, not respected, dominated. But what makes Daly's remark interesting and provocative is that it can be applied to an analysis of visual process. In this regard, she means to be differentiating the *character* of the 'background' from that of the 'foreground': what is in the foreground is what is in focus, under control, seen most clearly and distinctly, seen with greatest certainty; it corresponds to the functioning of a desire which is masculine, patriarchal, egocentric, anthropocentric, and logocentric. The visual foreground is what is mastered. By contrast, the background is not in sharp focus, not in control, not seen clearly and distinctly, not seen with such certainty; its dimensionality is not to be mastered, totally grasped and possessed. The essence of the background is that it demands that we *let it be*, and that, instead of knowing it, we sense it, feeling its presence as a whole. Thus, relative to our cultural tradition, Daly's associations are, I think, brilliantly accurate: it is the traditionally feminine mode of experiencing which has appreciated the background as background, and which our patriarchal culture has insisted on restricting to the channels of beam-like focus. As a culture, then, we need to work towards the valorisation and appreciation of the perceptual background; this is an aspect of the work we all need to do in order to integrate women into a society where the forms of oppression under which women have suffered are destroyed once and for all.

265 Nietzsche, *Thus Spake Zarathustra*, reprinted in Walter Kaufmann (ed.), *The Portable Nietzsche* (New York: Viking Press, 1954), p. 144. Bracketed words added. 'Women' are identified, here, with the earth and body archetypes; men are identified with the rule of the ego. Also see Susan

Rubin Suleiman (ed.), *The Female Body in Western Culture* (Cambridge: Harvard University Press, 1985) and Elaine Scarry, *The Body in Pain: The Making and Unmaking of the World* (New York: Oxford University Press, 1985).

266 Seyla Benhabib, *op. cit.*, p. 32.

267 Adrienne Rich, *Of Women Born: Motherhood as Experience and Institution* (New York: W. W. Norton, 1976), p. 95. Bracketed words added.

268 Jung, 'Women in Europe', *op. cit.*, pp. 65 and 75. I agree with him, so long as it is understood that women's psychology is not a pre-ordained fate, but an historical contingency which can be, and needs to be, changed. We may say that, in the history of our culture, the woman's psychology *has been* founded on the principle of Eros. But nothing in history makes this association, this manifestation of the archetype-complex, inevitable. Nor is it unchangeable.

269 Rich, *Of Woman Born*, p. 125. Bracketed word added. Also see an exciting paper by Andrea Nye, 'The Woman Clothed in the Sun: Julia Kristeva and the Escape from/to Language', forthcoming in the new feminist journal, *Signs*.

270 Benhabib, *op. cit.*, p. 3.

271 Jung, 'Women in Europe', *op. cit.*, p. 58.

272 Freud, *Three Essays on the Theory of Sexuality* (New York: Basic Books, 1975), p. 7. Italics added.

273 See James M. Robinson (ed.), *The Nag Hammadi Library* (San Francisco: Harper & Row, 1977), p. 462. Also see Elaine Pagles, 'The Gnostic Vision', *Parabola*, vol. 3, no. 4, pp. 6-9, and June Singer, *Androgyny: Towards a New Theory of Sexuality* (New York: Doubleday, 1979).

274 Arnold Mindell, *Dreambody: The Body's Role in Revealing the Self* (Santa Monica: Sigo Press, 1982), pp. 189-90.

275 Jung, *op. cit.*, p. 67.

276 Ralph Waldo Emerson, 'Education', in Mark van Doren (ed.), *The Portable Emerson* (New York: Viking Press, 1946), p. 269. Also see Foucault, 'The Eye of Power', in Colin Gordon (ed.), *Power/Knowledge: Selected Interviews and Other Writings, 1972-1977* (New York: Pantheon, 1980), pp. 146-165, and Susan Griffin, *Woman and Nature: The Roaring Inside Her* (New York: Harper & Row, 1978).

277 See Carol Gilligan, *In A Different Voice: Psychological Theory and Women's Development* (Cambridge: Harvard University Press, 1982), pp. 6 and 29. Bracketed words added. Also see Linda J. Nicholson, 'Women, Morality and History', *Social Research*, vol. 50, no. 3 (Autumn, 1983), pp. 514-36.

278 Nietzsche, *The Genealogy of Morals*, p. 255.

279 See the unpublished manuscript of Alison M Jaggar's *Feelings and Knowing: Emotion in Feminist Theory*, presented in part at 'Feminist Ways of Knowing', the second interdisciplinary seminar in Woman's Studies, held at Douglass College, Rutgers University, Fall, 1985, p. 31. Also see an extremely important paper by Naomi Scheman, 'Anger and the Politics of Naming', in McConnell-Ginet, Barker, and Furman (eds), *Women and Language in Literature and Society* (New York: Praeger, 1980).

280 Jaggar, *op. cit.*, p. 1.
281 *Ibid.*, p. 24.
282 *Ibid.*, p. 30.
293 *Ibid.*
284 *Ibid.*, p. 25. Also see Gendlin, 'A Philosophical Critique of the Concept of Narcissism', in Levin (ed.), *Pathologies of the Modern Self*, where this kind of analysis is developed in greater detail.
285 *Ibid.*, p. 28.
286 Gilligan, *op. cit.*, p. 4.
287 *Ibid.*, p. 23.
288 *Ibid.*, pp. 12-13.
289 *Ibid.*, p. 8. Bracketed words added.
290 *Ibid.*
291 *Ibid.*, p. 38. Bracketed words added.
292 *Ibid.*, p. 30.
293 *Ibid.*, p. 62.
294 *Ibid.*
295 *Ibid.*, p. 63.
296 *Ibid.*, p. 127. Also see p. 19.
297 Jerome Kagan, *The Nature of the Child* (New York: Basic Books, 1984), pp. 118-18.
298 *Ibid.*, p. 119.
299 *Ibid.*, p. 123.
300 See the fable which Heidegger quotes in *Being and Time*, p. 243.
301 Theodor Adorno, *Negative Dialectics* (New York: Seabury Press, 1973), p. 207.
302 Merleau-Ponty, 'The Concept of Nature', *Themes from the Lectures at the Collège de France, 1952-1960* (Evanston: Northwestern University Press, 1970), p. 82. Also see Calvin O. Schrag, *Radical Reflection and the Origin of the Human Sciences* (West Lafayette: Purdue University Press, 1980).
303 Herbert Marcuse, *Eros and Civilization: A Philosophical Inquiry into Freud* (New York: Vintage, 1962), pp. 34-5.
304 Pierre Bourdieu, *Distinction: A Social Critique of the Judgment of Taste* (Cambridge: Harvard University Press, 1984), p. 483.
305 Edward Sapir, 'The Unconscious Patterning of Behaviour in Society', D. G. Mandelbaum (ed.), *Selected Writings of Edward Sapir* (Berkeley: University of California Press, 1949), pp. 544-59.
306 Bourdieu, *op. cit.*, p. 474.
307 Nancy M. Henley, *Body Politics: Power, Sèx, and Nonverbal Communication* (Englewood Cliffs, New Jersey: Prentice-Hall, 1977), p. 3.
308 Foucault, *Power/Knowledge*, p. 97.
309 *Ibid*. Bourdieu, *op. cit.*, p. 471.
310 *Ibid.*, p. 193.
311 Henley, *op. cit.*, p. 155.
312 *Ibid.*, p. 142.
313 Bourdieu, *op. cit.*, p. 193.
314 Henley, *op. cit.*, p. 167.

315 *Ibid.*, p. 166.
316 Bourdieu, *op. cit.*, p. 208.
317 Henley, *op. cit.*, p. 89.
318 *Ibid.*, p. 131.
319 Bourdieu, *op. cit.*, p. 190.
320 Foucault, 'Why Study Power? The Question of the Subject', in Hubert Dreyfus and Paul Rabinow, *Michel Foucault: Beyond Structuralism and Hermeneutics* (Chicago: University of Chicago Press, 1982), p. 216.
321 Marcuse, *An Essay on Liberation* (Boston, Beacon Press, 1969), p. 53. Also see Max Horkheimer, 'Traditional and Critical Theory', in O'Connell *et al.* (eds), *Critical Theory: Selected Essays* (New York: Herder and Herder, 1972), p. 220; T. Adorno, *Negative Dialectics*, p. 52; and three works by Reiner Schürmann: 'Questioning the Foundation of Practical Philosophy', *Human Studies*, vol. 1 (1980), p. 357ff., 'Political Thinking in Heidegger', *Social Research*, vol. 45 (1978), p. 191ff., and *Heidegger et la question de l'agir* (Paris, 1982).
322 Adorno, *Minima Moralia*, p. 154.
323 See Naomi Scheman, 'Anger and the Politics of Naming', *op. cit.*, p. 185.
324 Jürgen Habermas, 'Questions Concerning the Theory of Power', in *The Philosophical Discourse of Modernity: Twelve Lectures* (Cambridge: Massachussetts Institute of Technology, 1987). As yet unpublished.
325 See Foucault's Howison Lecture on 'Truth and Subjectivity', delivered in Berkeley at the University of California on October 20, 1980. It has not been published. I am quoting from p. 7 of a transcript in circulation.
326 Schürmann, ' "What Can I Do?" in an Archaeological-Genealogical History', *Journal of Philosophy*, vol. 82, no. 3 (December, 1985), p. 541.
327 Schürmann, *op. cit.*, p. 544.
328 *Ibid.*
329 *Ibid.*, p. 546.
330 Michèle Najlis, 'Las viejas tribus', in *Nicaraguan Perspectives* (Berkeley: Nicaraguan Information Center, Fall-Winter, 1985-86), vol. 11, p. 33.
331 Foucault, 'Nietzsche, Genealogy, History', in Donald F. Bouchard (ed.), *Language, Countermemory, Practice* (Ithaca: Cornell University Press, 1977), p. 148. Also see Nancy Fraser, 'Foucault's Body-Language: A Post-Humanist Political Rhetoric', *Salmagundi*, no. 61 (Fall, 1983), pp. 55-70.
332 Foucault, *Power/Knowledge*, p. 186.
333 *Ibid.*, pp. 152-3.
334 See Habermas, 'Questions Concerning the Theory of Power', *op. cit.*
335 Foucault, *The Order of Things: An Archaeology of the Human Sciences* (New York: Pantheon Books, 1970), p. 13.
336 Heidegger, *What Is A Thing?* (Chicago: Henry Regnery Co., 1967), pp. 92-3.
337 Benhabib, *op. cit.*, p. 17.
338 Anthony Giddens, *The Constitution of Society* (Berkeley: University of California Press, 1984), p. 84.

339 Bourdieu, *op. cit.*, p. 192.
340 *Ibid.*, pp. 217-18.
341 *Ibid.*, p. 467.
342 *Ibid.*
343 *Ibid.*, p. 193.
344 *Ibid.*, p. 474.
345 *Ibid.*, p. 466.
346 *Ibid.*, p. 208.
347 *Ibid.*, p. 192.
348 Merleau-Ponty, *Phenomenology of Perception*, p. 146.
349 Bourdieu, *op. cit.*, p. 207.
350 Mindell, *Working with the Dreaming Body* (Boston and London: Routledge & Kegan Paul, 1986), p. 80.
351 *Ibid.*, p. 79.
352 *Ibid.*, p. 81.
353 Nietzsche, 'On the Use and Disadvantages of History for Life', *Untimely Meditations*, pp. 24-5.
354 Merleau-Ponty, 'The Child's Relations with Others', *op. cit.*, p. 118.
355 Heidegger and Fink, *Heraclitus Seminar 1966-1967* (University: University of Alabama Press, 1979), p. 24.
356 Edward Bellamy, *Looking Backward* (New York: New American Library, 1960), p. 215.
357 See Gilles Deleuze and Felix Guattari, *Anti-Oedipus: Capitalism and Schizophrenia* (Minneapolis: University of Minnesota Press, 1983).
358 Eugene Gendlin, 'Process Ethics and the Political Question', in *The Focusing Folio* (Chicago: The Focusing Institute, 1986), vol. 5, no. 2, pp. 69-87. For his precursors, see Carl Rogers, *Freedom to Learn* (Columbus, Ohio: Charles Merrill, 1969) and John Dewey, *Democracy and Education* (New York: Macmillan, The Free Press, 1966). Also see David M. Levin, 'Moral Education: The Body's Felt Sense of Value', in *The Body's Recollection of Being*, pp. 224-247. Gendlin has an answer to the question John O'Neill poses in his recently published book, *Five Bodies* (Ithaca: Cornell University Press, 1986), pp. 63 and 151.
359 Rogers, *Freedom to Learn*, p. 251.
360 Gendlin, 'Process Ethics and the Political Question', *op. cit.*, p. 81. In this quotation, and all those following, I have slightly revised the wording.
361 *Ibid.*, p. 83.
362 *Ibid.*, p. 84.
363 *Ibid.*, p. 69.
364 *Ibid.*
365 *Ibid.*, p. 70.
366 *Ibid.*, p. 77.
367 *Ibid.*, pp. 77-8.
368 *Ibid.*, p. 79.
369 *Ibid.*, p. 85.
370 See Gendlin, *Focusing* (New York: Bantam, 1981) and 'A Philosophical

Critique of the Concept of Narcissism', in Levin (ed.), *Pathologies of the Modern Self*.

371 Marcuse, *An Essay on Liberation*, p. 10.

372 Merleau-Ponty, 'The Child's Relations with Others', *op. cit.*, p. 146.

373 Kagan, *The Nature of the Child*, p. 130.

374 Merleau-Ponty, *Humanism and Terror*, p. xiv.

375 John H. Findlay, *Hegel: A Re-examination* (New York: Macmillan, 1962), p. 95. Italics added.

376 Merleau-Ponty, *Phenomenology of Perception*, p. 85.

377 *Ibid.*, p. 354.

378 G. W. F. Hegel, *The Phenomenology of Mind*, trans. J. B. Baillie (New York: Humanities Press, 1964), p. 339.

379 Edmund Burke, *Reflections on the Revolution in France*, ed. with an Introduction by Thomas H. Mahoney (Indianapolis: Bobbs-Merrill Co., 1955), p. 38. I am, of course, mischievously giving to Burke's metaphorical conception of a 'permanent body' a *literal* meaning alien to his own thinking. If we take his words in this sense, could we correlate the 'permanent body' with Merleau-Ponty's conception of the flesh – the human body in its primordial dismensionality? Such a correlation, however, would undermine the conservative politics Burke wants to advocate.

380 Foucault, *Power/Knowledge*, p. 190. Italics added.

381 Mary Daly, *Gyn/Ecology: The Metaethics of Radical Feminism*, pp. 415, 417. There is an etymological connection, but it is more indirect, more mediated than Daly suggests: the sense of *weaving* is related to *texere*, one of the interpretations of *ordiri*.

382 Merleau-Ponty, 'Reflection and Interrogation', *The Visible and the Invisible*, p. 49.

383 Merleau-Ponty, 'The Intertwining – The Chiasm', *op. cit.*, p. 264.

384 Merleau-Ponty, 'The Child's Relations with Others', *op. cit.*, pp. 118-19. What is between the quotes is a composite I put together.

385 Bourdieu, *op. cit.*, p. 474.

386 Kagan, *op. cit.*, p. 131.

387 Harold Searles, *Collected Papers on Schizophrenia and Related Subjects*, pp. 227-28. In psychiatry, Searles's clinical work with schizophrenics is universally respected and admired. And there are few whose theoretical work can be compared to his.

388 Terrence Des Pres, *The Survivor: An Anatomy of Life in the Death Camps* (New York: Oxford University Press, 1976), p. 199.

389 *Ibid.*, p. 192.

390 Thomas Hobbes, *The Leviathan* (New York: Bobbs-Merrill, 1958), p. 106.

391 Des Pres, *op. cit.*, p. 142.

392 *Ibid.*, p. 146.

393 *Ibid.*, p. 147.

394 *Ibid.*, pp. 198-199.

395 Jorge Semprun, *The Long Voyage* (New York: Grove Press; London: Weidenfeld & Nicolson, 1964), p. 205.

396 Des Pres, *op. cit.*, p. 199.
397 Konrad Lorenz, *On Aggression* (New York: Harcourt, Brace & World, 1966), p. 246.
398 Des Pres, *op. cit.*, p. 197.
399 *Ibid.*
400 Habermas, 'Moral Development and Ego Identity', in *Communication and the Evolution of Society* (Boston: Beacon Press, 1979), p. 81.
401 *Ibid.*
402 Merleau-Ponty, 'The Intertwining – The Chiasm', *op. cit.*, p. 139.
403 Hegel, *The Phenomenology of Mind*, p. 292.
404 Merleau-Ponty, 'The Intertwining – The Chiasm', *op. cit.*, p. 152.
405 Merleau-Ponty, 'The Child's Relations with Others', *op. cit.*, p. 107.
406 *Ibid.*, p. 119. Also see the *Phenomenology of Perception*, pp. 215-16, 330, 352-3.
407 *Ibid.*, p. 118.
408 Henley, *Body Politics*, p. 128.
409 Merleau-Ponty, 'The Child's Relations with Others', *op. cit.*, p. 119. Also see Joel Kovel, *The Age of Desire: Reflections of a Radical Psychoanalyst* (New York: Pantheon, 1981), pp. 63-64 and 233, for a discussion of the transhistorical body in relation to critical social theory. The transhistoricality of the body, i.e., the experiential dimension of a biological transhistoricality, makes possible an essential contribution to the Marxist critique of historical existence.
410 Habermas, 'Moral Development and Ego Identity', *op. cit.*, p. 88.
411 See Lawrence Kohlberg, 'Justice as Reversibility: The Claim to Moral Adequacy of a Highest Stage of Moral Judgment', in his *Essays On Moral Development*, vol. I: *The Philosophy of Moral Development* (New York: Harper & Row, 1981), p. 194.
412 See Shierry M. Weber, 'Individuation as Praxis', in Paul Breines (ed.), *Critical Interruptions: New Left Perspectives on Herbert Marcuse* (New York: Herder & Herder, 1970), pp. 22-59.
413 Benhabib, *op. cit.*, p. 8.
414 *Ibid.*, p. 18. Also see p. 20.
415 *Ibid.*, pp. 8-9.
416 Adorno, *Minima Moralia*, p. 103.
417 *Ibid.*
418 Heidegger, 'Poetically Man Dwells', *op. cit.*, pp. 218-19.
419 Des Pres, *op. cit.*, p. 208.
420 *Ibid.*
421 *Ibid.*
422 Merleau-Ponty, 'The Indirect Language', in *The Prose of the World* (Evanston: Northwestern University Press, 1973), p. 94. Also see pp. 81n and 83 on body, history and tradition.
423 Plato, *Meno*, 72, in Jowett, *The Dialogue of Plato*, vol. 1, p. 351. I have revised the translation after consulting the original Greek, substituting 'gaze' for 'eye' and 'steadied' for 'fixed'.

Chapter 3 Lighting: The Transformative Moment of Insight

1 See *G. W. F. Hegel: The Letters*, Clark Butler and Christine Seiler, trans. (Bloomington: Indiana University Press, 1984), p. 280. For me, our spiritual development begins when we are torn away from the normalized, consensually validated forms of experiencing into which we are fitted through the process of 'socialization' – torn away from 'concrete representations' in *this* sense. We must first 'die' to the sight and hearing lived by everyone-and-anyone, *das Man*, in order to begin to see and hear with 'fresh' organs of experience. In this sense, then, what spiritual growth calls for is precisely a *release* from the abstract forms of experiencing which are not truly and authentically our own (they lack the *concreteness* of *Gemeinigkeit*), because they have been imposed by the social system, have been taken in uncritically, and belong, therefore, to the consensus. What is authentic are the forms of experience which are truly concrete; what has not been made into one's own is what I'd like to call 'the abstract.'
2 Heidegger, *An Introduction to Metaphysics*, p. 37.
3 John Donne, *Devotions upon Emergent Occasions* (Ann Arbor: University of Michigan Press, 1959), p. 97. Quoted in Robert Jay Lifton, *In a Dark Time* (Cambridge: Harvard University Press, 1984), p. 140.
4 Fragment 64 is translated into English in Kathleen Freeman, *Ancilla to the Pre-Socratic Philosophers* (Cambridge: Harvard University Press, 1978), p. 29.
5 Hegel, *The Phenomenology of Mind*, J. B. Baillie, trans. (New York: Muirhead Philosophical Library, 1931), p. 490.
6 See *The Collected Poems of Theodore Roethke* (New York: Doubleday and Company; London: Faber & Faber, 1966), p. 239.
7 Rilke, 'Die Worte des Herrn an Johannes auf Patmos', *Gesammelte Gedichte* (Wiesbaden, 1962), p. 572. My own translation.
8 Heidegger, 'Hölderlins Erde und Himmel', *Erläuterungen zu Hölderlins Dichtung* (Frankfurt am Main: Vittorio Klostermann, 1963), p. 166. My own translation.
9 *Ibid.*
10 'Talavakara Upanishad', *The Upanishads*, translated by F. Max Muller (New York: Dover, 1962), Part I, pp. 151-2. Also see the 'Khandogya Upanishad' (p. 141): 'Now, where the sight has entered into the void, the open space, the black pupil of the eye, there is the *person* of the eye.'
11 Hegel, *The Phenomenology of Mind*, p. 701.
12 C. G. Jung, *Memories, Dreams, Reflections* (New York: Random House, Vintage Books, 1965), p. 269. Also see Jean Starobinski, *L'oeil vivant* (Paris: Gallimard, 1961).
13 Heidegger, *Being and Time*, p. 385.
14 See Heidegger, *Nietzsche*, vol. II: *The Eternal Recurrence of the Same*, p. 245. In David Farrell Krell's 'Analysis,' at the back of his translation, Krell comments on Heidegger's text (esp. pp. 41 and 56-7). The

passages in quotes are Heidegger's. On my reading, I would argue that what we have been *given* as our 'endowment' is a bodily *capacity* to see; and what we have been 'assigned' as a 'task' is therefore the *development* of this capacity: in particular, a development through which we become more open to experiencing the openness of Being as a whole.

15 Erich Neumann, *The Origins and History of Consciousness*, p. 323.

16 Neumann, *The Great Mother: An Analysis of the Archetype*, p. 212. The 'beginning' is Night. The 'uroboric' symbol of this condition refers to the holistic character of the experience: that it is an experience of Being as a whole.

17 *Ibid.*, p. 211. Wholeness and connectedness manifest the Feminine Principle of Wisdom.

18 *Ibid.*, p. 55n. Lightning is 'masculine,' given our cultural values, but it comes out of the womb of the Night; it also has a penetrating character.

19 Theodor Adorno, 'Über Mannheims Wissenssoziologie' (Frankfurt am Main: Adorno Estate, 1947), p. 4. Reprinted in Susan Buck-Morss, *The Origin of Negative Dialectics: Theodor W. Adorno, Walter Benjamin, and the Frankfurt Institute* (New York: Macmillan, Free Press, 1977), p. 84. My italics.

20 Norman Bryson, *Vision and Painting: The Logic of the Gaze* (New Haven: Yale University Press, 1983), pp. 121-2. For Bryson, the 'Gaze' enacts a metaphysics of presence: it imposes a normalizing, totalizing grid onto the world; it is eminently 'rational,' a source of rational order. It is also, we might add, in the service of the will to power. The 'Glance,' however, is inherently subversive: it values temporality more than the eternal, the fleeting more than the enduring, the disordered and anarchic more than the ordered and the legal, the unpredictable more than the predictable, the free more than the bound, the mobile more than the static, the quick more than the steady and the constant, the contingent more than the necessary. Bryson's account seems accurate as an analysis of our vision at this time; but I would not accept his description of the 'Gaze' as a *necessary* truth.

21 Neumann, *Amor and Psyche: The Psychic Development of the Feminine*, Ralph Mannheim, trans. (Princeton: Princeton University Press, 1973), p. 78.

22 Ralph Waldo Emerson, 'Circles,' in *Essays*, vol. I (New York: A. L. Burt), p. 309.

23 See *Letters of Rainer Maria Rilke, 1892-1910*, Jane B. Greene and M. D. Herter Norton, trans. (New York: W. W. Norton, 1945), p. 266. The letter was written by the poet to his wife, Clara, on March 8, 1907, from Villa Discopoli, Capri. Obviously, what Rilke means by 'gaze' is very different from what Bryson means: Rilke's 'gazing' has in fact many of the characteristics attributed by Bryson to the 'glance.' Foucault also distinguishes the 'gaze' from the 'glance.' See *The Birth of the Clinic*, pp. 121-2.

24 Arthur Deikman, 'Deautomatization and the Mystic Experience,' in Robert Ornstein (ed.), *The Nature of Human Consciousness* (San Francisco: W. H. Freeman, 1973), p. 228.

25 Lloyd Kaufman, *Perception: The World Transformed* (New York: Oxford University Press, 1979), p. 109.
26 Arnold Mindell, *Working with the Dreaming Body* (London and Boston: Routledge & Kegan Paul, 1985), p. 45.
27 Neumann, *The Great Mother*, p. 57. 'Luminous bodies' is a term which, for Neumann, refers only to planets, stars, moon, sun, comets and meteors, lightning flashes and clouds. But I want to refer to *all* beings of light, of enlightenment: beings such as the animals, and of course we ourselves, because eyes make visible, can bring into the light, can clarify. Neumann's sentence retains its truth even with this extension of his meaning. This is significant.
28 Konrad T. Preuss, *Die geistige Kultur der Naturvölker* (Leipzig: B. G. Teubner, 1914), p. 9. My translation.
29 Neumann, *The Great Mother*, p. 56.
30 *Ibid.*, p. 27.
31 *Ibid.*, p. 55.
32 *Ibid.*, pp. 179-180.
33 *Ibid.*, p. 223.
34 *Ibid.*, p. 226.
35 *Ibid.*
36 *Ibid.*, p. 57.
37 Nietzsche, *Beyond Good and Evil* (New York: Random House, 1966), p. 2.
38 Medard Boss, *I Dreamt Last Night* (New York: John Wyley & Sons: Halsted Press, 1977), p. 175.
39 On the dreambody, see, in particular, Eugene T. Gendlin, *Let Your Body Interpret Your Dreams* (Peru, Illinois: Chiron Publications, 1986) and Arnold Mindell, *Working With the Dreaming Body* (Routledge & Kegan Paul, 1985). Also see Nathan Schwartz-Salant, 'On the Subtle Body Concept in Clinical Practice,' *The Body in Analysis*, special issue of *Chiron*, vol. 1986; Sylvia Perera, 'Ceremonies of the Emerging Ego in Psychotherapy', *The Body in Analysis, Chiron*, 1986; Joan Chodrow, 'The Body as Symbol: Dance/Movement in Analysis', *Chiron*, 1986. To pursue the connection with the Tibetans, see Tarthang Tulku, *Space, Time, and Knowledge* (Berkeley: Dharma Publishing, 1977); Namkhai Norbu, *The Cycle of Day and Night* (Berkeley: Dzogchen Community in America, Zhang Zhung Editions, 1984); Chogyam Trungpa, *The Tibetan Book of the Dead: The Great Liberation Through Hearing in the Bardo* (Berkeley: Shambhala Publishing Co., 1975); and Herbert V. Guenther, *The Tantric View of Life* (Berkeley: Shambhala, 1976). Guenther's latest book, *The Matrix of Mystery* (Shambhala, 1985) is also to the point. I also wish to acknowledge the wisdom of the North American Indians concerning the dream vision and a vision of dreams, especially these: Lame Deer (with Richard Erdoes), *Lame Deer, Seeker of Visions* (New York: Simon & Schuster, 1972) and Black Elk (with John Neihart), *Black Elk Speaks* (New York: Simon & Schuster, 1972). I would also like to refer, marginally speaking, to Carlos Castaneda, *A Separate Reality* (New York: Simon & Schuster, 1971) and *Tales of Power* (New York: Simon & Schuster, 1974).

40 Stanley Corngold, 'The Question of the Self in Nietzsche during the Axial Period (1882-1888)', in Daniel O'Hara (ed.), *Why Nietzsche Now?* (Bloomington: Indiana University Press, 1985), p. 57.
41 Nietzsche, *Thus Spoke Zarathustra*, in W. Kaufmann (ed.), *The Portable Nietzsche* (New York: Viking Press, 1954), p. 146.
42 Nietzsche, *Beyond Good and Evil*, p. 20.
43 Boss, *op. cit*, p. 197.
44 *Ibid.*, p. 195.
45 Nietzsche, *Thus Spoke Zarathustra*, in Kaufmann (ed.), *The Portable Nietzsche*, p. 189.
46 Johann Wolfgang von Goethe, *Italian Journey, 1786-1788*, translated by W. H. Auden and Elizabeth Mayer (San Francisco: North Point Press, 1982), p. 46.
47 Nietzsche, *Thus Spoke Zarathustra*, in Kaufmann (ed.), *The Portable Nietzsche*, p. 144.
48 Walter Pater, *Miscellaneous Studies* (London: Macmillan, 1910), pp. 143-5. Quoted in Daniel O'Hara, 'The Prophet of Our Laughter: Or Nietzsche As – Educator?', in Daniel O'Hara (ed.), *Why Nietzsche Now?*, p. 15-16. Also see Nietzsche's remark on books and writings, shadows and light, in *The Joyful Wisdom* (New York: Frederick Ungar, 1960), Book II, p. 125. Here he touches on some neglected connections: matters of ancient provenance.
49 Heidegger and Fink, *Heraclitus Seminar, 1966/1967*, Charles Seibert, trans. (University: University of Alabama Press, 1979), p. 24.
50 *Ibid.*, p. 81.
51 Heidegger, 'The Turning', *The Question Concerning Technology and Other Essays*, pp. 41-2. Also see his *Addenda* to 'The Age of the World Picture,' *op. cit.*, p. 154, where he writes about the shadow: 'the shadow', he says, 'is a manifest, though impenetrable testimony to the concealed emitting of light.' This I take to be an *ontological* observation, based on, or anyway involved in, his experience with vision. In *The Genealogy of Morals*, Nietzsche writes, with Pindar no doubt in mind, that 'It is easy to tell a philosopher: he avoids . . . glare, and for this reason he avoids his own time and the "light" of its day. In this, he is like a shadow: the more the sun goes down, the larger he grows.' Curiously, he adds that the philosopher 'is fearful of being disturbed by lightning,' and then *likens the philosopher to the woman*, speaking of 'his "maternal" instinct, that secret love for what is growing in him' (pp. 245-6). But he puts the word, *maternal*, as we can see, *inside* quotation marks.
52 Heidegger, 'The Turning,' *op. cit.*, p. 37.
53 See Heidegger's deconstruction of the activity/passivity dualism in his essays on the history of metaphysics, published in *The End of Philosophy*.
54 Heidegger, 'The Turning,' *op. cit.*, p. 40.
55 Heidegger, 'The Word of Nietzsche: "God is dead" ', *op. cit.*, p. 108.
56 *Ibid.*, p. 72.
57 *Ibid.*, p. 44.

58 See Nietzsche, *The Will to Power*, Book III, pp. 288 and 294 on the reification we project through grammar, illustrated by the character of the lightning in our speech. Also see Benjamin Lee Whorf on this same point, same example, in *Language, Thought, and Reality* (Cambridge: MIT Press, 1956), p. 240.
59 Heidegger, 'The Turning,' *op. cit.*, p. 44. Italics added.
60 *Ibid.*, p. 45. Italics added. Heidegger also speaks of the 'glance of Being' in 'The Anaximander Fragment,' *Early Greek Thinking*, p. 27.
61 St. Augustine, *Confessions* (New York: E. P. Dutton, 1959), Book VII, p. 152.
62 Heidegger, 'The Turning,' *op. cit.*, p. 46.
63 *Ibid*. Italics added.
64 *Ibid.*, p. 47. Italics added.
65 *Ibid*. Italics added.
66 *Ibid*. Italics added.
67 Heidegger, *Nietzsche*, vol. II: *The Eternal Return of the Same*, p. 52.
68 Nietzsche, *The Will to Power*, p. 274.
69 Heidegger, 'The Turning,' *op. cit.*, p. 48. Italics added.
70 *Ibid.*, p. 49. Italics added.
71 The work to which I am alluding here – work with the dreambody and the body's felt sense – is more fully described and explained in Eugene Gendlin, *Focusing* (New York: Bantam, 1982). My own work is deeply indebted to the continual use of Gendlin's focusing process.
72 For more on the *homologein*, see my recently published book, a companion to this, *The Body's Recollection of Being*, pp. 93-109, 116-66, and 327-36
73 Emerson, 'The Oversoul,' *Essays*, vol. I, p. 281. Italics added. Also see the discussion of Pythagoras's experience of theoretical insight into the 'mathematics of being,' in Francis M. Cornford, 'The Harmony of the Spheres', *The Unwritten Philosophy and Other Essays*, edited by W. K. C. Guthrie (Cambridge: Cambridge University Press, 1967), pp. 25-7. Cornford argues that insight is always 'fused' with 'intense feeling,' that this feeling 'was an essential part of the original experience,' and that theory is born 'of the marriage of thought and feeling, in the fullness of an experience like that which I have imagined as felt in common by man and woman. . .'.
74 John Welwood, 'On Psychological Space,' *The Journal of Transpersonal Psychology*, vol. 9, no. 2 (1977), p. 101.
75 Ludwig Wittgenstein, *Tractatus Logico-Philosophicus* (London: Routledge & Kegan Paul, 1969), p. 147.
76 Rilke, *Letters 1910-1926* (New York: W. W. Norton, 1947), p. 375.
77 Erich Neumann, 'Creative Man and Transformation', *Art and the Creative Unconscious* (Princeton: Princeton University Press, 1959), p. 164.
78 Rilke, *Letters 1910-1926*, p. 376.
79 Heidegger, 'Letter on Humanism,' *op. cit.*, p. 237.
80 C. G. Jung, 'Aion: Phenomenology of the Self,' in *The Collected Works of C. G. Jung*, vol. 9, part 3; reprinted in Joseph Campbell (ed.), *The Portable Jung* (New York: Viking, 1971), p. 148.

81 The locution, 'Fury of Being,' comes from Herbert V. Guenther's latest work, *The Matrix of Mystery* (Berkeley: Shambhala, 1985).
82 Heidegger and Fink, *Heraclitus Seminar 1966/1967*, p. 128. Page references in what follows are to this text.
83 The translation is my own composition, based on Hermann Diels, *Die Fragmente der Vorsokratiker* (Berlin: Widemann, 1934) and Kathleen Freeman, *Ancilla to the Pre-Socratic Philosophers*, p. 26.
84 Chogyam Trungpa (with Francesca Fremantle), *The Tibetan Book of the Dead* (Berkeley: Shambhala, 1975), p. 41. Italics added. See Heidegger, 'Moira,' *Early Greek Thinking*, p. 100, where he says something startlingly similar to the Tibetan text: 'Ordinary perception certainly moves within the lightedness of what is present and sees what is shining out . . . in color; but [it] is dazzled by changes in color, . . . and pays no attention to the still light of the lighting that emanates from duality. . .' Also see Gerardo Reichel-Dolmatoff, 'The Loom of Life: A Kogi Principle of Integration,' *Journal of Latin American Lore* (vol. 4, no. 1, Summer, 1978), pp. 24-25, for a discussion of the experience of insight among the Kogi Indians of the Sierra Nevada de Santa Marta, in Colombia. I have spent time with these people myself.
85 Heidegger, 'Moira', *Early Greek Thinking*, pp. 96-100. Also see the 'Aletheia' essay, *op. cit.*, p. 122.
86 Medard Boss, *I Dreamt Last Night*, p. 47. Also see Jessie Taft, *The Dynamics of Therapy in Controlled Relationships* (New York: Dover, 1962), pp. 12-16. 97-9, 209n, and 282-7, on defenses, limits, boundaries, and panic.
87 James, M. Glass, *Delusion: Internal Dimensions of Political Life* (Chicago: University of Chicago, 1985), p. 80.
88 *Ibid.*, p. 90.
89 Heidegger, 'What is Metaphysics?', *Basic Writings*, p. 106.
90 Heidegger, 'Letter on Humanism,' *op. cit.*, p. 229. Also see p. 207.
91 *Ibid.*, p. 205.
92 *Ibid.*
93 Heidegger, 'What is Metaphysics?', *op. cit.*, p. 105.
94 Kierkegaard, *Concluding Unscientific Postscript* (Princeton: Princeton University Press, 1941), p. 79.
95 See my essay, 'Psychopathology in the Epoch of Nihilism,' included in David Michael Levin (ed.), *Pathologies of the Modern Self: Postmodern Studies on Narcissism, Schizophrenia, and Depression* (New York: New York University Press, 1987). The pathologies I discuss are: narcissism, schizophrenia, and depression. I see them as 'pathologies of the will' and as symptomatic of the nihilism raging in our time.
96 See *The Tibetan Book of the Dead*, p. 43. See Merleau-Ponty's remark on 'passive vision without gaze [*regard*], as in the case of dazzling light,' in *Phenomenology of Perception*, pp. 315-16. Concerning the ritual gesture of shielding one's eyes from the deity's radiance, see Otto Kern, *Die Religion der Griechen* (Berlin, 1926), vol. I, p. 25; also E. E. McCartney, 'The Blinding Radiance of the Divine Visage,' in *Classical Journal*, vol. 36 (1941), pp. 485-87, and Lillian B. Lawler, 'Blinding Radiance and the

Greek Dance,' *Classical Journal*, vol 37 (1941), pp. 94-6. In *Psychology and Religion: East and West, Collected Works*, vol. 11 (New York: Pantheon, 1958), Jung reports, of Bruder Klaus, that, on beholding the presence of divinity, he did not experience any great joy, but rather, 'Overcome with terror, he instantly turned his face away and fell to the ground.' (see p. 319.)

97 Heidegger, 'The Turning,' *op. cit.*, p. 47.

98 Heidegger quotes Periander at the closing of the *Heraclitus Seminar* (1966-1967) he gave with Fink.

99 Roger Bacon, *Opus Majus*; see David C. Lindberg, *John Pecham and the Science of Optics: Perspectiva Communis* (Madison: University of Wisconsin Press, 1970), p. 19, quoted in Samuel J. Edgerton, *The Renaissance Rediscovery of Linear Perspective* (New York: Harper & Row, Icon Edition, 1976), pp. 74-5.

100 Dante Alighieri, 'Purgatorio,' in John Ciardi (ed. and trans.), *The Divine Comedy* (New York: W. W. Norton, 1977), p. 30.

101 Heidegger, 'The End of Philosophy,' *Basic Writings*, p. 383. Also see 'Der Anfang der abendländischen Denkens', *Heraklit Gesamtausgabe*, vol. 55 (Frankfurt am Main: Vittorio Klostermann, 1979), p. 162, as well as the essay on the 'Logos' Fragment (Herakleitos, B50), in *Early Greek Thinking*, pp. 72-6, relating *Phusis* and light.

102 See Fink, *Heraclitus Seminar 1966/1967*, p. 5.

103 *Ibid.*

104 Merleau-Ponty, *Phenomenology of Perception*, p. 487.

105 Merleau-Ponty, *Le visible et l'invisible* (Paris: Gallimard, 1964), p. 195. My own translation.

106 *Service for the Day of Atonement* (New York: Hebrew Publishing Co., 1935), p. 279.

107 *Ibid.*, p. 307. In *The Ring and the Book* (New York: W. W. Norton, 1967), Robert Browning's meditation on the poet's vision, he speaks of 'that erect form, flashing brow, fulgurant eye.'

108 Heidegger, *Erläuterungen zu Hölderlins Dichtung*, p. 24.

109 Leibniz, 'On the Supersensible Element in Knowledge, and on the Immaterial in Nature' (Letter to Queen Sophie Charlotte of Prussia, 1702), in Philip Wiener (ed.), *Leibniz: Selections* (New York: Charles Scribner's Sons, 1951), p. 363.

110 Heidegger *Being and Time*, p. 387.

111 Reichel-Dolmatoff, *Amazonian Cosmos: The Sexual and Religious Symbolism of the Tukano Indians* (Chicago: University of Chicago Press, 1971), p. 126.

112 *Ibid.*, p. 127.

113 *Ibid.*, p. 137.

114 Heidegger, *Introduction to Metaphysics*, p. 130.

115 *Ibid.*, p. 52. I have revised the translation.

116 Hegel, *The Phenomenology of Mind*, p. 75.

117 Jung, *Memories, Dreams, Reflections* (New York: Pantheon, 1963), p. 268.

118 H. L. Martensen, *Jacob Behmen: His Life and Teaching: or Studies in Theosophy* (London: Hodder & Stoughton, 1855), p. 7.

119 Katsuki Sekida, *Zen Training: Methods and Philosophy* (New York: John Weatherhill, 1975), p. 135.
120 *Ibid.*, p. 42.
121 Heidegger, *Being and Time*, pp. 196-7.
122 Arthur Deikman, 'Deautomatization and the Mystic Experience,' in Robert Ornstein (ed.), *The Psychology of Consciousness*, p. 223.
123 *Ibid.*, p. 228.
124 John Welwood, 'Meditation and the Unconscious: A New Perspective,' *The Journal of Transpersonal Psychology*, vol. 9, no. 1 (1977), p. 17.
125 *Ibid.*, p. 20.
126 *Ibid.*, p. 18. Also see Herbert Guenther and Chogyam Trungpa, *The Dawn of Tantra* (Berkeley: Shambhala, 1975), p. 27.
127 Trungpa, *Cutting Through Spiritual Materialism* (Berkeley: Shambhala, 1973).
128 Welwood, *op. cit*, p. 18.
129 Nietzsche, *The Gay Science* (New York: Random House, 1974), p. 181. In the translation called *Joyful Wisdom* (New York: Frederick Ungar Publishing Co., 1960), p. 168.
130 Thomas J. J. Altizer, *The Descent into Hell: A Study of the Radical Reversal of the Christian Consciousness* (New York: Seabury Press, 1979), pp. 153-4.
131 Mark Taylor, *Erring: A Post-Modern A/Theology* (Chicago: University of Chicago Press, 1984), p. 9.
132 Heidegger, *Nietzsche*, vol. II: *The Eternal Recurrence of the Same*, p. 182.
133 Taylor, *op. cit*, p. 6.
134 J. Hillis Miller, *Poets of Reality: Six Twentieth Century Writers* (New York: Athaneum, 1969), p. 3.
135 On mirroring and the emergence of self, see Jacques Lacan, 'Some Reflections on the Ego,' *International Journal of Psycho-analysis*, vol. 34 (11953), pp. 11-17 and 'The Mirror Stage as Formative of the Function of the I as Revealed in Psychoanalytic Experience,' in *Écrits: A Selection*, Alan Sheridan, trans. (New York: W. W. Norton, 1977); Heinz Kohut, *The Restoration of the Self* (New York: International Universities Press, 1977); and Nathan Schwartz-Salant, *Narcissism and Character Transformation: The Psychology of Narcissistic Character Disorders* (Toronto: Inner City Books, 1982). On the 'weakness of the superego' in our time, and on the social consequences attributed to this 'weakness', see C. R. Badcock, *Madness and Modernity: A Study in Social Psychoanalysis* (Oxford: Basil Blackwell, 1983). Badcock's argument is highly speculative and rests on assumptions I would seriously dispute; nevertheless, it is a paradigm case of Freudian thinking, and is of some interest on that account.
136 Taylor, *op. cit.*, p. 20.
137 *Ibid.*, p. 30.
138 Heidegger, *Nietzsche*, vol. II: *The Eternal Recurrence of the Same*, p. 182.
139 *Ibid.*, p. 183.
140 *Ibid.*, p. 246.
141 *Ibid.*

142 Eugen Fink, *Heraclitus Seminar 1966/1967*, pp. 88-9.
143 David Farrell Krell, 'Analysis', in Heidegger, *Nietzsche*, vol. II: *The Eternal Recurrence of the Same*, p. 248.
144 *Ibid.*
145 Nietzsche, *Twilight of the Idols*, in Kaufmann (ed.), *The Portable Nietzsche*, p. 511.
146 Nietzsche, 'Thus Spoke Zarathustra,' *Ecce Homo*, in Oscar Levy (ed.), *The Complete Works of Friedrich Nietzsche*, vol. 16 (New York: Russell & Russell, 1964), pp. 101-2.
147 Heidegger, *Heraclitus Seminar 1966/1967*, p. 162.

Chapter 4 Truth

1 Heidegger, 'The Origin of the Work of Art', in *Basic Writings*, pp. 36-7.
2 Heidegger, *On the Essence of Truth*, English translation, excerpts reprinted in *Basic Writings*, p. 119.
3 Heidegger, *Der Satz vom Grund* (Pfullingen: Neske, 1957), p. 78.
4 *Ibid.* My own translation.
5 Heidegger, *Being and Time*, p. 268.
6 Heidegger, *Der Satz vom Grund*, p. 78.
7 Heidegger, *Being and Time*, p. 268.
8 See, for example, Heidegger, 'The End of Philosophy and the Task of Thinking', in *Basic Writings*, p. 389.
9 Heidegger, *Der Satz vom Grund*, p. 78. My own translation.
10 Roland Barthes, *Sur Racine* (Paris: Editions du Seuil, 1963), p. 33.
11 Heidegger, 'The Age of the World Picture', in *The Question Concerning Technology and Other Essays*, p. 154.
12 Sir Thomas Browne, *The Religio Medici*, in Ernest Rhys (ed.), *The Religio Medici and Other Essays* (New York: E. P. Dutton, 1928), p. 80.
13 Patricia Berry, *Echo's Subtle Body: Contributions to An Archetypal Psychology* (Dallas: Spring Publications, Inc., 1982), p. 189.
14 Pindar, 'Pythian Odes', in *The Odes of Pindar*, translated with an Introduction by John Sandys (Cambridge: Harvard University Press, Loeb Classical Library, 1968), Ode VIII, lines 92-100, pp. 268-9.
15 Sir Thomas Browne, *The Garden of Cyrus*, ch. 3, in *The Religio Medici and Other Essays*, pp. 219-20.
16 Henry David Thoreau, 'A Week on the Concord and Merrimack Rivers', in Carl Bode (ed.), *The Portable Thoreau* (New York: Viking Press, 1964), pp. 211-12.
17 See Samuel J. Todes, 'Shadows in Knowledge: Plato's Misunderstanding of Shadows and of Knowledge as Shadow-Free', in Don Ihde and James E. Edie (eds), *Dialogues in Phenomenology* (The Hague: Martinus Nijhoff, 1976).
18 William James, *A Pluralistic Universe*, p. 288, in an edition, edited by Ralph B. Perry, that binds it together with *Essays in Radical Empiricism* (New York: Longmans, Green & Co., 1958).
19 See Jacques Derrida, *Speech and Phenomena* (Evanston: Northwestern

University Press, 1974) and *La Dissémination* (Paris: Editions du Seuil, 1972), p. 272, where he writes of 'ombres noires sur fond blanc, profiles sans face', seeing very clearly the ontological similarity between shadows and writing. Also see his essay, 'White Mythology', in 'Mythology: Metaphor in the Text of Philosophy', *New Literary History*, vol. 6, no. 1 (Autumn, 1974), pp. 5-74.

20 Merleau-Ponty, 'Eye and Mind', in *The Primacy of Perception*, p. 166.

21 *Ibid.*, p. 167.

22 *Ibid.*, p. 164.

23 *Ibid.*, p. 166.

24 *Ibid.*, p. 165.

25 Heidegger, 'The Age of the World Picture', *op. cit.*, p. 154. The bracketed word is a word I added.

26 *Ibid.*, p. 136.

27 *Service for the New Year* (New York: Hebrew Publishing Co., 1935), p. 273.

28 Heidegger, 'The Turning', *op. cit.*, p. 45.

29 Heidegger, 'The End of Philosophy and the Task of Thinking', in *Basic Writings*, p. 384.

31 See an unpublished manuscript by Robert Romanyshyn, entitled 'Shadows and Reflections: Beauty and the Psychological Eye'. It is an important study.

32 Thoreau, 'A Week on the Concord and Merrimack Rivers', *op. cit.*, pp. 211-12.

33 James, *A Pluralistic Universe*, p. 289.

34 See Jacques Lacan, 'The Mirror Stage as Formative of the Function of the I as Revealed in Psychoanalytic Experience', in Alan Sheridan (ed.), *Écrits: A Selection* (New York: W. W. Norton, 1977), and Sheridan (ed.), *The Language of the Self* (New York: Delta Books, 1968). Also see Merleau-Ponty, 'The Child's Relations with Others', in *The Primacy of Perception*; Harry S. Sullivan, *The Interpersonal Theory of Psychiatry* (New York: W. W. Norton, 1953); Heinz Kohut, *The Restoration of the Self* (New York: International Universities Press, 1977); and Nathan Schwartz-Salant, *Narcissism and Character Transformation: The Psychology of Narcissistic Character Disorders* (Toronto: Inner City Books, 1982).

35 See John Sallis, *Delimitations: Phenomenology and the End of Philosophy* (Bloomington: Indiana University Press, 1986), p. 5.

36 Heidegger *Being and Time*, p. 266.

37 Heidegger, 'Poetically Man Dwells', in Hofstadter (ed.), *Poetry, Language, Thought*, p. 216.

38 Heidegger *Being and Time*, p. 266.

39 *Ibid.*, p. 267.

40 *Ibid.*, p. 197.

41 *Ibid.*, pp. 200-1.

42 On the question of univocity, see Heidegger's first lecture in *What Is Called Thinking?* and Herbert Marcuse, *One-Dimensional Man: Studies in the Ideology of Advanced Industrial Society* (Boston: Beacon Press, 1964), pp. 194-98. Univocity is a basic requirement of totalitarian power.

Ambiguity and polyphony are essentially subversive.

43 Heidegger, 'Science and Reflection', in *The Question Concerning Technology*, pp. 166-7.
44 Heidegger, 'The Turning', in *The Question Concerning Technology and Other Essays*, pp. 45, 48, 49.
45 Heidegger, 'The Origin of the Work of Art', *op. cit.*, p. 25. Italics added.
46 Harold Searles, *Collected Papers on Schizophrenia and Other Related Subjects*, p. 340. Italics and bracketed words added.
47 See Heidegger, 'Building, Dwelling, Thinking', in *Poetry, Language, Thought*, p. 150. Also see Marcuse, *Eros and Civilization*, p. 151: Marcuse extends the aletheic gaze of caring to the whole of nature.
48 Augustine, *On Free Choice of the Will* (New York: Bobbs-Merrill, 1964), Book II, p. 75. Also see p. 67.
49 Heidegger, 'Alētheia', in *Early Greek Thinking*, p. 122. Also see p. 123. Bracketed words have been added.
50 Fink and Heidegger, *Seminar on Heraclitus*, p. 154. Bracketed words added.
51 Heidegger, 'Poetically Man Dwells', *op. cit.*, p. 220.
52 On mandalas, see Rong-tha blo-bzang-dam-chos-rgya-mtsho, *The Creation of Mandalas*, illustrated by Don-'grub rdo-rje, in three volumes (New Delhi, 1971). I am grateful to Herbert Guenther for showing these books to me when I visited him. Also see Carl Jung, 'Concerning Mandala Symbolism', in *The Collected Works of Carl G. Jung*, vol. 9, Part 1: *The Archetypes and the Collective Unconscious* (Princeton: Princeton University Press, 1968).
53 Jung, *Memories, Dreams, Reflections*, p. 254.
54 Lame Deer and Richard Erdoes, *Lame Deer: Seeker of Visions* (New York: Simon & Schuster, 1972), pp. 108-9.
55 Gershom Scholem, *On the Kabbalah and Its Symbolism*, p. 115.
56 See Mircea Eliade's works on shamanism, e.g., *Shamanism: Archaic Techniques of Ecstasy* (Princeton: Princeton University Press, 1964) and James Hillman, *The Myth of Analysis* (New York: Harper & Row, 1978), p. 45.
57 Heidegger, 'Alētheia', *op. cit.*, p. 120.
58 *Ibid.*
59 *Ibid.*, p. 129.
60 Heidegger, 'Recollection in Metaphysics', in *The End of Philosophy*, p. 76.
61 Heidegger, *Being and Time*, p. 171.
62 Plato, *The Republic, The Dialogues of Plato*, vol. I, Jowett translation (New York: Random House, 1937), Book VII, 532, p. 792.
63 Emerson, 'Spiritual Laws', in *Essays*, vol. I, p. 156.
64 Heidegger and Fink, *Heracleitus Seminar*, p. 88. Also see p. 85.
65 Heidegger, 'Moira', in *Early Greek Thinking*, p. 97. Also see Eliade, *Patterns of Comparative Religions* (New York: Sheed & Ward, 1958), pp. 150-1.
66 Merleau-Ponty, 'Eye and Mind', in *The Primacy of Perception*, p. 186.
67 Herbert V. Guenther, 'On Spiritual Discipline', *Maitreya*, vol. 3

(Berkeley and London: Shambhala, 1972), pp. 29-34.
68 Merleau-Ponty, 'Eye and Mind', *op. cit.*, p. 166.
69 Heidegger, 'The Origin of the Work of Art', in *Poetry, Language, Thought*, p. 53.
70 Heidegger, 'The End of Philosophy and the Task of Thinking', in *Basic Writings*, p. 387.
71 Merleau-Ponty, *Phenomenology of Perception*, p. 425. Also see Heidegger *Being and Time*, p. 171.
72 Heidegger, *Der Satz vom Grund*, p. 78: 'der ekstatische Aufenthalt des Menschen in der Offenheit des Anwesens'.
73 Heidegger, 'The Anaximander Fragment', *Early Greek Thinking*, p. 38. For the German, see *Holzwege* (Frankfurt am Main: Vittorio Klostermann, 1950), p. 323.
74 Merleau-Ponty, *Phenomenology of Perception*, p. 68.
75 Heidegger, 'The Anaximander Fragment', *op. cit.*, p. 35. Italics added.
76 *Ibid.*, p. 36. Italics added. On 'beings as a whole', see also Heidegger, *Nietzsche*, vol. II: *The Eternal Recurrence of the Same*, pp. 52, 59, 136, and 187.
77 Heidegger, 'The Anaximander Fragment', *op. cit.*, p. 35.
78 See *The Letters of Rainer Maria Rilke, 1910-1926* (New York: W. W. Norton, 1947), p. 342. Also see the letter on p. 373.
79 Heyemeyohsts Storm, *Seven Arrows* (New York: Ballantine, 1973), p. 9.
80 Heidegger. 'The Anaximander Fragment', *op. cit.*, p. 36.
81 Adorno, *Gesammelte Schriften*, vol. 13: *Die Musikalischen Monographien: Wagner Mahler, Berg* (Frankfurt am Main: Suhrkamp, 1971), p. 350.
82 See Michel de Montaigne, *The Complete Essays* (Stanford: Stanford University Press, 1965), Book III, Essay 12, p. 804, and Blaise Pascal, *Pensées* (New York: E. P. Dutton, 1958), note 172, pp. 49-50.
83 Heidegger, 'The Anaximander Fragment', *op. cit.*, p. 35.
84 *Ibid.*, p. 36.
85 See Michael Heim, 'Translator's Introduction', in Heidegger, *The Metaphysical Foundations of Logic* (Bloomington: Indiana University Press, 1984), x-xi.
86 Merleau-Ponty, *Phenomenology of Perception*, p. 69. Italics added.
87 *Ibid.*, pp. 420-1.
88 John Sallis, 'Heidegger/Derrida: Presence', *The Journal of Philosophy*, vol. LXXXI, no. 10n (October, 1984), p. 597. This paper has been reprinted in his book, *Phenomenology and the End of Metaphysics* (Bloomington: Indiana University Press, 1986).
89 *Ibid.*
90 *Ibid.*, p. 598.
91 Heidegger. 'Hegel und die Griechen', in *Wegmarken* (Frankfurt am Main: Vittorio Klostermann, 1967), p. 272.
92 Heidegger, 'The End of Philosophy and the Task of Thinking', *op. cit.*, p. 385.
93 Heidegger, 'Letter on Humanism', in *Basic Writings*, p. 205.
94 See Marcuse, *One-Dimensional Man: Studies in the Ideology of Advanced Industrial Society*, pp. 124-98.

95 Heidegger, *Being and Time*, pp. 200-1.
96 *Ibid*. Also see p. 269.
97 See Erving and Mirriam Polster, *Gestalt Therapy Integrated* (New York: Random House, Vintage Books, 1974), pp. 132-7. Also see Tarthang Tulku, *Time, Space and Knowledge*, and his 'Foreword' to Moon and Randall (eds), *Dimensions of Thought* (Berkeley: Dharma Publishing, 1980), xxxviii.
98 Heidegger, *Discourse on Thinking*, p. 64.
99 Augustine, *On Free Choice of the Will*, Book II, p. 67.
100 Marcuse, *One-Dimensional Man*, p. 127.
101 Medard Boss, *Existential Foundations of Medicine and Psychology* (New York: Jason Aronson, 1979), p. 112. Boss's description is strikingly similar to the description of the gaze Nietzsche ascribes to Epicurus in *Joyful Wisdom*, Book I, p. 82 and to Heidegger's description of a vision that is marked by its 'clear serenity' and rejoices in the 'presence' of things, in 'Heimkunft/An die Verwandten', *Erläuterungen zu Hölderlins Dichtung*, p. 19. Also see Heidegger's discussion of the 'moment of vision' in his *Nietzsche*, vol. II, pp. 52, 59, and 187.
102 Heinz Kohut, *The Restoration of the Self* (New York: International Universities Press, 1977), p. 45.
103 Heidegger, 'The Anaximander Fragment', *op. cit.*, p. 39.
104 See Thoreau's journal entry for June 21, 1852, in Bradford Torrey and Francis Allen (eds), *The Journal of Henry David Thoreau*, vol. II (New York: Dover, 1906), p. 43.
105 Charles Olson, *Archaeologist of Morning* (New York: Grossman Publishers, 1973), p. 73. Merleau-Ponty examines our perception of colour in his *Phenomenology of Perception*, pp. 153, 209-14, 227-9, 234, and 304-313, and on pp. 131-32 of 'The Intertwining – The Chiasm'. Heidegger, perhaps more surprisingly, writes about *aisthēsis*, the seeing of colours, in *Being and Time*, p. 57, in his essay on Parmenides, 'Moira (Parmenides VIII, 34-41)', in *Early Greek Thinking*, pp. 93, 96, and 100, and in his interpretation of Nietzsche's references to colour in *Thus Spake Zarathustra*: for this, see his *Nietzsche*, vol. II: *The Eternal Recurrence of the Same*, pp. 51, 131-2. In *Joyful Wisdom*, Book III, pp. 184-185, Nietzsche reflected on the cultural meanings that saturate our perception of colour and suggested that we can see in colour the evidence of historical change. Also see William James, 'The Place of Affectional Facts in a World of Pure Experience', *Essays in Radical Empiricism*, pp. 137-54; Herbert V. Guenther, *Philosophy and Psychology of the Abhidharma*, pp. 137-41; and Gerardo Reichel-Dolmatoff, *Amazonian Cosmos: The Sexual and Religious Symbolism of the Tukano Indians*, pp. 104-8, 122-7, and 137-8.
106 Merleau-Ponty, 'Eye and Mind', *op. cit.*, p. 166.
107 Wallace Stevens, *The Collected Poems of Wallace Stevens* (New York: Alfred E. Knopf, 1961), p. 203.
108 Heidegger, 'The Question Concerning Technology', *op. cit.*, p. 34.
109 Heidegger, 'Andenken', in *Erläuterungen zu Hölderlins Dichtung*, p. 134.
110 Heidegger, 'The Origin of the Work of Art', *op. cit.*, pp. 81, 83. Also see

his *Discourse on Thinking*, p. 65; *What Is Called Thinking?*, p. 19; *Der Satz vom Grund*, p. 102; and Nietzsche, vol. I: *The Will to Power as Art*, pp. 196-7.

111 See Vladimir Nabokov, *Lectures on Russian Literature* (New York: Harcourt, Brace, Jovanovich, 1981), p. 141. Also see p. 310. The usual word for truth is *pravda*. *Istina* seems to carry, for Nabokov, a meaning strikingly close to the meaning Heidegger ascribes to *alētheia*.

112 Merleau-Ponty, *Phenomenology of Perception*, p. 377.

113 Heidegger, 'The Anaximander Fragment', *op. cit.*, p. 36.

114 Merleau-Ponty, *Phenomenology of Perception*, pp. 308-9.

115 John Dewey, *Art as Experience* (New York: G. P. Putnam's Sons, 1958), p. 232. Also see Adorno, *Minima Moralia*, p. 48 on 'the hurrying eye' and pp. 227-8, on the conflict between the 'useful' and the 'colourful' in our present political economy.

116 Heidegger, *Nietzsche*, vol. IV: *Nihilism*, p. 245.

117 Dewey, *Experience and Nature* (Illinois: Open Court, 1929), p. 40.

Appendix

1 Augustine, *The Confessions* (New York: E. P. Dutton, 1950), Book 10, p. 258.

2 Vincent Scully, *The Earth, The Temple, and the Gods: Greek Sacred Architecture* (New York: Frederick A. Praeger, 1969), pp. 74-6. I have made a composite of separate textual passages.

3 R. Gordon Wasson, Albert Hofmann, and Carl A. P. Ruck, *The Road to Eleusis: Unveiling the Secret of the Mysteries* (New York and London: Harcourt, Brace, Jovanovich, 1978), pp. 84, 49, 36-7. A composite of passages I have made, ordered as the page references indicate.

4 For the biography of Ayo Khandro, se Tsultrim Allione, *Women of Wisdom* (New York and London: Routledge & Kegan Paul, 1984), pp. 236-57.

5 See volumes I and II of the transcripts that record the lectures given by Namkhai Norbu during recent retreats in the United States. These volumes are being published by the Dzogchen Community of America. Concerning a vision which 'lets be what is' (*snang-ba cog-bzhag*), see vol. I, pp. 75-7, 81-2, 137-41, 183-5, 259, 274-81, 289-90, 302-5. Also see p. 189 on reintegration with the elements of phenomenal vision. The body of light is discussed on pp. 374-8, and the Dark Retreat on pp. 139-40.

6 See Herbert V. Guenther , *Philosophy and Psychology in the Abhidharma* (Berkeley: Shambhala, 1976) and *Tibetan Buddhism in Western Perspective* (Berkeley: Dharma Publishing, 1977). Also see Chogyam Trungpa, *Glimpses of Abhidharma* (Boulder: Vajradhatu, 1975).

7 See Herbert V. Guenther, *Buddhist Philosophy in Theory and Practice* (Berkeley: Shambhala, 1971) and Tarthang Tulku, *Time, Space and Knowledge* (Berkeley: Dharma Publishing, 1980.)

8 Heidegger, 'Moira (Parmenides VIII, 34-41)', in David F. Krell and

Frank Capuzzi (translators and editors), *Early Greek Thinking* (New York: Harper & Row, 1975), p. 100.

9 See Faber Birren, *Color Psychology and Color Therapy* (Secaucus, New Jersey: The Citadel Press, 1961). Birren emphasizes (p. 132) that every part of the body is sensitive to luminous radiation and that, in this sense, we see with our entire body. He cites experiments which indicate that colours can be 'seen' even without seeing them by way of the eyes. Thus, for example, there is a substantive increase in blood sugar under the action of red light. My capacity to 'see' my own body moving in the dark is briefly touched on by Maurice Merleau-Ponty in *Phenomenology of Perception* (Routledge & Kegan Paul, 1962) and discussed at greater length in 'The Intertwining – The Chiasm' and his 'Working Notes', published in *The Visible and the Invisible* (Evanston: Northwestern University Press, 1969). In the latter, this capacity is characterized as a 'narcissism' of the flesh. The flesh is radiant, luminous; it can see and touch itself. Through the proprioceptive interiority of the sensorimotor schema, the moving body seems able to see itself and make itself be seen – even in the most complete darkness.

Bibliography

Adorno, T., *Die Musikalische Monographien: Wagner, Mahler, Berg*, 'Gesammelte Schriften', vol. 13, Frankfurt am Main, Suhrkamp, 1971.
Adorno, T., *Negative Dialectics*, New York, Seabury Press, 1973.
Adorno, T., *Minima Moralia: Reflections from Damaged Life*, London, New Left Books, 1974, Verso Edition, 1978.
Adorno, T., *The Positivist Dispute in German Sociology*, London, Heinemann, 1981.
Allione, T., *Women of Wisdom*, Boston and London, Routledge & Kegan Paul, 1984.
Altizer, T., *The Descent into Hell: A Study of the Radical Reversal of the Christian Consciousness*, New York, Seabury, 1979.
Althusser, L., *Lire le Capital*, vol. 1, Paris, Maspero, 1968.
Augustine, *Confessions*, New York, Dutton, 1959.
Augustine, *On Free Choice of the Will*, New York, Bobbs-Merrill, 1964.
Badcock, C., *Madness and Modernity: A Study in Social Psychoanalysis*, Oxford, Basil Blackwell, 1983.
Balmary, M., *Psychoanalyzing Psychoanalysis: Freud and the Hidden Fault of the Father*, Baltimore, Johns Hopkins University Press, 1982.
Barker, F., *The Tremulous Private Body: Essays on Subjection*, New York, Methuen, 1984.
Barthes, R., *Sur Racine*, Paris, Editions du Seuil, 1963.
Beckett, S., *Endgame*, New York, Grove Press, 1958.
Bellamy, E., *Looking Backward*, New York, New American Library, 1960.
Benhabib, S., 'The Generalized and the Concrete Other: Toward a Feminist Critique of Substitutionality Universalism', in E. Kittay and D. Meyers, ed., *Proceedings of the Women and Moral Theory Conference*, New Jersey, Rowman and Allenheld, 1986.
Benoit, H., *The Supreme Doctrine*, New York, Viking, 1955.
Bergson, H., *The Two Sources of Morality and Religion*, Greenwood, New Jersey, 1935.
Berkeley, G., *An Essay Towards a New Theory of Vision*, London, Dent, 1934.
Berry, P., *Echo's Subtle Body: Contributions to an Archetypal Psychology*, Dallas, Spring Publications, 1982.
Blake, W., *The Marriage of Heaven and Hell*, New York and London, Oxford University Press, 1975.
Blake, W., *The Works of William Blake*, ed. G. Bentley, New York and London, Oxford University Press, 1979.

BIBLIOGRAPHY

Blythe, R., *Akenfield: Portrait of an English Village*, New York, Dell, 1969.

Boss, M., *Erinnerung an Martin Heidegger*, Pfullingen, Neske, 1977.

Boss, M., *I Dreamt Last Night*, New York, Wiley and Sons, 1977.

Boss, M., *The Existential Foundations of Medicine and Psychology*, New York, Aronson, 1979.

Bourdieu, P., *Distinction: A Social Critique of the Judgment of Taste*, Cambridge, Harvard University Press, 1984.

Breines, P., ed., *Critical Interruptions: New Left Perspectives on Herbert Marcuse*, New York, Herder & Herder, 1970.

Brown, J., ed., *The Sacred Pipe: Black Elk's Account of the Seven Rites of the Oglala Sioux*, New York, Penguin, 1971.

Brown, N., *Life Against Death: The Psychoanalytic Meaning of History*, Middletown, Wesleyan University Press, 1959.

Browne, T., *The Religio Medici and Other Writings*, London, Dent & Sons, New York, Dutton, 1928.

Browning, R., *The Ring and the Book*, New York, Norton, 1967.

Bryson, N., *Vision and Painting: The Logic of the Gaze*, New Haven, Yale University Press, 1983.

Buck-Morss, S., *The Origin of Negative Dialectics: Theodor Adorno, Walter Benjamin and the Frankfurt Institute*, New York, Macmillan, 1977.

Burke, E., *Reflections on the Revolution in France*, Indianapolis, Bobbs-Merrill, 1955.

Carroll, L., *Through the Looking Glass*, New York, Random House, 1946.

Castaneda, C., *A Separate Reality*, New York, Simon & Schuster, 1971.

Castaneda, C., *Tales of Power*, New York, Simon & Schuster, 1974.

Cavell, S., *Must We Mean What We Say?*, New York, Charles Scribner's Sons, 1969.

Cavell, S., *The Claim of Reason*, New York and London, Oxford University Press, 1979.

Cooley, C., *Human Nature and the Social Order*, New Brunswick, New Jersey, Transaction Books, 1983.

Cornford, F., *From Religion to Philosophy*, New York, Harper & Row, 1957.

Cornford, F., *The Unwritten Philosophy and Other Essays*, Cambridge, Cambridge University Press, 1967.

Corngold, S., 'The Question of the Self in Nietzsche during the Axial Period (1882-1888)', in D. O'Hara, ed., *Why Nietzsche Now?*, Bloomington, Indiana University Press, 1985.

Daly, M., *Gyn/Ecology: The Metaethics of Radical Feminism*, Beacon Press, 1978.

Dante, *The Divine Comedy*, New York, Norton, 1977.

Deikman, A., 'Deautomatization and the Mystical Experience', in R. Ornstein, ed., *The Nature of Human Consciousness*, San Francisco, W. H. Freeman, 1973.

Deleuze, G. and Guattari, F., *Anti-Oedipus: Capitalism and Schizophrenia*, Minneapolis, University of Minnesota Press, 1983.

Derrida, J., *La Dissémination*, Paris, Editions du Seuil, 1972.

Derrida, J., *Speech and Phenomena*, Evanston, Northwestern University Press, 1974.

Derrida, J., 'White Mythology', *New Literary History*, vol. 6, no. 1, Autumn, 1974.

Derrida, J., 'Interview', in R. Kearny, ed., *Dialogues with Contemporary Continental Thinkers: The Phenomenological Heritage*, Dover, Manchester University Press, 1984.

Descartes, R., *The Philosophical Works of Descartes*, ed. E. Haldane and G. Ross, New York, Dover Publications, 1955.

Des Pres, T., *The Survivor: An Anatomy of Life in the Death Camps*, New York and London, Oxford University Press, 1976.

Dewey, J., *Experience and Nature*, Chicago, Open Court, 1929.

Dewey, J., *Art as Experience*, New York, G. P. Putnam's Sons, 1958.

Dewey, J., *Democracy and Education*, New York, Macmillan, 1966.

Diner, H., *Mothers and Amazons: The First Feminist History of Culture*, New York, Doubleday, 1973.

Donne, J., *Devotions upon Emergent Occasions*, Ann Arbor, University of Michigan, 1959.

Dreyfus, H. and Rabinow, P., *Michel Foucault: Beyond Structuralism and Hermeneutics*, Chicago, University of Chicago Press, 1983.

Dufrenne, M., *The Phenomenology of Aesthetic Experience*, Evanston, Northwestern University Press, 1973.

Edgerton, S., *The Renaissance Rediscovery of Linear Perspective*, New York, Harper & Row, 1976.

Eliade, M., *Patterns of Comparative Religions*, New York, Sheed & Ward, 1958.

Emerson, R., *Essays*, vols. 1 and 2, New York, A. Burt, 1936.

Emerson, R., *The Portable Emerson*, ed. M. van Doren, New York, Viking, 1946.

Fink, E. and Heidegger, M., *Heraclitus Seminar, 1966/1967*, University, Alabama, University of Alabama Press, 1979.

Findlay, J., *Hegel: A Re-Examination*, New York, Macmillan, 1962.

Fischer, M. and Marcus, G., *Anthropology as Cultural Critique: An Experimental Movement in the Human Sciences*, Chicago, University of Chicago Press, 1986.

Foucault, M., *The Order of Things: An Archaeology of the Human Sciences*, New York, Random House, 1973.

Foucault, M., *The Birth of the Clinic: An Archaeology of Medical Perception*, New York, Random House, 1975.

Foucault, M., *Language, Counter-Memory, Practice: Selected Essays and Interviews*, ed. C. Gordon, Ithaca, Cornell University Press, 1977.

Foucault, M., *Discipline and Punish: The Birth of the Prison*, London, Allen Lane, 1977, New York, Pantheon, 1978.

Foucault, M., *Power/Knowledge: Selected Interviews and Other Writings, 1971/1977*, ed. C. Gordon, New York, Pantheon, 1980.

Foucault, M., *Le Souci de Soi*, Paris, Gallimard, 1984.

Foucault, M., *L'Usage des Plaisirs*, Paris, Gallimard, 1984.

Fraser, N., 'Foucault's Body-Language: A Post-Humanist Political Rhetoric', *Salmagundi*, no. 61, Fall, 1983.

Freeman, K., ed., *Ancilla to the Pre-Socratic Philosophers*, Cambridge, Harvard University Press, 1978.
Freud, S., 'The Psycho-Analytic View of Psychogenic Disturbances of Vision', in *The Complete Works of Sigmund Freud*, Standard Edition, vol. 11, London, Hogarth Press, 1953.
Freud, S., 'On Narcissism', in *The Complete Works of Sigmund Freud*, Standard Edition, vol. 14, London, Hogarth Press, 1953.
Freud, S., *The Ego and the Id*, New York, Norton, 1960.
Freud, S., *Civilization and Its Discontents*, New York, Doubleday. 1962.
Freud, S., *Inhibition, Symptom and Anxiety*, New York, Norton, 1963.
Freud, S., *Three Essays on the Theory of Sexuality*, New York, Basic Books, 1975.
Freud, S., *The Origins of Psycho-Analysis: Letters to William Fliess*, ed. M. Bonaparte, A. Freud, E. Kris, New York, Basic Books, 1977.
Frye, N., *Fearful Symmetry: A Study of William Blake*, Princeton, Princeton University Press, 1947.
Frye, N., *The Anatomy of Criticism*, Princeton, Princeton University Press, 1957.
Gendlin, E., 'Analysis', in M. Heidegger, *What Is A Thing?*, Chicago, Henry Regnery, 1967.
Gendlin, E., 'Experiential Phenomenology', in M. Natanson, ed., *Phenomenology and the Social Sciences*, Evanston, Northwestern University Press, 1973.
Gendlin, E., 'Experiential Psychotherapy', in R. Corsini, ed., *Current Psychotherapies*, Itasca, Illinois, F. E. Peacock, 1973.
Gendlin, E., '*Befindlichkeit*: Heidegger and the Philosophy of Psychology', *Review of Existential Psychology and Psychiatry*, vol. 16, nos. 1-3, 1978-1979.
Gendlin, E., 'Non-Logical Moves and Nature Metaphors', *Analecta Husserliana*, vol. 19, Dordrecht, D. Reidel Publishing, 1985.
Gendlin, E., 'Process Ethics and the Political Question', *The Focusing Folio*, Chicago, the Focusing Institute, vol. 5, no. 2, 1986.
Gendlin, E., *Let Your Body Interpret Your Dreams*, Peru, Illinois, Chiron Publications, 1986.
Gendlin, E., 'A Philosophical Critique of the Concept of Narcissism', in D. Levin, ed., *Pathologies of the Modern Self: Postmodern Studies on Narcissism, Schizophrenia, and Depression*, New York, New York University Press, 1987.
Geuss, R., *The Idea of Critical Theory: Habermas and the French School*, New York and Cambridge, Cambridge University Press, 1981.
Giddens, A., *The Constitution of Society*, Berkeley, University of California Press, 1984.
Gilligan, C., *In a Different Voice: Psychological Theory and Women's Development*, Cambridge, Harvard University Press, 1982.
Giorgi, A., *Psychology as a Human Science*, New York, Harper & Row, 1970.
Glass, J., *Delusion: Internal Dimensions of Political Life*, Chicago, University of Chicago Press, 1985.
Goethe, J., *Italian Journey, 1786-1788*, San Francisco, North Point Press, 1982.

Goodman, P., et al., *Gestalt Therapy: Excitement and Growth in the Human Personality*, New York, Delta, 1951.
Griffin, S., *Woman and Nature: The Roaring Inside Her*, New York, Harper & Row, 1978.
Guattari, F. and Deleuze, G., *Anti-Oedipus: Capitalism and Schizophrenia*, Minneapolis, University of Minnesota Press, 1983.
Guenther, H., *Buddhist Philosophy in Theory and Practice*, Baltimore and Harmondsworth, Penguin, 1971.
Guenther, H., 'On Spiritual Discipline', *Maitreya*, vol. 3, Berkeley and London, Shambhala, 1972.
Guenther, H., *The Royal Song of Saraha*, Berkeley and London, Shambhala, 1973.
Guenther, H., *Philosophy and Psychology in the Abhidharma*, Berkeley and London, Shambhala, 1974.
Guenther, H. and Trungpa, C., *The Dawn of Tantra*, Berkeley and London, Shambhala, 1975.
Guenther, H., *The Tantric View of Life*, Berkeley and London, Shambhala, 1976.
Guenther, 'Commentary', in Longchen-pa, *Kindly Bent to Ease Us*, 3 vols., Berkeley, Dharma Publishing, 1976.
Guenther, H., *Tibetan Buddhism in Western Perspective*, Berkeley, Dharma Publishing, 1977.
Guenther, H., *The Matrix of Mystery*, Berkeley and London, Shambhala, 1985.
Gurwitsch, A., *Studies in Phenomenology and Psychology*, Evanston, Northwestern University Press, 1966.
Habermas, J., *Communication and the Evolution of Society*, Boston, Beacon Press, 1979.
Habermas, J., *The Philosophical Discourse of Modernity: Twelve Lectures*, Cambridge, M.I.T. Press, 1987.
Heaton, J., *The Eye: Phenomenology and Psychology of Function and Disorder*, London, Tavistock, 1968.
Hegel, G. W. F., *The Letters*, ed. C. Butler and C. Seiler, Bloomington, Indiana University Press, 1984.
Heidegger, M., *Vom Wesen der Wahrheit*, Frankfurt am Main, Klostermann, 1943.
Heidegger, M., *Der Satz vom Grund*, Pfullingen, Neske, 1957.
Heidegger, M., *Introduction to Metaphysics*, New York, Doubleday, 1961.
Heidegger, M., *Being and Time*, New York, Harper & Row, 1962.
Heidegger, M., *Holzwege*, Frankfurt am Main, Klostermann, 1963.
Heidegger, M., *Was Ist Metaphysik?*, Frankfurt am Main, Klostermann, 1965.
Heidegger, M., *Discourse on Thinking*, New York, Harper & Row, 1966.
Heidegger, M., *Wegmarken*, Frankfurt am Main, Klostermann, 1967.
Heidegger, M., *What Is Called Thinking?*, New York, Harper & Row, 1968.
Heidegger, M., *Erläuterungen zu Hölderlins Dichtung*, Frankfurt am Main, Klostermann, 1971.
Heidegger, M., *Poetry, Language, Thought*, ed. A. Hofstadter, New York, Harper & Row, 1971.

Heidegger, M., *The End of Philosophy*, ed. J. Stambaugh, New York, Harper & Row, 1973.
Heidegger, M., *Early Greek Thinking*, eds. F. Capuzzi and D. Krell, New York, Harper & Row, 1975.
Heidegger, M., *Basic Writings*, ed. D. Krell, New York, Harper & Row, 1977.
Heidegger, M., *The Question Concerning Technology and Other Essays*, ed. W. Lovitt, New York, Harper & Row, 1977.
Heidegger, M., *Nietzsche*, vol. 1: *The Will to Power as Art*, New York, Harper & Row, 1979.
Heidegger, M., and Fink, E., *Heraclitus Seminar 1966/1967*, University, University of Alabama Press, 1979.
Heidegger, M., *Heraklit*, 'Gesamtausgabe', vol. 55, Frankfurt am Main, Klostermann, 1979.
Heidegger, M., *The Basic Problems of Phenomenology*, Bloomington, Indiana University Press, 1982.
Heidegger, M., *Nietzsche*, vol. 4:*Nihilism*, New York, Harper & Row, 1982.
Heidegger, M., *Nietzsche*, vol. 2: *The Eternal Recurrence of the Same*, New York, Harper & Row, 1984.
Heim, M., 'Translator's Introduction', in M. Heidegger, *The Metaphysical Foundations of Logic*, Bloomington, Indiana University Press, 1984.
Henley, N., *Body Politics: Power, Sex, and Non-verbal Communication*, Englewood Cliffs, New Jersey, Prentice-Hall, 1977.
Henry, J., *Pathways to Madness*, New York, Harper & Row, 1973.
Hillman, J., *The Myth of Analysis*, New York, Harper & Row, 1978.
Hobbes, T., *The Leviathan*, New York, Bobbs-Merrill, 1958.
Hofman, A., *et al.*, *The Road to Eleusis: Unveiling the Secret of the Mysteries*, New York and London: Harcourt, Brace & Jovanovich, 1978.
Homer, *The Odyssey*, Harmondsworth and New York, Penguin, 1950.
Horkheimer, M., 'Tradition and Critical Theory', in O'Connell *et al.*, eds., *Critical Theory: Selected Essays*, New York, Herder & Herder, 1972.
Horkheimer, M., *Dawn and Decline: Notes 1926-1931 and 1950-1969*, New York, Seabury Press, 1978.
Hoy, D., *The Critical Circle: Literature and History in Contemporary Hermeneutics*, Berkeley, University of California Press, 1978.
Husserl, E., *Ideas: General Introduction to Pure Phenomenology*, vol. 1, New York, Macmillan, 1931.
Husserl, E., *The Crisis of European Sciences and Transcendental Phenomenology*, Evanston, Northwestern University Press, 1970.
James, W., *The Principles of Psychology*, New York, Henry Holt, 1890.
James, W., *A Pluralistic Universe*, New York, Longmans, Green & Co., 1958.
James, W., *Essays in Radical Empiricism*, New York, Longmans, Green & Co., 1958.
James, W., *The Writings of William James*, ed. J. McDermott, New York, Random House, 1968.
Jonas, H., 'The Nobility of Sight: A Study in the Phenomenology of the Senses', in S. Spicker, ed., *Philosophy of the Body*, Chicago, Quadrangle, 1970.

Jung, C. G., *Relations between the Ego and the Unconscious*, 'The Collected Works of Carl G. Jung', vol. 7, Princeton, Princeton University Press, 1953.
Jung, C. G., 'Psychological Commentary on "The Tibetan Book of Great Liberation" ', *Psychology and Religion: West and East*, 'The Collected Works of Carl G. Jung', vol. 11, Princeton, Princeton University Press, 1954.
Jung, C. G., *Two Essays on Analytical Psychology*, New York, World Publishing Co., 1956.
Jung, C. G., 'The Concept of the Libido', *Symbols of Transformation*, 'The Collected Works of Carl G. Jung', vol. 5, New York, Pantheon, 1956.
Jung, C. G., *Memories, Dreams, Reflections*, New York, Random House, 1965.
Jung, C. G., 'Consciousness, Unconscious and Individuation', *The Archetypes and the Collective Unconscious*, 'The Collected Works of C. G. Jung', vol. 9, Part 1, Princeton, Princeton University Press, 1968.
Jung, C. G., 'Concerning Mandala Symbolism', *The Archetypes and the Collective Unconscious*, 'The Collected Works of Carl G. Jung', vol. 9, Part 1, Princeton, Princeton University Press, 1968.
Jung, C. G., *Civilization in Transition*, 'The Collected Works of Carl G. Jung', vol. 10, Princeton, Princeton University Press, 1970.
Jung, C. G., 'Aion: Phenomenology of the Self', in J. Campbell, ed., *The Portable Jung*, New York, Viking, 1971.
Jung, C. G., *The Portable Jung*, ed. J. Campbell, New York, Viking, 1971.
Jung, C. G., 'On the Nature of the Psyche', *The Structure and Dynamics of the Psyche*, 'The Collected Works of Carl G. Jung', vol. 8, Princeton, Princeton University Press, 1975.
Jung, C. G., *Aspects of the Feminine*, ed. W. McGuire, Princeton, Princeton University Press, 1982.
Kagan, J., *The Nature of the Child*, New York, Basic Books, 1984.
Kant, I., *The Critique of Practical Reason*, New York, Bobbs-Merrill, 1956.
Kaufman, L., *Perception: The World Transformed*, New York and London, Oxford University Press, 1979.
Kehr, D., 'The End of the Line: A Review of Claude Lanzmann's "Shoah" ', *Chicago Magazine*, January, 1986.
Kierkegaard, S., *Concluding Unscientific Postscript*, Princeton, Princeton University Press, 1941.
Kohlberg, L., *Essays on Moral Development*, vol. 1: *The Philosophy of Moral Development*, New York, Harper & Row, 1981.
Köhler, W., *Gestalt Psychology*, New York, Mentor Books, 1947.
Köhler, W., *Dynamics in Psychology*, New York, Washington Square Press, 1965.
Kohut, H., *The Restoration of the Self*, New York, International Universities Press, 1977.
Kovel, J., *The Age of Desire: Reflections of a Radical Psychoanalyst*, New York, Pantheon, 1982.
Lacan, J., *The Language of the Self*, ed. A. Sheridan, New York, Delta, 1968.
Lacan, J., *Le Séminaire, I, Les Écrits Techniques de Freud*, Paris, Editions du Seuil, 1975.
Lacan, J., *Ècrits: A Selection*, ed. A. Sheridan, New York, Norton, 1977.

Laing, R., *The Divided Self: An Existential Study in Society and Madness*, Harmondsworth and Baltimore, Penguin, 1965.
Lame Deer, *Lame Deer: Seeker of Visions*, ed. Richard Erdoes, New York, Simon and Schuster, 1972.
Lasch, C., *The Culture of Narcissism: American Life in an Age of Diminishing Expectations*, New York, Norton, 1979.
Leder, D., 'Medicine and Paradigms of Embodiment', *Journal of Medicine and Philosophy*, vol. 9, 1984.
Lefebvre, H., *Le Vie Quotidienne dans le Monde Moderne*, Paris, Gallimard, 1968.
Leibniz, G., *Selections*, ed. Philip Wiener, New York, Charles Scribner's Sons, 1951.
Leibniz, G., *Monadology*, in G. Parkinson, ed., *Leibniz: Philosophical Writings*, London, J. M. Dent & Sons, 1973.
Levi, R., *Survival in Auschwitz: The Nazi Assault on Humanity*, New York, Collier-Macmillan, 1961.
Levin, D., 'The Poetic Function in Phenomenological Discourse', in W. McBride and C. Schrag, eds., *Phenomenology in a Pluralistic Context: Selected Studies in Phenomenological and Existential Philosophy*, Albany, State University of New York, 1983.
Levin, D., *The Body's Recollection of Being*, London and Boston, Routledge & Kegan Paul, 1986.
Levin, D., 'Psychopathology in the Epoch of Nihilism', in D. Levin, ed., *Pathologies of the Modern Self: Postmodern Studies on Narcissism, Schizophrenia, and Depression*, New York, New York University Press, 1987.
Levin, R., 'Cancer and the Self: How Illness Constellates Meaning', in D. Levin, ed., *Pathologies of the Modern Self: Postmodern Studies on Narcissism, Schizophrenia, and Depression*, New York, New York University Press, 1987.
Lévi-Strauss, C., *Tristes Tropiques*, New York, Atheneum, 1964.
Lévy-Bruhl, L., *The Soul of the Primitive*, London, George Allen & Unwin, 1965.
Lindberg, D., *John Pecham and the Science of Optics: Perspectiva Communis*, Madison, University of Wisconsin, 1970.
Lindberg, D., *Theories of Vision from Al-kindi to Kepler*, Chicago, University of Chicago Press, 1976.
Lindberg, D., and Steneck, N., 'The Sense of Vision and the Origins of Modern Science', in A. Debus, ed., *Science, Medicine and Society in the Renaissance: Essays to Honour Walter Pagel*, New York, Science History Publications, 1981.
Lifton, R., *The Life of the Self: Towards a New Psychology*, New York, Basic Books, 1983.
Lifton, R., *In a Dark Time*, Cambridge, Harvard University Press, 1984.
Lingis, A., 'The Elemental Background', in J. Edie, ed., *New Essays in Phenomenology*, Chicago, Quadrangle Books, 1969.
Longchenpa, 'Now That I Come to Die', *Crystal Mirror*, vol. 5, Berkeley, Dharma Publishing, 1977.

Lorenz, K., *On Aggression*, New York, Harcourt, Brace & World, 1966.
Lyotard, J., *Discours Figure*, Paris, Klincksieck, 1972.
Mallarmé, S., *Oeuvres Complètes*, ed. H. Mondor and G. Jean-Aubry, Paris, Gallimard, 1945.
Marcus, G. and Fischer, M., *Anthropology as Cultural Critique: An Experimental Moment in the Human Sciences*, Chicago, University of Chicago Press, 1986.
Marcuse, H., *Eros and Civilization: A Philosophical Inquiry into Freud*, New York, Vintage, 1962.
Marcuse, H., *One-Dimensional Man: Studies in the Ideology of Advanced Industrial Society*, Boston, Beacon Press, 1964.
Marcuse, H., *An Essay on Liberation*, Boston, Beacon Press, 1969.
Martensen, H., *Jacob Behmen: His Life and Teaching: Or Studies in Theosophy*, London, Hodder and Stoughton, 1855.
Meister Eckhart, *Meister Eckhart: An Introduction to the Study of His Works, with an Anthology of His Sermons*, ed. J. Clark, London, Nelson & Sons, 1957.
Merleau-Ponty, M., *Phenomenology of Perception*, New York and London, Routledge & Kegan Paul, 1962.
Merleau-Ponty, M., *The Primacy of Perception*, ed. J. Edie, Evanston, Northwestern University Press, 1964.
Merleau-Ponty, M., *Sense and Non-Sense*, ed. H. Dreyfus and P. Dreyfus, Evanston, Northwestern University Press, 1964.
Merleau-Ponty, M., *Signs*, ed. R. McCleary, Evanston, Northwestern University Press, 1964.
Merleau-Ponty, M., *Le Visible et l'Invisible*, Paris, Gallimard, 1964.
Merleau-Ponty, M., *The Visible and the Invisible*, ed. C. Lefort, Evanston, Northwestern University Press, 1968.
Merleau-Ponty, M., *Humanism and Terror*, Boston, Beacon Press, 1969.
Merleau-Ponty, M., *Themes from the Lectures at the Collège de France*, Evanston, Northwestern University Press, 1970.
Merleau-Ponty, M., *The Prose of the World*, Evanston, Northwestern University Press, 1973.
Miller, H., *Poets of Reality: Six Twentieth Century Writers*, New York, Atheneum, 1969.
Mindell, A., *Dreambody: The Body's Role in Revealing the Self*, Santa Monica, Sigo Press, 1982.
Mindell, A., *Working with the Dreaming Body*, Boston and London, Routledge & Kegan Paul, 1985.
Montaigne, M., *The Complete Essays*, ed., D. Frame, Stanford, Stanford University Press, 1965.
Muller, J., and Richardson, R., *Lacan and Language*, New York, International Universities Press, 1982.
Murdoch, I., *The Sovereignty of Good*, New York and Harmondsworth, Penguin, 1970.
Nabokov, V., *Lectures on Russian Literature*, ed. D. Bowers and M. Fredson, New York, Harcourt, Brace & Jovanovich, 1981.
Nagarjuna, *Mulamadhyamikakarika*, ed. K. Inada, Tokyo, Hojuseido Press, 1970.

Najlis, Michèle, 'Las viejas tribus', *Nicaraguan Perspectives*, vol. 11, Berkeley, Nicaraguan Information Center, Fall-Winter, 1985-86.
Neumann, E., *The Origins and History of Consciousness*, Princeton, Princeton University Press, 1970.
Neumann, E., *Art and The Creative Unconscious*, Princeton, Princeton University Press, 1971.
Neumann, E., *The Great Mother: An Analysis of the Archetype*, Princeton, Princeton University Press, 1972.
Neumann, E., *Amor and Psyche: The Psychic Development of the Feminine*, Princeton, Princeton University Press, 1973.
Nicolson, L., 'Women, Morality and History', *Social Research*, vol. 50, no. 3, Autumn, 1983.
Nietzsche, F., *The Portable Nietzsche*, ed. W. Kaufmann, New York, Viking, 1954.
Nietzsche, F., *The Birth of Tragedy*, New York, Doubleday, 1956.
Nietzsche, F., *The Genealogy of Morals*, New York, Doubleday, 1956.
Nietzsche, F., *Joyful Wisdom*, New York, Frederick Ungar, 1960.
Nietzsche, F., *Ecce Homo*, 'The Complete Works of Friedrich Nietzsche', vol. 17, ed. O. Levy, New York, Russell and Russell, 1964.
Nietzsche, F., *Beyond Good and Evil*, New York, Random House, 1966.
Nietzsche, F., *The Will to Power*, New York, Random House, 1967.
Nietzsche, F., *Twilight of the Idols*, New York and Harmondsworth, Penguin, 1968.
Nietzsche, F., *Daybreak: Thoughts on the Prejudices of Morality*, Cambridge, Cambridge University Press, 1982.
Nietzsche, F., *Untimely Meditations*, Cambridge, Cambridge University Press, 1983.
Norbu Namkhai, *The Mirror: Advice on Presence and Awareness*, Arcidosso, Italy, Shang Shung Editions, 1983.
Norbu Namkhai, *The Cycle of Day and Night*, Berkeley, Zhang Zhung Editions, 1985.
Norbu Namkhai, *Retreats in the United States*, vols. 1 and 2, Conway, Massachussetts, Dzogchen Community of America, 1985, 1986.
Olson, C., *Archaeologist of Morning*, New York, Grossman, 1973.
O'Neill, J., *Sociology as a Skin Trade: Essays towards a Reflexive Sociology*, London, Heinemann Educational Books, 1972.
O'Neill, J., *Five Bodies: The Human Shape of Modern Society*, Ithaca, Cornell University Press, 1985.
Paris, J., *L'espace et le regard*, Paris, Editions du Seuil, 1965.
Paris, J., *Painting and Linguistics*, Pittsburgh, Carnegie-Mellon University, 1975.
Pascal, B., *Pensées*, New York, Dutton, 1958.
Pater, W., *Miscellaneous Studies*, London, Macmillan, 1910.
Perls, F. *et al.*, *Gestalt Therapy: Excitement and Growth in the Human Personality*, New York, Delta, 1951.
Perls, F., 'Four Lectures', in J. Fagan and I. Shepherd, eds., *Gestalt Therapy Now*, New York, Harper & Row, 1970.
Philippi, D., ed., *Songs of Gods, Songs of Humans: The Epic Tradition of the*

Ainu, San Francisco, North Point Press, 1982.
Pindar, *The Odes of Pindar*, ed. J. Sandys, Cambridge, Harvard University Press, 1968.
Plato, *The Republic*, in B. Jowett, ed., 'The Dialogue of Plato', vol. 1, New York, Random House, 1937.
Poetzl, O. *et al.*, 'Preconscious Stimulation in Dreams, Associations and Images', *Psychological Issues*, vol. 2, 1960.
Polster, E. and Polster, M., *Gestalt Therapy Integrated*, New York, Random House, 1974.
Preuss, K., *Die geistige Kultur der Naturvölker*, Leipzig, Teubner, 1914.
Rabinow, P., ed., *The Foucault Reader*, New York, Pantheon, 1984.
Rabinow, P. and Dreyfus, H., *Michel Foucault: Beyond Structuralism and Hermeneutics*, Chicago, University of Chicago Press, 1983.
Rawls, J., *A Theory of Justice*, Cambridge, Harvard University Press, 1971.
Reichel-Dolmatoff, G., *Amazonian Cosmos: The Sexual and Religious Symbolism of the Tukano Indians*, Chicago, University of Chicago Press, 1971.
Rich, A., *Of Woman Born: Motherhood as Experience and Institution*, New York, Norton, 1976.
Richardson, W. and Muller, J., *Lacan and Language*, New York, International Universities Press, 1982.
Rifkin, J., *Declaration of a Heretic*, London and New York, Routledge & Kegan Paul, 1971.
Rilke, R., *The Letters of Rainer Maria Rilke, 1892-1910*, ed. J. Greene and M. Herter-Norton, New York, Norton, 1945.
Rilke, R., *The Letters of Rainer Maria Rilke, 1910-1926*, ed. J. Green and M. Herter-Norton, New York, Norton, 1947.
Rilke, R., *The Notebooks of Malte Laurids Brigge*, New York, Norton, 1949.
Rilke, R., *Gesammelte Gedichte*, Frankfurt am Main, Insel-Verlag, 1962.
Robinson, J., ed., *The Nag Hammadi Library*, New York, Harper & Row, 1977.
Roethke, T., *The Collected Poems*, New York, Doubleday, London, Faber & Faber, 1966.
Rogers, C., *On Becoming a Person*, Boston, Houghton Mifflin, 1961.
Rogers, C., *Freedom to Learn*, Columbus, Ohio, Charles Merrill Publishing, 1969.
Romanyshyn, R., 'Copernicus and the Beginning of Modern Science', *Journal of Phenomenological Psychology*, vol. 3, no. 2, Spring,1973.
Romanyshyn, R., 'Science and Reality: Metaphors of Experience and Experience as Metaphorical', in R. Valle and R. von Eckhartsberg, eds., *Metaphors of Consciousness*, New York, Plenum, 1980.
Romanyshyn, R., *Psychological Life: From Science to Metaphor*, Austin, University of Texas Press, 1982.
Romanyshyn, R., 'The Despotic Eye', in D. Kruger, Ed., *The Changing Reality of Modern Man: Essays in Honor of J. H. van den Berg*, Cape Town, Juta, 1984.
Rong-tha blo-bzang-dam-chos-rgya-mtsho, *The Creation of Mandalas*, New Delhi, 1971.
Rosanes-Berrett, M., 'Gestalt Therapy as an Adjunct Treatment for Some

Visual Problems', in J. Fagan and I. Shepherd, eds., *Gestalt Therapy Now*, New York, Harper & Row, 1970.
Ruck, C. *et al.*, *The Road to Eleusis: Unveiling the Secret of the Mysteries*, New York and London, Harcourt, Brace & Jovanovich, 1978.
Sallis, J., 'Heidegger/Derrida: Presence', *Journal of Philosophy*, vol. 81, no. 10, October, 1984.
Sallis, J., *Delimitations: Phenomenology and the End of Metaphysics*, Bloomington, Indiana University Press, 1986.
Santideva, *Entering the Path of Enlightenment*, ed. M. Matics, New York, Macmillan, 1970.
Sapir, E., 'The Unconscious Patterning of Behaviour in Society', in D. Mandelbaum, ed., *Selected Writings of Edward Sapir*, Berkeley, University of California Press, 1949.
Sartre, J. P., *Being and Nothingness*, New York, Philosophical Library, 1956, London, Methuen, 1957.
Sartre, J. P., *The Transcendence of the Ego*, New York, Noonday Press, 1957.
Scarry, E., *The Body in Pain: The Making and Unmaking of the World*, New York and London, Oxford University Press, 1985.
Schelling, F., *Of Human Freedom*, Chicago, Open Court, 1936.
Scheman, N., 'Anger and the Politics of Naming', in McConnell-Ginet *et al.*, eds., *Women and Language in Literature and Society*, New York, Praeger, 1980.
Scholem, G., *The Kabbalah and Its Symbolism*, New York, Schocken, 1969.
Schrag, C., *Radical Reflection and the Origin of the Human Sciences*, West Lafayette, Purdue University Press, 1980.
Schroedinger, *Mind and Matter*, London, Cambridge University Press, 1969.
Schürmann, R., 'Political Thinking in Heidegger', *Social Research*, vol. 45, 1978.
Schürmann, R., 'Questioning the Foundations of Practical Philosophy', *Human Studies*, vol. 1, 1980.
Schürmann, R., ' "What Can I Do?" in an Archaeological-Genealogical History', *Journal of Philosophy*, vol. -82, no. 3, December, 1985.
Schürmann, R., *Heidegger on Being and Acting: From Principles to Anarchy*, Bloomington, Indiana University Press, 1986.
Schwartz-Salant, N., *Narcissism and Character Transformation: The Psychology of Narcissistic Character Disorders*, Toronto, Inner City Books, 1982.
Schwartz-Salant, N., 'On the Subtle Body Concept in Clinical Practice', in *Chiron*, Peru, Illinois, Chiron Publications, vol. 3, 1986.
Scully, V., *The Earth, the Temple and the Gods: Greek Sacred Architecture*, New York, Praeger, 1969.
Searles, H., *Collected Papers on Schizophrenia and Related Subjects*, New York, International Universities Press, 1965.
Sekida, Katsuki, *Zen Training: Methods and Philosophy*, New York, John Weatherhill, 1975.
Service for the New Year: Rosh Hashanah, New York, Hebrew Publishing Co., 1935.
Service for the Day of Atonement: Yom Kippur, New York, Hebrew Publishing Co., 1935.

Shapiro, K., *Bodily Reflective Modes: A Phenomenological Method for Psychology*, Durham, Duke University Press, 1985.
Singer, J., *Androgyny: Towards a New Theory of Sexuality*, New York, Doubleday, 1979.
Sontag, S., *On Photography*, New York, Dell, 1977.
Starobinski, J., *L'Oeuil Vivant*, Paris, Gallimard, 1961.
Stevens, W., *The Collected Poems of Wallace Stevens*, New York, Alfred Knopf, 1961.
Storm, H., *Seven Arrows*, New York, Ballantine, 1973.
Straus, E., *The Primary World of the Senses*, London and New York, Macmillan, 1963.
Straus, E., 'Born to See, Bound to Behold: Reflections on the Function of the Upright Posture in the Esthetic Attitude', in S. Spicker, ed., *Philosophy of the Body*, New York, Quadrangle, 1970.
Suleiman, S., ed., *The Female Body in Western Culture*, Cambridge Harvard University Press, 1985.
Sullivan, H., *The Interpersonal Theory of Psychiatry*, New York, Norton, 1953.
Suzuki, D., *Studies in Zen*, New York, Dell, 1955.
Suzuki, D., *The Zen Doctrine of No Mind*, New York, Samuel Weiser, 1973.
Taylor, M., *Erring: A Postmodern A/Theology*, Chicago, University of Chicago Press, 1984.
Thoreau, H., *The Journal of Henry David Thoreau*, vol. 2, ed. B. Torrey and F. Allen, New York, Dover, 1906.
Thoreau, H., *The Portable Thoreau*, ed. Carl C. Bode, New York, Viking, 1964.
Tarthang Tulku, 'The Life and Liberation of Padmasambhava', *Crystal Mirror*, vol. 4, Berkeley, Dharma Publishing, 1975.
Tarthang Tulku, *Gesture of Balance*, Berkeley, Dharma Publishing, 1977.
Tarthang Tulku, *Time, Space and Knowledge*, Berkeley, Dharma Publishing, 1977.
Tarthang Tulku, *Kum Nye Relaxation*, 2 vols., Berkeley, Dharma Publishing, 1978.
Todes, S., 'Shadows in Knowledge: Plato's Misunderstanding of Shadows and of Knowledge as Shadow-Free', in D. Ihde and J. Edie, eds., *Dialogues in Phenomenology*, The Hague, M. Nijhoff, 1976.
Trungpa, C., *Cutting Through Spiritual Materialism*, Boulder, Shambhala, 1973.
Trungpa, C., *The Tibetan Book of the Dead: The Great Liberation Through Hearing in the Bardo*, Berkeley, Shambhala, 1975.
Trungpa, C. and Guenther, H., *The Dawn of Tantra*, Berkeley, Shambhala, 1975.
Trungpa, C., *Glimpses of Abhidharma*, Boulder, Vajradhatu, 1975.
Turnbull, C., *The Forest People: A Study of the Pygmies of the Congo*, Simon & Schuster, 1961.
Turner, B., *The Body and Society: Explorations in Social Theory*, Oxford and New York, Basil Blackwell, 1984.
The Upanishads, 2 vols., trans. M. Muller, New York, Dover, 1962.
Vico, G., *The New Science*, Ithaca, Cornell University Press, 1970.

Wallen, R., 'Gestalt Therapy and Gestalt Psychology', in J. Fagan and I. Shepherd, eds., *Gestalt Therapy Now*, New York, Harper & Row, 1970.

Wasson, G. et al., *The Road to Eleusis: Unveiling the Secret of the Mysteries*, New York and London, Harcourt, Brace & Jovanovich, 1978.

Weber, S., 'Individuation as Praxis', in P. Breines, ed., *Critical Interruptions: New Left Perspectives on Herbert Marcuse*, New York, Herder & Herder, 1970.

Welwood, J., 'Exploring Mind: Form, Emptiness and Beyond', *Journal of Transpersonal Psychology*, vol. 8, no. 2, 1976.

Welwood, J., 'On Psychological Space', *Journal of Transpersonal Pscyhology*, vol. 9, no. 2, 1977.

Welwood, J., 'Meditation and the Unconscious: A New Perspective', *Journal of Transpersonal Psychology*, vol. 9, no. 1, 1977.

Welwood, J., 'Principles of Inner Work: Psychological and Spiritual', *Journal of Transpersonal Psychology*, vol. 6, no. 1, 1984.

Werner, H., *Comparative Psychology of Mental Development*, New York, International Universities Press, 1957.

Wheelwright, P., trans., *The Presocratics*, New York, Odyssey, 1966.

Whorf, B., *Language, Thought and Reality*, Cambridge, M.I.T. Press, 1956.

Wiesel, E., *Night*, New York, Hill & Wang, 1960.

Wilber, K., *The Spectrum of Consciousness*, Wheaton, Illinois Theosophical Publishing Co., 1979.

Winnicott, D., *Playing and Reality*, London, Tavistock, 1971.

Wittgenstein, L., *Philosophical Investigations*, trans. G. E. M. Anscombe, New York and London, Macmillan, 1953.

Wittgenstein, L., *Tractatus Logico-Philosophicus*, London, Routledge & Kegan Paul, 1969.

Wordsworth, W., *The Prelude*, ed. J. Maxwell, New York and Harmondsworth, Penguin, 1972.

Wyschogrod, E., *Spirit in Ashes: Hegel, Heidegger and Man-Made Mass Death*, New Haven, Yale University Press, 1985.

Yeshé Tsogyal, *The Life and Liberation of Padmasambhava*, Berkeley, Dharma Publishing, 1978.

Index